T0211510

Electron Paramagnetic Resonance Spectroscopy

Patrick Bertrand

Electron Paramagnetic Resonance Spectroscopy

Applications

Patrick Bertrand
Aix-Marseille University
Marseille, France

This translation has been supported by **UGA Éditions**, publishing house of Université Grenoble Alpes and the **région Auvergne-Rhône-Alpes**. https://www.uga-editions.com/.

ISBN 978-3-030-39670-1   ISBN 978-3-030-39668-8   (eBook)
https://doi.org/10.1007/978-3-030-39668-8

Translated, revised and adapted from "La Spectroscopie de Résonance Paramagnétique Electronique, Vol. I: Applications", P. Bertrand, EDP Sciences, 2014.

Grenoble Sciences

The French version of this book has been selected by "Grenoble Sciences". "Grenoble Sciences" directed by Professor Jean Bornarel, was between 1990 and 2017 an expertising and labelling centre for scientific works, with a national accreditation in France. Its purpose was to select the most original high standard projects with the help of anonymous referees, then submit them to reading comittees that interacted with the authors to improve the quality of the manuscripts as long as necessary. Finally, an adequate scientific publisher was entrusted to publish the selected works worldwide.

About this Book

This book is translated, revised and adapted from *La spectroscopie de résonance paramagnétique éléctronique – Applications* edited by Patrick Bertrand, EDP Sciences, Grenoble Sciences Series, 2014, ISBN 978-2-7598-1191-5.

The translation from original French version has been performed by: Maighread Gallagher, scientific editor, TWS Editing.

The reading committee of the French version included the following members:

▷ C. Autret-Lambert, lecturer at Tours University

▷ A.-L. Barra, CNRS research director, Grenoble

▷ G. Chouteau, retired professor at Grenoble Alpes University

▷ C. Duboc, CNRS research director, Grenoble

▷ Y. Li, CNRS research engineer, Paris

▷ M. Orio, CNRS researcher, Lille

▷ H. Vezin, CNRS research director, Lille

The authors

Thierry ALLARD - CNRS research director; Institut de Minéralogie, de Physique des Matériaux et de Cosmochimie (IMPMC) (Institute of Mineralogy and Physics of Condensed Media), Sorbonne Universities, UPMC Paris VI University, CNRS UMR 7590, IRD UMR 206, MNHN, Paris.

Jean Jacques BAHAIN - professor at the Muséum National d'Histoire Naturelle (French National Museum of Natural History), Prehistory Department, UMR 7194, CNRS & MNHN, Paris.

Etienne BALAN - IRD researcher; Institut de Minéralogie, de Physique des Matériaux et de Cosmochimie (IMPMC) (Institute of Mineralogy and Physics of Condensed Media), Sorbonne Universities, UPMC Paris VI University, CNRS UMR 7590, IRD UMR 206, MNHN, Paris.

Valérie BELLE - professor at Aix-Marseille University; Laboratoire de Bioénergétique et Ingénierie des Protéines ((Laboratory of Bioenergetics and Protein Engineering), UMR 7281, Institute of Microbiology of the Mediterranean, CNRS & Aix-Marseille University, Marseille.

Elie BELORIZKY - professor at Grenoble Alpes University; Laboratoire Interdisciplinaire de Physique (Interdisciplinary Laboratory of Physics), CNRS & Grenoble Alpes University, Grenoble.

Patrick BERTRAND - honorary professor at Aix-Marseille University; Laboratoire de Bioénergétique et Ingénierie des Protéines (Laboratory of Bioenergetics and Protein Engineering), UMR 7281, Institute of Microbiology of the Mediterranean, CNRS & Aix-Marseille University, Marseille.

Laurent BINET - lecturer at the École Nationale Supérieure de Chimie (National Superior School of Chemistry); Institut de Recherche de Chimie-ParisTech (Chemistry-ParisTech Research Institute), UMR 8247 CNRS, Chemistry-ParisTech, Paris.

George CALAS - professor at Pierre & Marie Curie University; Institut de Minéralogie, de Physique des Matériaux et de Cosmochimie (IMPMC) (Institute of Mineralogy and Physics of Condensed Media), Sorbonne Universities, UPMC Paris VI University, CNRS UMR 7590, IRD UMR 206, MNHN, Paris.

Jean-Louis CANTIN - lecturer at Pierre & Marie Curie University, Sorbonne Universities; Institut des NanoSciences de Paris (Paris Institute of NanoSciences), UMR 7588, CNRS & Paris VI University, Paris.

Jean-Michel DOLO - CEA research engineer; Institut d'Imagerie Biomédicale (I2BM) (Institute of Biomedical Imaging), French Atomic and Alternative Energy Commission (CEA), Orsay.

Pierre DORLET - CNRS researcher; Laboratoire Stress Oxydant et Détoxication (Laboratory of Oxidative Stress and Detoxification), UMR 8221, CNRS & CEA/iBiTec-S/SB²SM, Gif-sur-Yvette.

Christophe FALGUÈRES - CNRS research director; Muséum National d'Histoire Naturelle (French National Museum of Natural History), Prehistory Department, UMR 7194, CNRS & MNHN, Paris.

André FOURNEL - honorary lecturer at Aix-Marseille University; Laboratoire de Bioénergétique et Ingénierie des Protéines (Laboratory of Bioenergetics and Protein Engineering), UMR 7281, Institute of Microbiology of the Mediterranean, CNRS & Aix-Marseille University, Marseille.

Pascal H. FRIES - CEA research engineer, Reconnaissance Ionique et Chimie de Coordination, Chimie Inorganique et Biologique (Ion Recognition and Coordination Chemistry, Inorganic and Biological Chemistry) UMR E3, Grenoble Alpes University, CEA/DSM/INAC/SCIB, Grenoble.

Emmanuel FRITSCH - IRD research director; Institut de Minéralogie, de Physique des Matériaux et de Cosmochimie (IMPMC) (Institute of Mineralogy and Physics of Condensed Media); Sorbonne Universities, UPMC Paris VI University, CNRS UMR 7590, IRD UMR 206, MNHN, Paris.

Serge GAMBARELLI - CEA research engineer; Laboratoire de Résonance Magnétique (Laboratory of Magnetic Resonance), Department of Inorganic and Biological Chemistry, DSM/ INAC, UMR E3 CEA, FRE 3200 CNRS & Grenoble Alpes University, Grenoble.

Tristan GARCIA - CEA research engineer; LIST, Laboratoire National Henri Becquerel (Henri Becquerel National Laboratory), Gif-sur-Yvette.

François GENDRON - lecturer at Pierre & Marie Curie University (Paris VI), Sorbonne Universities; Institut des NanoSciences de Paris (Paris Institute of NanoSciences), UMR 7588, CNRS & Paris VI University, Paris.

Didier GOURIER - professor at the École Nationale Supérieure de Chimie (National Superior School of Chemistry); Institut de Recherche de Chimie-ParisTech (Chemistry-ParisTech Research Institute), UMR 8247 CNRS, Chemistry-ParisTech, Paris.

Stéphane GRIMALDI - lecturer at Aix-Marseille University; Laboratoire de Bioénergétique et Ingénierie des Protéines ((Laboratory of Bioenergetics and Protein Engineering), UMR 7281, Institute of Microbiology of the Mediterranean, CNRS & Aix-Marseille University, Marseille.

Christelle HUREAU - CNRS researcher; Laboratoire de Chimie de Coordination (Laboratory of Coordination Chemistry), UPR 8241, CNRS, Toulouse.

Robert LAURICELLA - honorary lecturer at Aix-Marseille University and CNRS associate tenure scientist; Institut de Chimie Radicalaire (Institute of Radical Chemistry) UMR 7273, Marseille.

Vincent MAUREL - CEA research engineer; Laboratoire de Résonance Magnétique (Laboratory of Magnetic Resonance), Department of Inorganic and Biological Chemistry, DSM/ INAC, UMR E3 CEA, FRE 3200 CNRS & Grenoble Alpes University, Grenoble.

Guillaume MORIN - CNRS research director; Institut de Minéralogie, de Physique des Matériaux et de Cosmochimie (IMPMC) (Institute of Mineralogy and Physics of Condensed Media), Sorbonne Universities, UPMC Paris VI University, CNRS UMR 7590, IRD UMR 206, MNHN, Paris.

Brigitte PÉPIN-DONAT - CNRS research director; Laboratoire Structures et Propriétés d'Architectures Moléculaires (Structures and Properties of Molecular Architectures Laboratory), DSM/INAC/SPrAM, UMR 5819, CNRS, CEA & Grenoble Alpes University, CEA, Grenoble.

Jérôme POULENARD - professor at Savoie University; Laboratoire Environnement Dynamique et Territoires de la Montagne (Laboratory of Environment Dynamics and Territories of the Mountain), UMR 5204, CNRS & University of Savoie, Le Bourget-du-Lac.

François TROMPIER - tenure scientist at the Institut de Radioprotection et de Sûreté nucléaire (Institute of Radioloprotection and Nuclear Safety), External dosimetry Department, Laboratory of dosimetry for ionising radiation, Fontenay-aux Roses.

Béatrice TUCCIO - lecturer at Aix-Marseille University; Institut de Chimie Radicalaire (Institute of Radical Chemistry), UMR 7273, CNRS & Aix-Marseille University, Marseille.

Philippe TUREK - professor at Strasbourg University; Institut de chimie (Chemistry Institute), UMR 7177, CNRS & Strasbourg University, Strasbourg.

Hans Jürgen VON BARDELEBEN - CNRS emeritus research director; Institut des NanoSciences de Paris (Paris Institute of NanoSciences), UMR 7588, CNRS & Paris VI University, Paris.

Collaborators

Arnaud FABIEN - CNRS researcher; Laboratoire Environnement Dynamique et Territoires de la Montagne (Laboratory of Environment Dynamics and Territories of the Mountain), UMR 5204, CNRS & University of Savoie, Le Bourget-du-Lac.

Thibaut BLONDEL[1] - PhD candidate at the University of Avignon and the Vaucluse; UMR Environnement Méditerranéen et Modélisation des Agro-Hydrosystèmes (Modelling Agricultural and hydrological systems in the mediterranean environment), Laboratory of Hydrogeology, Avignon, INRA & University of Avignon and the Vaucluse, Avignon.

Jean-Jacques DELANNOY - professor at Savoie University; Laboratoire Environnement Dynamique et Territoires de la Montagne (Laboratory of Environment Dynamics and Territories of the Mountain), UMR 5204, CNRS & University of Savoie, Le Bourget-du-Lac.

Jean-Marcel DORIOZ - INRA research director; UMR Centre Alpin de Recherche sur les Réseaux Trophiques et Ecosystèmes Limniques (Alpine Research Centre on Trophic Networks and Limnic Ecosystems), INRA & University of Savoie, Le Bourget-du-Lac.

---

[1]   Position held during his/her contribution to writing this book.

Yves Dudal - INRA researcher; UMR Biogeochimie du Sol et de la Rhizosphère (Biogeochemistry of the Soil and of the Rhizosphere), INRA & Montpellier SupAgro, Montpellier.

Christophe Emblanch - lecturer at the University of Avignon and the Vaucluse; UMR Environnement Méditerranéen et Modélisation des Agro-Hydrosystèmes (Modelling Agricultural and hydrological systems in the mediterranean environment), Laboratory of Hydrogeology, Avignon, INRA & University of Avignon and the Vaucluse, Avignon.

Bernard Fanget - research engineer at Savoie University; Laboratoire Environnement Dynamique et Territoires de la Montagne (Laboratory of Environment Dynamics and Territories of the Mountain), UMR 5204, CNRS & University of Savoie, Le Bourget-du-Lac.

Charline Giguet-Covex - post-doctoral fellow at Savoie University; Laboratoire Environnement Dynamique et Territoires de la Montagne (Laboratory of Environment Dynamics and Territories of the Mountain), UMR 5204, CNRS & University of Savoie, Le Bourget-du-Lac.

Christian Lombard - CEA engineer; Laboratoire Structures et Propriétés d'Architectures Moléculaires (Structures and Properties of Molecular Architectures Laboratory), DSM/INAC/SPrAM, UMR 5819, CNRS, CEA & Grenoble Alpes University, CEA, Grenoble.

Cécile Miège - IRSTEA research engineer; Laboratoire de chimie des milieux aquatiques (Chemistry of aquatic media laboratory), UR QELY, IRSTEA, Lyon.

Yves Perrette - CNRS researcher; Laboratoire Environnement Dynamique et Territoires de la Montagne (Laboratory of Environment Dynamics and Territories of the Mountain), UMR 5204, CNRS & University of Savoie, Le Bourget-du-Lac.

Myriam Protière [1] - PhD candidate at Grenoble Alpes University; Laboratoire Structures et Propriétés d'Architectures Moléculaires (Structures and Properties of Molecular Architectures Laboratory), DSM/INAC/SPrAM, UMR 5819, CNRS, CEA & Grenoble Alpes University, CEA, Grenoble.

Preface

In a first volume devoted to the *Foundations* of electron paramagnetic resonance (EPR) spectroscopy, hereafter called "Volume 1", we showed how the shape and intensity of the spectrum are determined by the nature, the arrangement and the number of paramagnetic centres present in the sample. These results were obtained using the tools of quantum mechanics, the abstract nature of which can hinder some beginners and slow down their learning of this technique. However, it must be admitted that most EPR applications do not explicitly require the use of these tools. It is sufficient for example to measure the amplitude of the spectrum to track the evolution of the concentration of a substance or to simply simulate a composite spectrum to analyse a mixture. It is slightly more difficult to identify a molecule from the parameters of the spin Hamiltonian or to determine the energy of excited states from the temperature dependence of the signal intensity, but once again it is not necessary to resort to the general formalism. Beginners will learn EPR more eagerly and more easily if they know that it can be used at several levels and that the information sought is often obtained without returning to the basic principles. The aim of this second volume is to illustrate the different levels of use of EPR through examples of applications. Readers who have only a basic knowledge will readily understand the easiest levels and will progressively become familiar with the more difficult ones by relying on the contents of the volume presenting the foundations [Volume 1]. Back and forth reading between the two volumes should therefore improve and accelerate learning of the concepts and methods of EPR by first-year Master's students in Physics, Physical Chemistry, Chemistry, and also Biophysics and Biochemistry.

Another novelty of this volume is the diversity of applications which are dealt with. To fully appreciate it, the aim of the first EPR studies must be recalled. The first experiments were performed on solutions of Mn^{2+} and Cu^{2+}, independently in Kazan and Pittsburg [Zavoisky, 1945; Cummerow and Halliday,

1946]. They preceded, by a few months, the first experiments in *nuclear magnetic resonance* performed at Stanford [Bloch *et al.*, 1946] and Harvard [Purcell *et al.*, 1946]. The study by EPR of Cr^{3+} alum at the Clarendon Laboratory in Oxford [Bagguley and Griffith, 1947] marked the start of an impressive series of studies devoted to transition ion and rare earths complexes [Bleaney, 1992]. It was also in Oxford that the first *ferromagnetic resonance* experiments were performed on thin films of iron, cobalt and nickel [Griffith, 1946], the interpretation of which was presented shortly afterwards by Kittel [Kittel, 1947; Kittel, 1948]. All these studies allowed experimental devices to be optimised and to develop the theoretical bases of the new spectroscopies, but they were only of interest to physicists. The invention of the *spin Hamiltonian* marked a decisive step in the generalisation of EPR [Abragam and Pryce, 1951]. Indeed, this remarkable tool allows the experimenter to extract the most relevant parameters from the spectrum and to let theoreticians focus on interpreting them... From this point on, progress was rapid. For example, most of the concepts necessary to analyse spectra produced by free radicals in solution had already been developed by 1957. Spectra for gases such as NO, NO_2, O_2 had been correctly interpreted as early as 1950–1954 and the study of radicals produced by the irradiation of solid matrices had started [Fraenkel, 1957]. EPR was found to be a promising technique to test models of chemical bonds. By comparison, experiments devoted to biology appeared relatively anecdotal. They involved, for example, detection of radicals in cancer tissue, chloroplasts under illumination and cigarette smoke [Commoner *et al.*, 1954; Sogo *et al.*, 1957; Lyons *et al.*, 1958] or the elucidation of the orientation of the plane of the haeme in a monocrystal of myoglobin [Bennett *et al.*, 1957]. Despite these incursions into fields pertaining to chemistry and biology, EPR remained mainly a physicists' field until the start of the 1960s: most of the communications presented at the first international EPR spectroscopy conference (Jerusalem, July 1962) were based on transition and rare earths ions in solids [Low, 1963].

Fifty years later, EPR is certainly one of the spectroscopic techniques for which the applications are the most diverse. They currently relate to sectors that the first experimentalists would have had difficulty imagining, related to physics and chemistry but also Earth sciences, social sciences, agricultural sciences, environmental science, biology and health.

The subjects dealt with in this volume clearly illustrate the diversity of current applications of EPR. They use this technique at very varied levels and we have presented them in increasing order of difficulty. Here is a foretaste of their contents:

Chapter 1 - Natural or artificial irradiation of certain solid materials generates paramagnetic centres, free radicals, defects or ions, which are stable enough to be identified and quantified by EPR. The irradiation dose can be assessed from the intensity of the spectrum, which opens opportunities for numerous applications such as dosimetry of radiation used in radiotherapy or to sterilise food, determination of the dose received by individuals during chronic exposure or radiological accidents, or even to date archaeological samples.

Chapter 2 - Natural organic matter in soils contains stable semiquinone radicals, the nature of which depends on their origin and their maturation state. The relative proportions of these radicals, deduced from simulations of the EPR spectrum, constitute a signature which can be used to differentiate soils and even their horizons. It allows us to monitor the transfer of organic matter in a drainage basin from the soil to the hydrological systems, from the systems to their spillways and right up to the natural recorders (sediments, speleothems). The power of this tracing method is clearly illustrated by the determination of the catchment area and the residence time for drinking water in karst formations.

Chapter 3 - Numerous chemical reactions occurring in solution produce highly reactive radicals with a very short lifetime which prevents their direct study by conventional EPR spectroscopy. They can be detected by causing them to react with trap molecules chosen so as to obtain relatively stable radical adducts, the spectrum of which allows us to identify the original radical without ambiguity and to thus determine the mechanism of the initial reaction. This method is frequently used to identify the radicals generated by oxidative stress, but also to monitor polymerisation and photochemical reactions.

Chapter 4 - Neurodegenerative diseases (Alzheimer's, Creutzfeldt-Jacob, Parkinson's, etc.) are characterised by the deposit of protein aggregates in the intracellular medium in the brain. They start by conformational changes of partially disordered peptides or proteins which have a high affinity for transition ions, in particular copper ions. EPR reveals that, in the presence of these ions, these flexible macromolecules fold in different ways to ensure the 4-ligand equatorial coordination which stabilises Cu(II) complexes. These ions could play an essential role in the aggregation processes and in the phenomena of oxidative stress associated with these diseases.

Chapter 5 - Clay minerals from the kaolinite group contain impurities (transition ions) and point defects (radicals). Their study by EPR provides information on the physicochemical conditions in which these minerals were formed. In addition, by measuring the intensity of the spectrum for certain defects created by natural irradiation, it is possible to date the kaolinites and thus determine the timescale of the alteration processes to which these minerals were subjected. These data are very useful to understand the mechanisms of the chemical evolution of continental surfaces and to reconstitute historical climates.

Chapter 6 - The catalytic activity of redox enzymes relies on the existence of an active site and redox centres. These are organic groups and transition ion complexes which can generally be prepared in a paramagnetic state. Their study by EPR, combined with theoretical chemical models, provides detailed information on their structure and redox properties, but also on the catalytic mechanism of the enzymes. It thus contributes to the development of bio catalysts or synthetic "bio-inspired" catalysts which can be distinguished from their industrial homologues by the absence of rare metals and their non-toxicity.

Chapter 7 - The oldest carbonaceous matter in the solar system, aged at 4.5 billion years old, can be found trapped in meteorites. On Earth, the oldest carbonaceous matter of biological origin is fossilised in cherts dated at 3.5 billion years old. The EPR study of samples of this primitive carbonaceous matter provides information on the nature, the environment and the means by which carbon-based radicals were formed. This information is precious to develop scenarios for the formation of organic matter in the solar system and the emergence of life on Earth. The same methods could be used to analyse samples collected on Mars.

Chapter 8 - When a nitroxide radical is bound to a protein in solution, the shape of its spectrum and its relaxation properties depend on its interactions with its environment. These "paramagnetic probes" can be used to study structural transitions in proteins, which are very important phenomena which are currently poorly understood. By labelling an appropriately selected site, it is possible, for example, to monitor conformational changes in an enzyme induced by its interaction with its physiological partners. It is also possible to draw a "mobility map" to identify the domain involved in folding of an "intrinsically disordered protein" induced by its interaction with another protein.

Chapter 9 - The nature (ferro- or antiferromagnetic) and magnitude of exchange interactions between nitroxide radicals connected by organic linkers are determined by topological rules that EPR can reveal: the spectrum for a dilute solution of molecules containing pairs and triads of radicals recorded at room temperature reveals intramolecular interactions, whereas the spin of the ground state and the exchange parameters can be deduced from the temperature-dependence of the total intensity, used here as a measure of the magnetisation. This novel "molecular magnetism" approach is used in the context of current developments in nanosciences.

Chapter 10 - There exist several methods to record the spectrum of short-lived paramagnetic species. The temporal resolution of "stopped flow" and "freeze quench" techniques is around 10^{-3} s. Time resolved techniques based on in situ generation of species, direct detection and repeating the experiment for several values of the magnetic field can reduce this to 10^{-7} s. These techniques are used in particular to study photochemical reactions. The shape and intensity of the spectra produced by radical intermediates are determined by the CIDEP effect and its analysis allows us to distinguish between "radical pair" and "triplet state" mechanisms.

Chapter 11 - In Magnetic Resonance Imaging (MRI), complexes of Gd^{3+} are very often used as contrast agents to locally accelerate relaxation of water protons. The theoretical models developed to quantitatively describe this phenomenon involve parameters which determine the spin-lattice relaxation for the complexes. Their values can be obtained by simulating a series of EPR spectra recorded at different frequencies over a range of temperatures. These studies reveal the factors determining the efficacy of the complexes and can be used to optimise their structure.

Chapter 12 - Ferromagnetic resonance (FMR) uses the same equipment as EPR to study strong interactions between spins characterising crystallised ferromagnetic materials. The parameters describing these very anisotropic interactions are deduced from the variation in the position of the resonance line as a function of the direction of the applied magnetic field. This technique is illustrated by experiments performed on several types of nano structured materials: Fe films deposited by epitaxy on GaAs, thin layers of semiconductors doped with Mn^{2+} and ferrofluids. Its applications relate to magnetic recording, spintronics, and biomedical fields.

Several chapters in this volume include *complements* which need not be read to understand the main body of the text, but which may be of interest to more demanding readers. Some applications use results from techniques which are not dealt with in volume 1 presenting the foundations [Volume 1]. These techniques are presented in complements to the chapters and in *independent appendices* which can be found at the end of the present volume:

▷ complement 1 and 2 to chapter 7 describe the principles of quadrature phase detection and EPR imaging,

▷ appendix 1 introduces the Bloch equations and pulsed methods,

▷ appendix 2 presents pulsed EPR and describes ESEEM, HYSCORE and PELDOR experiments,

▷ appendix 3 covers continuous wave ENDOR spectroscopy,

▷ appendix 4 describes the general properties of peptides and protein, complex macromolecules studied in several chapters.

It was not possible to treat several important applications of EPR in this volume, such as Dynamic Nuclear Polarisation (DNP) [Abragam and Goldman, 1976; Barnes *et al.*, 2008], molecular conductors [Coulon and Clérac, 2004], oximetry [Ahmad and Kuppusamy, 2010], molecular magnetism of transition ions and rare earths (see the introduction to chapter 9).

This volume should be of interest to those curious about the applications of spectroscopic techniques, and teachers will find ample material to illustrate their lessons. The two complementary volumes devoted to the *Foundations* and *Applications* constitute a "treatise on EPR" in which the technical elements are enriched by the chapter describing FMR and by the appendices and complements in the present volume. To our knowledge, there is no equivalent of this collection.

Anatole Abragam left us in June 2011 when this collection was maturing. He played a major role in the development of magnetic spectroscopies both through his scientific production and his personal commitment. Thanks to their quality and depth of analysis, the treatises that he wrote over 40 years ago remain reference works. Anatole Abragam's scientific rigour was combined with an unshakeable sense of humour which clearly emerge in the subtitles of the chapters of his book on nuclear magnetism [Abragam, 1983]. The door of his office bore a "*Smoking, no hydrogen*" sign, and in his laboratory he had posted this essential recommendation: "before definitively consigning quantum mechanics to the bin, check the fuses one last time" [Abragam, 1999]. We dedicate this work to him.

Patrick Bertrand

References

ABRAGAM A. (1983) *The Principles of Nuclear Magnetism*, Clarendon Press, Oxford.

ABRAGAM A. & PRYCE M.H.L. (1951) *Proceedings of the Royal Society London A* **205**: 135-153.

ABRAGAM A. & GOLDMAN M. (1976) *Reports on Progress in Physics* **41**: 395-467.

ABRAGAM A. (1999) *Time Reversal, an autobiography*, Oxford University Press, Oxford.

AHMAD R. & KUPPUSAMY P. (2010) *Chemical Review* **110**: 3212-3236.

BAGGULEY D.M.S & GRIFFITH J.H.E. (1947) *Nature* **160**: 532-533.

BARNES A.B. *et al.* (2008) *Applied Magnetic Resonance* **34**: 237-263.

BENNETT J.E. *et al. (1957) Proceedings of the Royal Society London* **A 240**: 67-82.

BLEANEY B. (1992) *Applied Magnetic Resonance* **3**: 927-946.

BLOCH F. *et al.* (1946) *Physical Review* **70**: 474-485.

COMMONER B. *et al.* (1954) *Nature* **174**: 689-691.

COULON C. & CLÉRAC R. (2004) *Chemical Review* **104**: 5655-5687.

CUMMEROW R.L. & HALLIDAY D. (1946) *Physical Review* **70**: 433-433.

FRAENKEL G.K. (1957) *Annals of the New York Academy of Sciences* **67**: 546-569.

GRIFFITHS J.H.E. (1946) *Nature* **158**: 670-671.

KITTEL C. (1947) *Physical Review* **71**: 270-271.

KITTEL C. (1948) *Physical Review* **73**: 155-161.

LOW W. (1963) *Paramagnetic Resonance*, Academic Press, New York.

LYONS M.T. *et al.* (1958) *Nature* **182**: 1003-1004.

PURCELL E.M. *et al.* (1946) *Physical Review* **69**: 37-38.

SOGO P.B. *et al.* (1957) *Proceedings of the National Academy of Sciences of the USA* **43**: 387-393.

Volume 1: BERTRAND P. (2020) *Electron Paramagnetic Resonance Spectroscopy - Fundamentals*, Springer, Heidelberg.

ZAVOISKY E.J. (1945) *Journal of Physics* (U.S.S.R.) **9**: 211-216.

Contents

Chapter 1

Dosimetry
of ionising radiation

Bahain J.J.[a] , Dolo J.M.[b] , Falguères C.[a], Garcia T.[c], Trompier F.[d]

[a] *French National Museum of Natural History,*
Prehistory Department, UMR 7194, Paris.

[b] *French Atomic and Alternative Energy Commission (CEA),*
Institute of Biomedical Imaging, Orsay.

[c] *French Atomic and Alternative Energy Commission (CEA),*
Henri Becquerel National Laboratory, Gif-sur-Yvette.

[d] *Institute of Radioprotection and Nuclear Safety,*
External dosimetry Department, Fontenay-aux Roses.

1.1 - Introduction

Dosimetry of ionising radiation aims to quantify the energy received by a sample of matter (inert matter or living organism) as a result of its interaction with ionising radiation (photons, electrons, protons, neutrons, α particles, etc.). The advantages of EPR in this field are obvious, and its use was first proposed in the 1950s. Indeed, the interaction of Ionising Radiation (IR) with matter generates paramagnetic entities by excitation and ionisation of the atoms, and by breaking the bonds between atoms. These entities, free radicals, defects or ions, can be detected by conventional EPR spectrometry when their lifetime is sufficiently long. As the number of radio-induced paramagnetic species is directly linked to the absorbed dose [1], it can be determined by measuring the intensity of the spectrum. In EPR dosimetry, the absolute number of radio-induced radicals or paramagnetic defects is not measured. Instead, the peak-to-peak amplitude A_{pp}

---

[1] The dose is the energy absorbed per unit of mass of the material being studied. It is expressed in grays [Gy]; 1 Gy = 1 J kg^{-1}.

© Springer Nature Switzerland AG 2020
P. Bertrand, *Electron Paramagnetic Resonance Spectroscopy,*
https://doi.org/10.1007/978-3-030-39668-8_1

for a structure of the signal is measured, and the absorbed dose can be deduced by applying one of the following methods:

▷ One or more calibration curves is created from reference samples with a similar nature to those to be assayed. These samples are irradiated with known doses and the A_{pp} amplitude of their spectrum can be used to establish a curve linking A_{pp} to the dose. When creating the calibration curve, it is important to use a dose range covering all of the doses to be assessed, as extrapolation can be inaccurate due to saturation of the radio-induced signal beyond a certain dose for some materials. The nature of the radiation used to irradiate the reference samples is also very important. Thus, for a given dose, the number of radio-induced species can vary considerably with the type of radiation (photon, electron, neutron, etc.) and the dose rate [in Gy s^{-1}]. It is therefore preferable that the characteristics of the calibration beam correspond as closely as possible to those of the radiation to which the samples assayed were exposed. It is not always easy to obtain a source of radiation with the desired characteristics, and it is sometimes necessary to apply correction coefficients. Calibration curves are used with materials presenting little variability, compatible with the desired accuracy of the dose measurement, such as alanine and dental enamel.

▷ When the sample to be assayed presents broad variability (bone tissues, minerals), the notion of reference sample is no longer valid. In these cases, the "additive dose" method is used, whereby the relationship between A_{pp} and the dose for *the sample under investigation* is established. The sample is irradiated at known doses, to allow its coefficient of sensitivity $\Delta A_{pp}/\Delta$dose to be determined, taking matrix effects into account. This coefficient is used to calculate an absorbed dose from the amplitude of the signal recorded before the post-irradiation process.

Since the earliest experiments performed on materials such as glass or alanine [Combrisson and Uebersfeld, 1954; Gordy *et al.*, 1955], EPR dosimetry has been used in a wide range of applications:

▷ dating of archaeological samples based on the accumulated dose of natural irradiation,

▷ estimation of the dose received by individuals during chronic and/or past exposure, from biopsies of dental enamel,

▷ estimation of the dose received during recent radiological accidents from biopsies taken from victims or samples of materials worn by them or present in their immediate environment,

▷ calibration and control of radioactive sources, generally using alanine as standard,

▷ identification of irradiated food products.

This chapter does not aim to be exhaustive, but to use some examples to illustrate the variety of EPR applications in the field of dosimetry of ionising radiation.

1.2 - Archaeological dating

1.2.1 - Principle of the method

EPR dating involves quantification of the paramagnetic centres created in a mineral through exposure to *natural radioactivity*. The minerals or biominerals tested are generally embedded in a sediment characterised by a very low level of natural radioactivity, produced by uranium (U), thorium (Th) and potassium (K) contained in some grains. The minerals absorb part of the radiation energy and thus constitute natural dosimeters. The total dose absorbed by the sample throughout its lifetime, known as its palaeodose, is expressed in grays, and can be written:

$$D(T) = \int_0^T d_a(t)\, dt \qquad [1.1]$$

where d_a is the dose rate (or annual dose) in Gy year^{-1}, i.e., the dose absorbed by the sample over a year, and T is the duration of exposure in years, i.e., the age of the sample. When the dose rate is constant, the sample's age is simply given by

$$\text{age} = D(T)/d_a \qquad [1.2]$$

EPR spectrometry can be used to estimate the sample's palaeodose, and the dose rate is calculated from the quantities of radioactive elements (U, Th and its descendents, K) in the sample and the sediment surrounding it.

To be useful in dating, an EPR signal must meet a certain number of conditions: the signal produced by the mineral studied must be radiosensitive; its initial intensity must, if possible, be null; the signal must not be contaminated by other signals; and it must be sufficiently heat-stable to allow dating over a geological timescale. Figure 1.1 shows the EPR spectra produced by the minerals most commonly used in archaeological dating: carbonates of speleothems (stalagmites) and corals, dental enamel and quartz.

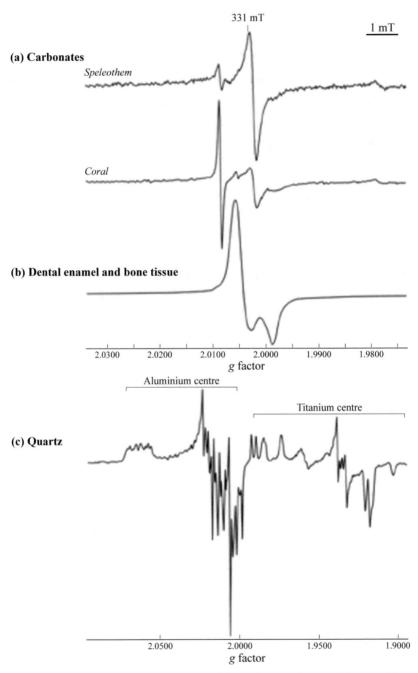

Figure 1.1 - EPR spectra produced by carbonates (cave calcite, corals), dental enamel and quartz in the $g = 2$ region.

(a) Carbonates. Experimental conditions: room temperature, power 5 mW, modulation amplitude 0.1 mT. The carbonate spectra present three radiosensitive lines at $g = 2.0057$; 2.0036; 2.0007 attributed, respectively, to SO_2^-, SO_3^- and CO_2^- radicals associated with a water molecule.

(b) Dental enamel and bone tissue. Experimental conditions: room temperature, power 10 mW, modulation amplitude 0.1 mT. The hydroxyapatite signal used in dating is axial with $g_\perp = 2.0018$; $g_{//} = 1.9977$. It results from the superposition of several signals attributed to carbonated centres (see section 1.3.1).

(c) Quartz. Experimental conditions: $T = 100$ K, power 5 mW, modulation amplitude 0.1 mT. The *aluminium* centre $[AlO_4]^0$ is characterised by $g = 2.0602$; 2.0085; 2.0019 and a hyperfine structure due to the Al nucleus ($I = \frac{5}{2}$). The *titanium* $[TiO_4]^0$ centres are associated with a compensatory cation (Li^+, Na^+, H^+) and their spectra present a superhyperfine structure due to the cation's nuclear spin ($\frac{3}{2}$ for Li^+ and Na^+, $\frac{1}{2}$ for H^+).

EPR dating can be applied to a wide variety of samples over long time periods. In particular, it can be used to determine chronological reference points for the Lower Pleistocene and the start of the Middle Pleistocene, a period between around 2.0 and 0.5 Ma (figure 1.2).

This method is therefore of great importance, in particular to date samples from the limestone regions of western Europe. For more recent periods, it can be applied in combination with the uranium series (U-Th) method on the same support (dental enamel, mollusc shell, etc.) and the results compared to those obtained by other methods, such as luminescence (thermoluminescence, optically stimulated luminescence) and carbon 14 dating. For recent epochs, the field of application is limited by the radio-sensitivity of the samples.

For more ancient epochs, the number of paramagnetic defects available and their lifetime – if it is too short – are limiting factors which vary depending on the type of sample. Ages of around a million years have been determined from dental enamel and quartz, whereas marine carbonates only appear to be compatible with dating to around a few hundred thousand years.

A summary of the data relating to EPR dating can be found in [Falguères and Bahain, 2002; Grün, 2006]. Application of this method to dating of minerals contained in soils is described in chapter 5 of this volume.

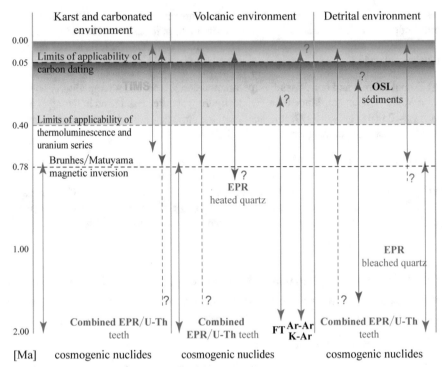

Figure 1.2 - Comparison of the domains of application of the EPR method
and the main methods used to date samples from the Quaternary period, depending
on the different types of environment of the prehistoric sites.
TIMS = **U-Th** method by thermo-ionisation mass spectrometry;
FT = fission track method; **Ar-Ar**/**K-Ar** = argon-argon and potassium-argon methods;
OSL = optically stimulated luminescence; **U-Th** = uranium-series method.

1.2.2 - Determining the equivalent dose (palaeodose)

The palaeodose is determined by the additive dose method (section 1.1). Several
aliquots of the sample to be dated are artificially irradiated with increasing doses
of beta or gamma radiation, and their EPR spectra are recorded. The variation
in signal intensity as a function of the dose constitutes the sensitivity curve for
the sample, and its extrapolation to an intensity of zero gives the palaeodose
(figure 1.3a).

The sensitivity curve is obtained by exposure to a single type of radiation.
However, in nature the samples are exposed to complex radiation which includes several types. For this reason, this method does not directly supply the
palaeodose, but what is known as an "equivalent dose" D_E. The sensitivity
curve is generally well described by the following equation:

$$I = I_{\infty}[1 - e^{-\mu(D_{art} + D_E)}]$$ [1.3]

where I is the intensity of the EPR signal of a sample artificially irradiated with the dose D_{art} [Gy], I_{∞} is the saturation intensity and μ is the sensitivity coefficient [Yokoyama *et al.*, 1985]. Other equations may also be used [Duval *et al.*, 2009].

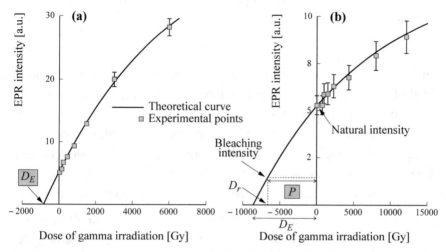

Figure 1.3 - Sensitivity curves and calculation of the equivalent dose considered as the palaeodose. **(a)** fossilised dental enamel, **(b)** quartz. The dose due to bleaching must be subtracted from the total dose to obtain the archaeological dose. D_E = equivalent dose; D_r = residual dose; P = palaeodose (or archaeological dose) with $D_E = D_r + P$. The continuous line corresponds to calculations performed using equation [1.3].

The quartz grains present in river terraces are often used for archaeological dating. Before sedimenting, these grains were "bleached" by UV radiation from sunlight during their transport by wind and water. This bleaching generates a residual EPR signal which must be subtracted when dating the terrace. To do so, in the laboratory an artificial bleaching experiment is performed on the "natural" aliquot (not artificially irradiated) using lamps which reproduce part of the solar spectrum (figure 1.3b) [Voinchet *et al.*, 2004].

1.2.3 - Assessing the dose rate

The annual dose, which is the result of all the ionising radiation to which the sample is exposed over a year, varies depending on the nature of the dose integrator (bone, dental enamel, minerals, sediments). It depends on the concentration and distribution of the radioactive elements in the sample and its environment, and on the intensity of the cosmic radiation. It can be written as follows:

$$d_a = kd_\alpha + d_\beta + d_\gamma + d_{cos} \qquad [1.4]$$

where d_α, d_β, d_γ and d_{cos} represent the annual doses of α, β, γ and cosmic radiation received by the sample, and k is a factor that takes into account the lower efficacy of α radiation to create trapped electrons. The value of this factor varies depending on the nature of the support. In practice, it is convenient to express the annual dose as a function of the origin of the radiation:

$$d_a = d_{external} + d_{internal} \qquad [1.5]$$

The dose due to the radiation external to the sample is measured either *in situ* with the help of a portable gamma spectrometer or a thermoluminescent dosimeter, or in the laboratory on the corresponding sediments. The dose due to the internal radioactivity is *calculated* from the amounts of radioactive elements contained in the sample, as measured by gamma spectrometry in the laboratory. It is rarely constant over time as numerous samples from the quaternary period present radioactive imbalances which must be taken into consideration.

The distribution of uranium in teeth is heterogeneous due to differences in chemical composition of the constituent tissues (enamel, dentine, cement). The uranium content can vary from less than one ppm in well-preserved enamel to more than 100 ppm in altered dentine from the same tooth. When dating fossil dental enamel from samples containing sufficient uranium, the uranium series can be used to calculate variations in annual dose rates. This *post-mortem* enrichment is linked to the difference in solubility between the different valence states of uranium [Gascoyne, 1982]. A number of models have been proposed to describe this phenomenon, such as the early uptake [Bischoff and Rosenbauer, 1981] and linear uptake [Ikeya, 1982] uranium incorporation models.

Grün and collaborators proposed combining U-Th and EPR data when calculating the age T of fossil samples. They described how the samples' uranium content changes over time by using the $U(t)$ function below [Grün *et al.*, 1988]:

$$U(t) = U_0(t/T)^{p+1} \qquad [1.6]$$

where U_0 is the uranium content measured, and p is the diffusion parameter for uranium, determined simultaneously from the EPR and U-Th data. The function $U(t)$ is illustrated in figure 1.4 for a few values of p.

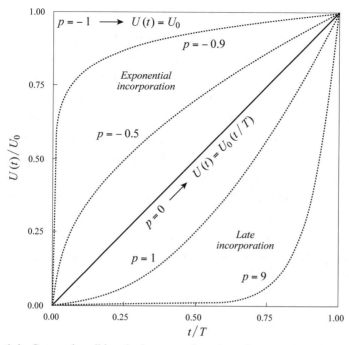

Figure 1.4 - Curves describing the incorporation of uranium over long time periods, as calculated using equation [1.6]. [From Grün *et al.*, 1988]

1.2.4 - Examples of archaeological and geological dating

◇ Visogliano paleolithic site, Italy

Among the sites in western Europe, the Visogliano deposit, located in the karst around Trieste, Italy, is considered by pre-historians as a reference sequence for the Middle Pleistocene, the period between −780,000 and −130,000 years. This deposit contained fossil human remains, attributed to *Homo heidelbergensis*, prehistoric tools and animal bones [Abbazzi *et al.*, 2000]. This site contains a large number of well-defined successive archaeological layers from which it was possible to establish a chronostratigraphy so as to situate this sequence in a broader context at the level of western Europe. Teeth from large herbivores from the deposit were dated by the combined EPR/U-Th method [Falguères *et al.*, 2008]. The oldest layers were dated at between −480,000 and −360,000 years (figure 1.5). As these lower levels contained most of the human remains, the results give a minimum age of −350,000 years for Visogliano Man.

◇ *Fossil river formations of the Creuse valley, France*

When the bones and teeth are altered or absent from a site, quartz is often the only support which can be dated by EPR in a non-volcanic environment [Falguères and Bahain, 2002]. This was the case, for example, in the Creuse valley, which presents a system of alluvial terraces covering the whole Quaternary period, i.e., around the last two-million years (figure 1.6) [Despriée *et al.*, 2006]. Numerous samples were collected from the different terraces and the ages determined by EPR were used to propose, for the first time, an interesting and solid chronological frame for this region. Several prehistoric sites were discovered in the fluvial formations in this valley. One of the most ancient deposits was found at Pont-de-Lavaud, an open-air site consisting of deposits on a riverbank. It is fossilised in a very high fluvial terrace located more than 100 m above the current level of the Creuse, and was dated at around 1 million years.

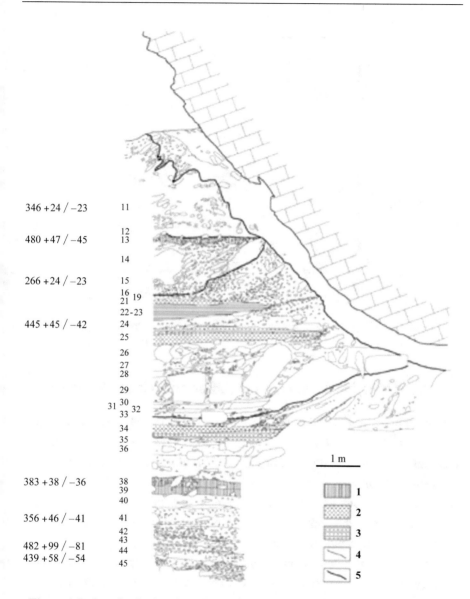

Figure 1.5 - Longitudinal profile of the Visogliano site showing how the main geoarcheological levels, numbered from 11 to 45 are arranged.The ages determined by the EPR/U-Th method are expressed in ka on the left. **1**: occupied soils, **2**: lœss levels, **3**: altered lœss levels,**4**: limits of lithological units, **5**: main discontinuities.

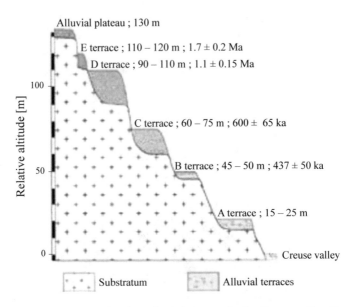

Alluvial plateau ; 130 m

E terrace ; 110 – 120 m ; 1.7 ± 0.2 Ma

D terrace ; 90 – 110 m ; 1.1 ± 0.15 Ma

C terrace ; 60 – 75 m ; 600 ± 65 ka

B terrace ; 45 – 50 m ; 437 ± 50 ka

A terrace ; 15 – 25 m

Creuse valley

Substratum   Alluvial terraces

Figure 1.6 - EPR dating of various fluvial terraces in the Creuse valley, France. The results cover almost the whole Quaternary period, i.e., around 2 million years.

1.3 - Retrospective dosimetry and radiation accident dosimetry

In this context, the aim of the study is to make a *post hoc* estimation of the doses of Ionising Radiation (IR) received by individuals in case of accident or chronic external exposure [2]. In the case of past and/or chronic exposure, it is mainly a question of comparing the estimated doses and the effects observed (cancer and non-cancer, cataract, fibrosis, infarct, etc.) to better understand the effects of IR on health in the context of epidemiological or radiobiological studies. With recent acute accidental irradiation, it is a question of helping to diagnose victims to allow healthcare teams to determine the most appropriate therapeutic strategy or to identify people who were exposed during a major accident.

EPR spectroscopy is one of the techniques used in retrospective dosimetry. Indeed, as the circumstances of the exposure are very varied and generally poorly known, a single technique cannot always reliably estimate the received dose. It is therefore often necessary to apply a multi-technique approach (cytogenetics, luminescence, numerical simulation) to better determine the received doses [Alexander *et al.*,

---

[2]   The source of radiation is outside the organism, unlike with 'internal' exposure, which is induced by inhalation or ingestion of radioactive products.

2007; Simon *et al.*, 2007; Trompier *et al.*, 2010; Ainsbury *et al.*, 2011]. In the case of severe irradiation, the dose is generally very heterogeneously deposited and the EPR approach can be used to determine its distribution in the organism.

1.3.1 - Chronic and/or past exposure

During chronic exposure, individuals are exposed to relatively low doses, but on a permanent or semi-permanent basis over relatively long periods, from a few months to several decades. It could be the result of exposure to "reinforced" natural radiation due to radon[3] or exposure linked to radioactive contamination of the environment (Chernobyl, Mayak, radium industry[4]) or of contaminated materials used to build housing, as was the case in Taiwan[5]. Assessing the received dose during past exposure can improve what we know about the long-term effects of IR. Thus, the whole modern radioprotection system, for example the definition of the statutory exposure limits, is based in large part on the effects observed on survivors of the Nagasaki and Hiroshima bombings. Retrospective dosimetry techniques have, in particular, been used to estimate the doses received during these two explosions.

Although in some cases the main parameters of the exposure can be assessed, and the dose estimated by applying models to this data, it is always preferable to validate these approaches through direct measurements of the dose received by exposed persons or by materials contained in their environment or on the site of the accident. EPR spectra for biopsies of dental enamel is a reference method in this field. Like bone tissue, dental enamel is mainly composed of hydroxyapatite crystals $(Ca)_{10}(PO_4)_6(OH)_2$ (96 % of the mass), but it is better crystallised and contains much less water (3 %) and organic matter (1 %). It has numerous advantages over bone from the point of view of measuring the dose by EPR:

▷ the signal induced by irradiation is specific for the interaction with the IR,

▷ some components of the radio-induced signal are extremely stable over time (half-life around 10^7 years), which is rare,

---

[3] Radon is a rare gas produced by thorium 232 and uranium 238 decay, it is naturally present in the earth's crust, mainly in granitic rocks. Its isotopes are radioactive.

[4] After the discovery of radium by Marie Curie, an industry was developed to produce radium from uranium ore for a range of applications: luminescent paints, cosmetics, energy drinks, etc. Some sites where this activity was performed remain contaminated to this day.

[5] Iron contaminated with cobalt 60, a radioactive element, was used in the construction of housing in Taiwan. It resulted in chronic exposure of the residents over several years.

▷ paramagnetic species are extensively produced, which makes it possible to detect relatively low doses. The limit of detection is around 50 mGy for 100 mg, whereas the annual dose received by the enamel in normal conditions is around a few mGy,

▷ the variation in signal intensity as a function of the dose can be considered linear and the inter-sample variability is relatively weak (standard deviation around 10 %), which makes it possible to use pre-established calibration curves.

Most radio-induced species are either carbonate groups ($CO_2^{\bullet-}$, $CO_3^{\bullet-}$, $CO^{\bullet-}$, $CO_3^{\bullet3-}$, etc.), phosphate groups ($PO_4^{\bullet2-}$), or oxygen derivatives ($O_3^{\bullet-}$, $O^{\bullet-}$, etc.). But not all are heat-stable. For example, the $CO_3^{\bullet-}$ radical disappears at room temperature a few weeks after irradiation. Figure 1.7 shows the EPR spectrum for a sample of dental enamel which received a dose of 500 mGy and the simulation obtained by superposing the radio-induced signal on the "native" signal (present before irradiation).

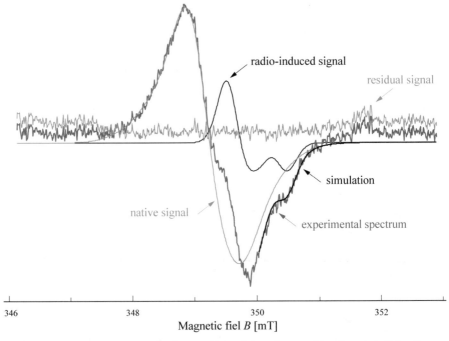

Figure 1.7 - X-band spectrum for a sample of dental enamel irradiated at 500 mGy. Experimental conditions: room temperature, microwave power 2 mW. The simulation shows that it results from the superposition of a radio-induced component and a "native" component. The "residual signal" is the difference between the experimental spectrum and the simulation.

Numerous studies have been performed to identify the various components of the radio-induced signal and the associated paramagnetic species [Fattibene and Callens, 2010]. At X-band, the spectrum has the appearance of an axial signal characterised by $g_{//} = 2.0018$ and $g_\perp = 1.9971$, but in reality it includes a dominant rhombic component and an axial component. Today, it is relatively well established that these two components are due to $CO_2^{\bullet-}$ radicals located in two different environments [Vanhaelewyn *et al.*, 2002]. The "native" signal present before irradiation, characterised by $g = 2.0045$ and a peak-to-peak width $\Delta B = 0.8$ mT, is due to organic matter present in the enamel. It disappears when this matter is eliminated. The various components of this signal and the associated paramagnetic species have not yet been clearly identified.

The two methods described in section 1.1 can be used to estimate the dose received by the dental enamel. The additive dose method is generally considered more accurate [Wieser *et al.*, 2005]. However, it involves destruction of the initial information and it is time-consuming to perform as well as requiring ready access to irradiation devices. It is thus more appropriate to use a pre-established calibration curve when studying a large number of samples. The advantages and disadvantages of the two methods are discussed in [IAEA, 2002] and [Fattibene and Callens, 2010].

Several parasitic effects can introduce bias when determining the intensity of the radio-induced signal and the associated dose. For example, drilling the teeth induces a heat-stable "mechanical" signal, characterised by $g = 2.002$ and $\Delta B = 0.8$ mT (figure 1.8).

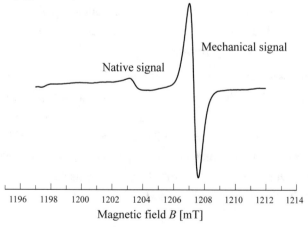

Figure 1.8 - Q-band EPR spectrum for a sample of mechanically-stressed calcified tissue. Experimental conditions: room temperature, power 1 mW.

This signal is due to radicals formed by the mechanical stress, mainly located at the surface of samples. It can be subtracted from the spectrum by simulation or suppressed by treating the sample with dilute acetic acid to uniformly eliminate the surface layer of the enamel. UV radiation and irradiation produce identical radicals, which can lead to overestimation of the received dose [Romanyukha *et al.*, 1996]. To avoid this potential pitfall, samples from the front faces of the canines and incisors are generally not used.

In contaminated areas, doses linked to the ingestion or inhalation of radioactive particles are added to the external irradiation. In some cases, radioactive contaminants can significantly increase the dose deposited in dental enamel, and as a result the dose estimated in the enamel will no longer be the external dose due to emissions from radioactive deposits in the environment. This is the case of strontium 90, a beta emitter, with chemical properties similar to those of calcium, which preferentially binds to dental enamel or dentine (the internal part of the tooth) as the teeth grow. Strontium generates localised dose deposits due to the short trajectory of beta radiation in these tissues. These localised deposits can result in significant local doses. As a result, the doses measured in the enamel must be interpreted carefully and additional measurements and calculations are often required [Simon *et al.*, 2007]. This type of contamination occurred, for example, in the Techa river basin in Russia.

Most applications of dosing using dental enamel relate to exposure to photonic radiation. In the case of victims of the Hiroshima and Nagasaki bombings, a neutronic component must be added [Ikeya, 1993]. The characteristics of the radio-induced signals are identical in both cases, but at equal dose, the intensity of the EPR signal induced by neutrons only represents a few % of that of the signal induced by photons [Trompier *et al.*, 2006].

In conclusion, using EPR to measure the dose in samples of dental enamel is recognised as a valid technique to estimate received doses linked to external exposure, whether recent or past, high or low. Nevertheless, there remain a certain number of open or unresolved questions (influence of UV, origin of signals), the resolution of which could ultimately improve the performance of this method. A very detailed literature review on this topic, covering the last 30 years, can be found in [Fattibene and Callens, 2010].

1.3.2 - Radiation accident dosimetry and emergency situations

Radiological accidents [6] result in a wide range of situations as they can involve a variety of types of radiation (photon, electron, neutron, etc.), energies (from a few meV to a hundred MeV) and accidental circumstances (duration, topology, number of victims). Such accidents can occur in industrial installations, during manipulation of sources or particle generators, in a nuclear setting (e.g. criticality accidents), during the recovery or theft of "orphan sources" [7], malicious use of radioactive substances, or in a medical context, mainly during radiotherapy and interventional radiology. These accidents sometimes lead to extreme exposure, and can have a significant impact on victims' health, occasionally requiring amputations and even causing death. It is therefore important to diagnose exposure as precisely as possible to better adapt the treatment offered to victims of irradiation. Several techniques can be implemented to estimate effects on the body as a whole, otherwise, the distribution of the dose can be determined for the whole body or for specific organs of interest, such as haematopoietic centres [8] and the skin. In complement to the analysis of clinical symptoms (vomiting, rash, necrosis, aplasia), accident reconstruction techniques or biological or physical retrospective dosimetry are used. There is no "standard" or ideal method. These techniques are often complementary and that selected will depend on the type of accident (localised or whole-body irradiation), the type of radiation involved, the time since irradiation, access to the site of the accident and/or to patients for biopsy biodosimetry, and information on the circumstances and parameters of the accident.

Estimation of the received dose by EPR has the advantage of not requiring information on the circumstances of the accident [9], which is generally partial and imprecise. Another advantage is that it can estimate the received dose at several points in the victims' bodies, as most irradiation is very heterogeneous or localised and produces very strong dose gradients. In this case, the mean whole-body dose, estimated on blood samples by cytogenetic techniques, provides no information on the damage incurred locally by the body.

---

[6] Here, we only mention accidents linked to external exposure as approaches to estimate exposure following ingestion or inhalation have yet to be described.

[7] Orphan sources are radioactive sources which have been abandoned or left unattended, and sometimes without any particular identification.

[8] Red bone marrow, responsible for the production of blood cells.

[9] Unlike accident reconstruction techniques, whether experimental or digital.

Materials eligible for this EPR dosing must present an exploitable EPR signal, if possible characteristic and measurable several weeks or several months after the accident, from dose-levels greater than around a few Gy. Although the potentially lethal whole-body dose is around 4 to 6 Gy, doses of up to several thousand Gy involving small volumes with very heterogeneous irradiation may not endanger the victims' lives in the short-term [Huet *et al.*, 2007]. Numerous materials are likely to present the required properties, but they must be present during the irradiation and recoverable for analysis. In the case of accidents for which dosimetric analyses have been published, *calcified tissues* (dental enamel and bone) were the main samples used [Trompier *et al.*, 2007; Clairand *et al.*, 2006]. These materials present good sensitivity to IR and give a stable EPR signal. Although the sensitivity of bone tissues is 10- to 20-fold lower than that of dental enamel, it remains sufficient to perform measurements, particularly when biopsies of bone tissue (generally a few mg) are taken near to the most irradiated zones, where the doses are high. The distribution of the bone tissues throughout the body means that samples can be obtained whatever the configuration of the irradiation and the location of the dose deposit. The dosimetric signal given by bone tissues has the same characteristics as that of dental enamel. However, the bone is living material which completely renews itself every 7–8 years, as a result the signal decreases over time. Nevertheless, once extracted, the signal is as stable as that for dental enamel. In radiation accident dosimetry, where the dose must be measured very precisely, the additive dose method is preferred due to the significant variability in dose-sensitivity observed with bone tissues.

To limit the invasiveness of the sampling, measurements can be performed *in vivo* on exposed teeth or bone tissues located near to the surface [Swartz *et al.*, 2006; Zdravkova *et al.*, 2004]; alternatively, micro biopsies (1–2 mg) can be used, taking advantage of the greater sensitivity of EPR at higher frequencies (K- and Q-bands) [Romanuykha *et al.,* 2007; Gomez *et al.*, 2011; Trompier *et al.*, 2010]. To perform *in vivo* measurements, surface antennae and a low-frequency (0.3 to 2 GHz) spectrometer are used, making it possible to study dense samples which may contain a significant proportion of water without causing heating of the tissues. Sensitivity with this technique is lower than at X-band, with a limit of detection of around 1 to 2 Gy for *in vivo* measurements at L-band on molars or pre-molars, whereas it is around 0.5 Gy at Q-band with 2–3 mg of enamel. These two techniques have yet to be used in real cases of accidents.

They appear promising, but numerous technical and methodological obstacles remain to be overcome.

Other materials present on the victims or in their environment can be used in EPR dosimetry. For example, sucrose can be used [Nakajima, 1994; Hütt *et al.*, 1996; Kai *et al.*, 1990; Shirashi *et al.*, 2002] as can soda-lime glass from watches [Wu *et al.*, 1998]. Figure 1.9 shows the EPR spectra for irradiated sucrose and soda-lime glass samples, the latter was taken from a liquid-crystal screen from a mobile telephone [Trompier *et al.*, 2011].

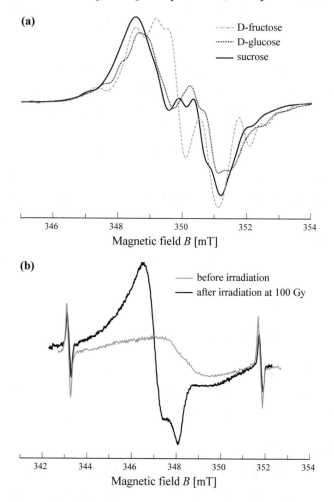

Figure 1.9 - X-band EPR spectra for various irradiated samples.
(a) Samples of sucrose and the two monosaccharides composing it were irradiated at 100 Gy and their spectra were recorded. **(b)** Soda-lime glass, before and after irradiation at 100 Gy. Spectra were recorded at room temperature at 2 mW.

A recent study identified up to six radio-induced radicals in sucrose [De Cooman *et al.*, 2008]. In the case of soda-lime glass, photons produce a signal which is generally attributed to $O^{•-}$ ions, but neutrons, α particles and photons at very high doses (greater than a kGy) produce E' ($O \equiv Si^{•}$) centres [Ikeya, 1993]. Sugars in general and some food additives (sorbitol, ascorbic acid, aspartame, etc.) can be used in EPR dosimetry [Hervé *et al.*, 2006]. Other potentially informative materials (polymers, cotton, wool, leather, table salt, nail fragments, hair clippings, etc.) have never been used in practice, either because they are not widely available, or because of unstable signals, or indeed because the relevant methods have yet to be validated. A complete description can be found in [Trompier *et al.* 2009a].

Among all these materials, nail fragments are by far the most interesting, particularly in cases of irradiation of the hands or feet, which is common during accidents. However, the method to estimate the received dose has not yet been established as the radio-induced signal is difficult to distinguish from signals induced by mechanical stress [Trompier *et al.*, 2009b].

In conclusion, EPR is frequently used to estimate the doses received during accidental exposure. The most frequently used materials are calcified tissues, sugars and glass. Other materials are regularly reviewed and particular efforts are being made to develop methods for use with common materials (glass from liquid-crystal screens and polymers from plastic cases, keyboards and screens from mobile phone), to allow EPR dosimetry to be used as a method to sort populations when major accidents occur.

1.4 - A reference method: Alanine/EPR dosimetry

In metrology for ionising radiation, several techniques are used to measure the absorbed dose: calorimetry, ionometry, thermoluminescence, chemical dosimetry. EPR dosimetry of alanine is one of these reference techniques.

1.4.1- Description of the method

Alanine is an amino acid of formula $H_2N-CH(CH_3)-COOH$. L–α–alanine in a polycrystalline state is used in dosimetry. Samples are conditioned as powders, films or pellets (figure 1.10).

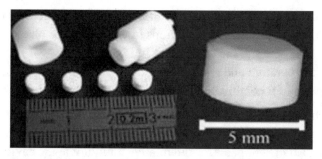

Figure 1.10 - Alanine dosimeter containing several pellets and its container.

Radiolysis of polycrystalline alanine is a complex process which leads to the formation of several stable free radicals which generate the EPR spectrum presented in figure 1.11.

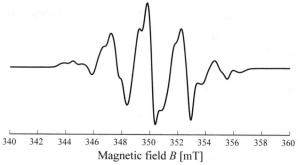

Magnetic field B [mT]

Figure 1.11 - X-band EPR spectrum obtained by radiolysis of alanine in the solid phase (powder). Experimental conditions: power: 2 mW, modulation: frequency 100 kHz, amplitude 0.3 mT.

This spectrum was simulated by superposing two major components attributed to the $H_3C-{}^\bullet CH-COOH$ (55 %) and $H_3C-{}^\bullet C(NH_3^+)-COO^-$ (35 %) radicals and a less-well-defined minor component [Hydari *et al.*, 2002; Malinen *et al.*, 2003]. However, it should be noted that spin trapping experiments (see chapter 3 in this volume) of radicals created by radiolysis of polycrystalline alanine only detected two species, in different proportions to those indicated above [Raffi *et al.*, 2008].

In dosimetry, the amplitude of the central peak on the spectrum is measured. The relation between this amplitude and the dose received by the sample depends on numerous factors that can be classified in two categories [Dolo *et al.*, 1998; Dolo *et al.*, 2005]:

▷ The first relates to the parameters influencing the physico-chemical reaction and the environmental conditions in which it takes place:
  - the dose rate and the duration of irradiation,
  - the temperature and relative humidity,
  - the rate of recombination of the radical species created.

▷ The second relates to the conditions in which the EPR spectrum is recorded, i.e., the acquisition parameters per se (power, modulation amplitude, etc.) but also the temperature and relative humidity in the cavity and, naturally, the mass of the sample.

In dose metrology, all these parameters are precisely defined. In particular, dosimeters must be designed to avoid radical mobility and water absorption. The quantitative control of the EPR spectrum for alanine has made it a reference method for the dosimetry of high doses (10^2 to 10^5 Gy) [Regulla *et al.*, 1982] but also for lower doses, such as those used in radiotherapy (2 to 50 Gy) [Garcia *et al.*, 2009]. It can be used to measure doses with an uncertainty of less than around 2 %.

1.4.2 - Research and development in radiotherapy

For a number of years, research into alanine-EPR dosimetry has focused on increasing the sensitivity of the method for the range of doses used in radiotherapy (2–50 Gy). Several axes of development are currently under investigation in metrology laboratories, in particular to improve the detector, the acquisition parameters and exploitation of EPR spectra [Garcia *et al.*, 2009].

Every year, around 200,000 patients in France and 1 million in Europe are treated by radiotherapy, whatever the technique. The new systems which emerged over the last few years use several beams of small dimension, the orientation of which varies during each treatment session. This is the case, for example with tomotherapy-type systems Cyberknife®, arctherapy and Gamma-Knife. As the irradiation conditions are very different to those defined in the reference radiotherapy protocols, specific metrological adaptations must be made. Alanine dosimeters can be very useful in this field thanks to their small size and their specificities (non-destructive measurements, integrating dosimeter affected neither by the dose rate nor by the energy) which make their use very flexible. They can thus be used to calibrate beams or to verify the dose delivered by irradiating the dosimeters in the conditions used to treat patients.

When using photons, the reference conditions for accelerator calibration are a 10×10 cm^2 irradiation area, with the dosimeter placed in a vat of water measuring $30 \times 30 \times 30$ cm^3 at 10 cm depth and at a distance of 1 m from the source of radiation. During a tomotherapy session, the source of radiation moves around the patient in a ring. The ring has a radius of 85 cm and the surface area has a maximal field of 40×5 cm^2. In this case, the reference protocol cannot be used. To calibrate this type of radiation source by alanine EPR, a series of dosimeters must first be irradiated in the reference conditions using a beam with a known dose rate. This first step can be used to establish a calibration curve, which gives the intensity of the EPR signal as a function of the dose. This curve is linear over the range of doses used in radiotherapy (figure 1.12).

Dosimeters are then placed on the tomotherapy apparatus, irradiated and analysed by EPR, and the received dose is deduced from the reference curve by applying corrections linked to the geometry of the irradiation device. In this case, it is essential to consider the area of the field, the source-detector distance, the depth in water. For the area of the field, numerical simulations are required. For the source-detector distance and the depth in water, the correction simply involves the $1/d^2$ variation and the value of the absorption coefficients, respectively [Perichon *et al.*, 2011].

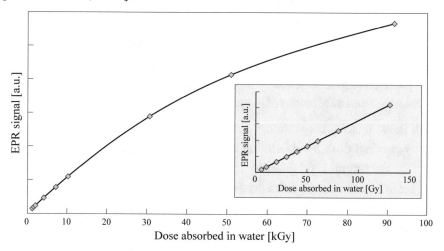

Figure 1.12 - Examples of alanine calibration curves for various dose ranges.

Electron beam radiotherapy is used to treat tumours at a low depth. In this case, EPR alanine dosimetry can also be used, on the condition that a calibration curve is established using an electron beam. Indeed, for the same irradiation dose in water, the EPR signal produced by an electron beam is somewhat weaker than that produced by a photon beam. The calibration curve is linear over the 8–22 MeV energy range used in radiotherapy.

Proton therapy is used to treat tumours located close to very sensitive organs, particularly in the head. Indeed, proton beams can destroy cancerous cells at a well-defined depth within the body without affecting neighbouring healthy cells. Alanine EPR dosimetry can also be used with this technique, but few studies have been devoted to this subject.

1.4.3 - Sterilisation by irradiation

Alanine EPR dosimetry is a reference technique for industrial irradiation [ISO / ASTM 51607, 2004]. Irradiation is used for sterilisation, to improve hygiene and to increase the shelf-life of a range of food and health products [IAEA, 1999]. This procedure generates low heat and provides better sterilisation than traditional techniques, some of which have been shown to be dangerous to consumers, such as fumigation with ethylene oxide. Irradiation is used in numerous activity sectors, and is associated with strict regulatory controls:

 ▷ medicine: sterilisation of single-use materials (syringes, catheters, prostheses), food for immunodepressed patients, or blood products,

 ▷ pharmacy: sterilisation of some innovative biotechnological and /or thermosensitive drugs, in particular amino-acid-based treatments,

 ▷ food and agriculture: disinfection of foods contaminated due to human manipulation (e.g. mechanically separated meats) or deinfestation of naturally contaminated substances (e.g. spices).

In all these sectors, the statutory authorisations indicate characteristic product-by-product dose levels aiming to guarantee a degree of sterility varying from the maximum level (sterility in the medical sense of the term) to simple "hygienic" measures. As these doses are in the range of a few tens of grays to several kilograys, alanine EPR dosimetry is particularly well adapted to their measurement.

1.5 - Identifying irradiated food products

The quality of irradiated food products is guaranteed by the requirements to ensure traceability of the received doses and labelling. The term retained for the use of this procedure is *ionisation*, and in Europe a logo must be displayed on any ionised product, even if only some of the ingredients were ionised (figure 1.13).

Figure 1.13 - Logo which must be present on all ionised food products.

In this context, the legislation has two main objectives. A standard authorises the treatment of food products by ionisation with, for each authorised product, an indication of the doses to be applied. Another standard authorises the ionised products and their circulation, with an obligation of clear labelling for consumer's information. These statutory obligations have been associated with control methods to combat fraud [Raffi, 2003]. The first objective is to detect irradiated foods first by applying a qualitative approach, then a quantitative approach which aims to assess whether the doses delivered conformed to the prescriptions. Several European standards have defined protocols for use with a range of products. EPR was recognised as appropriate for these requirements and it is used to this end for:

▷ meat and fish bones, by measuring radicals of hydroxyapatite,

▷ strawberries, pistachios and paprika, measuring the characteristic signal produced by cellulose,

▷ dried fruits (figs, papaya, raisins), measuring the signal from semiquinone radicals produced from crystallised sugars.

These protocols are reliable enough to simply and rapidly guarantee proof that the conditions of application were in line with the statutory prescriptions. They are often extended to other products, but only as part of routine verifications to detect unexpected product behaviour.

1.6 - Conclusion

Measuring the dose of ionising radiation absorbed by matter based on EPR quantification of paramagnetic species is used in a wide range of fields: archaeological dating, study of the effects of radiation on the body, management of radiological accidents, metrology, detection of irradiated food products. This variety of applications is possible thanks to the sensitivity of EPR, the wide range of exploitable signals and the ease with which their amplitude can be measured. *Alanine*/EPR dosimetry can be used to estimate doses over a very broad range, from a few tens of mGy to the MGy, it is non-destructive and can be applied to a wide range of materials. It thus offers many possibilities in dosimetry and recent technical developments in instrumentation should increase its use in this field.

References

ABBAZZI L. *et al.* (2000) *Journal of Archaeological Science* **27**: 1173-1186.

AINSBURY E.A. *et al.* (2011) *Radiation Protection Dosimetry* **147**: 573-592.

ALEXANDER G.A. *et al.* (2007) *Radiation Measurements* **42**: 972-996.

BISCHOFF J.L. & ROSENBAUER R.J. (1981) *Science* **213**: 1003-1005.

CLAIRAND I. *et al.* (2006) *Radiation Protection Dosimetry* **120**: 500-505.

COMBRISSON J. & UEBERSFELD J. (1954) *Comptes Rendus Hebdomadaires de l'Académie des Sciences* **238**: 572-574.

DE COOMAN H. *et al.* (2008) *Journal of Physical Chemistry B* **112**: 7298-7307.

DESPRIÉE J. *et al.* (2006) *Comptes Rendus Palevol* **5**: 821-828.

DOLO J.-M., RAFFI J. & PICCERELLE P. (2005) *Revue Française de Métrologie* **4**: 5-15.

DOLO J.-M., PICHOT E. & FEAUGAS V. (1998) *Applied Magnetic Resonance* **15**: 269-277.

DOLO J.-M. & FEAUGAS V. (2005) *Applied Radiation and Isotopes* **62**: 273-280.

DUVAL M. *et al.* (2009) *Radiation Measurements* **44**: 477-482.

FALGUÈRES C. & BAHAIN J.J. (2002) « La datation par résonance paramagnétique électronique (RPE) » *in Géologie de la Préhistoire, Méthodes, Techniques, Applications* Miskovsky, J.-C., ed., Géopré.

FALGUÈRES C. *et al.* (2008) *Quaternary Geochronology* **3**: 390-398.

FATTIBENE P. & CALLENS F. (2010) *Applied Radiation and Isotopes* **68**: 2033-2116.

GARCIA T. *et al.* (2009) *Radiation Physics and Chemistry* **78**: 782-790.

GASCOYNE M. (1982) « Geochemistry of the actinides and their daughters » in *Uranium series disequilibrium: applications to environmental problems*, Ivanovich M. and Harmon R.S. eds, Oxford University Press.

GÓMEZ J.A. *et al.* (2011) *Radiation Measurements* **46**: 754-759.

GORDY W., ARD W. & SHIELDS H. (1955) *Proceedings of the National Academy of Sciences of the USA* **41**: 983-996.

GRÜN R., SCHWARCZ H.P. & CHADAM J. (1988) *Nuclear Tracks* **14**: 237-241.

GRÜN R. (2006) *Yearbook of Physical Anthropology* **49**: 2-48.

HERVE ML. *et al.* (2006) *Radiation Protection Dosimetry* **120**: 205-209.

HUET C. *et al.* (2007) *Radioprotection* **42**: 489-500.

HÜTT G., BRODSKI L. & POLYAKOV V. (1996) *Applied Radiation and Isotopes* **47**: 1329-1334.

HYDARI M. *et al.* (2002) *Journal of Physical Chemistry A* **106**: 8971-8977.

IAEA (International Atomic Energy Agency) (1999) Proceedings of the IAEA International Symposium on Techniques for High Dose Dosimetry in Industry, Agriculture and Medecine, IAEA TECDOC 1070, Vienna.

IAEA (International Atomic Energy Agency) (2002) IAEA TECDOC 1331, Vienna.

IKEYA M. (1982) *Japanese Journal of Applied Physics* **22**: 763-765.

IKEYA M. (1993) *New Applications of Electron Spin Resonance-Dating, Dosimetry and Microscopy*. World Scientific, Singapore.

ISO/ASTM (2004) n°51607.

KAI A., IKEYA M. & MIKI T. (1990) *Radiation Protection Dosimetry* **34**: 307-310.

MALINEN E. *et al.* (2003) *Radiation Research* **159**: 23-32.

NAKAJIMA T. (1994) *Applied Radiation and Isotopes* **45**: 113-120.

PÉRICHON N. *et al.* (2011) *Medical Physics* **38**: 1168-1177.

RAFFI J. *et al.* (2008) *Spectrochimica Acta Part A*, **69**: 904-910.

RAFFI J. (2003) *Techniques de l'ingénieur*, IN 12.

REGULLA D.F. & DEFFNER U. (1982) *International Journal of Applied Radiation and Isotopes* **33**: 1101-1114.

ROMANYUKHA A.A., WIESER A. & REGULLA D.F. (1996) *Radiation Protection Dosimetry* **65**: 389-392.

ROMANYUKHA A. *et al.* (2007) *Health Physics* **93**: 631-635.

SHIRAISHI K. *et al.* (2002) *Advances in ESR Applications* **18**: 207-209.

SIMON S.L. *et al.* (2007) *Radiation Measurements* **42**: 948-971.

SWARTZ H.M. *et al.* (2006) *Radiation Protection Dosimetry* **120**: 163-170.

TROMPIER F. *et al.* (2006) *Radiation Protection Dosimetry* **120**: 191-196.

TROMPIER F. *et al.* (2007) *Radiation Measurements* **42**: 1025-1028.

TROMPIER F. *et al.* (2009) (a) *Annal of Istuto Superiore di Sanità* **45**: 287-296.

TROMPIER F. *et al.* (2009) (b) *Radiation Measurements* **44**: 6-10.

TROMPIER F. *et al.* (2010) *Radiation Protection Dosimetry* **144**: 571-574.

TROMPIER F. *et al.* (2011) *Radiation Measurements* **46**: 827-831.

VANHAELEWYN, G.C.A.M. *et al.* (2002) *Radiation Research* **158**: 615-625.

VOINCHET P. *et al.* (2004) *Quaternaire* **15**: 135-141.

YOKOYAMA Y., FALGUÈRES C. & QUAEGEBEUR J.P. (1985) *Nuclear Tracks* **10**: 921-928.

WIESER A. *et al.* (2005) *Applied Radiation and Isotopes* **62**: 163-171.

WU K. *et al.* (1998) *Radiation Protection Dosimetry* **77**: 65-67.

ZDRAVKOVA M. *et al.* (2004) *Physics in Medicine and Biology* **49**: 2891-2898.

Chapter 2

Tracing natural organic matter at the scale of drainage basins

Pépin-Donat B. [a], Poulenard J. [e], Blondel T. [c], Lombard C. [a], Protière M. [a], Dudal Y. [d], Perrette Y. [e], Fanget B. [e], Miège C. [f], Delannoy J.-J. [e], Dorioz J.-M. [b], Emblanch C. [c], Arnaud F. [e], Giguet-Covex C. [e]

[a] *UMR SPrAM (CEA-CNRS-UJF), INAC, CEA-Grenoble, Grenoble.*

[b] *UMR CARRTEL, (University of Savoie-INRA), Le Bourget-du-Lac.*

[c] *UMR EMMAH, Laboratory of Hydrogeology of Avignon (INRA-University of Avignon), Avignon.*

[d] *UMR BSR (INRA-SupAgro), Montpellier.*

[e] *UMR EDYTEM (CNRS-University of Savoie), Le Bourget-du-Lac.*

[f] *UR QELY (IRSTEA), Cemagref, Lyon.*

2.1 - Introduction

Organic matter from plants and animals is transformed over time by two types of reaction: mineralisation and humification. These processes produce antagonistic effects: production of greenhouse gases for the first [Lugo and Brown, 1993; Raich and Potter, 1995], and synthesis of complex organic systems [Kelleher and Simpson, 2006] allowing carbon storage on Earth for the second [Lal and Bruce,1999]. Complex organic systems resulting from humification make up what is commonly known as soil Natural Organic Matter (NOM) which is the *source* of terrestrial NOM. NOM has an extremely complex structure, it is made up of a wide variety of more or less aromatic molecules and macromolecules assembled in supramolecular architectures. To further add to the complexity, the structural properties of NOM depend on numerous parameters including,

© Springer Nature Switzerland AG 2020
P. Bertrand, *Electron Paramagnetic Resonance Spectroscopy*,
https://doi.org/10.1007/978-3-030-39668-8_2

among others, the nature of the waste from which it is produced, the mineral compounds making up the rock on which it undergoes decomposition (lithological factors), the climatic conditions, the pH conditions of the medium in which it is found, conditions linked to human activity, its depth of burial. In other words, not only is the structure of NOM extremely complex, in addition it depends on the location of its sampling site, on the surface or at a depth. In practice, we can consider that there are as many structures of organic matter as there are samples collected!

It is nevertheless crucial to characterise the flow and reactivity of natural organic matter. Indeed, NOM is the largest carbon reservoir on Earth, with a mass estimated at 1500×10^9 tons [Batjes, 1996]. It plays a major role in the nutrient cycle within terrestrial ecosystems and between terrestrial and aquatic ecosystems due to its partial solubility. It can interact with numerous toxic organic and mineral substances [Jerzykiewicz, 2004]. Its supramolecular structure, linked to its amphiphilic nature, could explain some critical soil degradation [Poulenard *et al.*, 2004; Szajdak *et al.*, 2003]. To manage ecosystems, it is therefore necessary to better understand the dynamics and reactivity of NOM and how they depend on climate, plant coverage, hydrodynamic regimes, lithology and human activity. To be complete, NOM should be studied at various scales, from the molecular to the scale of the Earth as a whole. However, as drainage basins represent a scale covering both terrestrial life and human management, we chose to work at this intermediate scale.

As the structure of NOM is highly complex and variable it is absolutely impossible to determine it, and because of this, reactive properties and flow must be studied without reference to its structure. This type of study would be impossible with classical methods. However, NOM can be studied using "tracing" methods. These methods do not attempt to precisely determine the structures of the organic systems studied but use a fingerprint linked to a tracer to identify major sub-structural types present in the overall system. The tracers exploited must be ubiquitous, stable over time and space, capable of distinguishing between relatively similar sub-structures (i.e., be discriminative), and easy to detect. The markers currently used do not combine all these characteristics. They can be classified in two major categories: "analytical" markers (atoms or molecules), these are very discriminative but impossible to monitor over large space and time scales [Peters, 1993; Simoneit, 2002; Jacob *et al.*, 2008; Blyth *et al.*, 2008] and spectroscopic markers (IR, UV-Vis., 3D fluorescence), which are easy to track over large spatial scales but are not very discriminative [Stedmon *et al.*, 2003; Sierra *et al.*, 2005]. Existing tracing methods therefore cannot be used to study the structure, reactivity and dynamics of NOM over extended spatial and time scales.

EPR tracing involves the use of new tracers: semiquinone-type free radicals. These radicals are generated and trapped within aromatic structures in NOM during the humification process. This process mainly consists in oxidation of lignin followed by a loss of aliphatic and carboxylic groups to produce polyphenol units. These units are subsequently oxidised to produce quinone units, passing through semiquinone intermediates. Semiquinone free radicals are ubiquitous, stable and readily detectable by EPR. They have been extensively used since the 1960s to characterise the degree of humification of NOM [Steelink and Tollin, 1967; Senesi, 1990; Nickel-Pépin-Donat *et al.*,1990; Yabuta *et al.*, 2005; Saab and Martin-Neto, 2008]. Schnitzer suggested that they could be used as tracers as early as 1972 [Schnitzer and Khan, 1972], but the idea was never followed through or implemented.

In this chapter, we first show that the EPR spectrum of semiquinone radicals differentiate between NOM from neighbouring soils and even from different levels within the same soil. We then show that tracing of these radicals by EPR can be used to monitor the transfer of NOM from soils in a drainage basin to its hydrological system, from this system to its spillway, and even down to their natural recorders such as sediments and speleothems (stalagmites). We tested the efficacy of this method on three clearly differentiated drainage basins and compared the results to those given by independent studies applying other techniques. The results show that this tracing method has great potential as a tool to help with management of drainage basins. Numerous applications can now be imagined, and as an example we show how it can help to determine the catchment area and the residence time for drinking water in karst formations.

2.2 - Transformation and transfer of natural organic matter

On Earth, the different forms of natural organic matter can be classified as a function of their ages:

    ▷ recent NOM, present in soils and spring waters. Their age varies between 1 year and a few thousand years,

▷ NOM from natural recorders such as sediments, peat, stalagmites, ice. Their age ranges from several hundred years to several hundred thousand years,

▷ fossil NOM in coal and petrol, which is between a few hundred thousand and several billion years old,

▷ extraterrestrial NOM that is found in meteorites.

In this chapter we are interested in the first two categories. EPR also provides information on the structural and reactive properties of fossil NOM [Nickel-Pépin-Donat *et al.*, 1990; Pépin-Donat and Brunel, 1994] and on extraterrestrial organic matter. Chapter 7 in this volume describes how EPR contributes to the study of primitive, terrestrial and extraterrestrial, carbonaceous matter in the context of the search for the origins of life.

The variety of NOM precursors, which depends on the type of vegetation and the presence of agricultural or pastureland activities, results in diverse ecosystems (figure 2.1a). The progressive incorporation of these precursors into soils and their interaction with the products of the alteration of rocks lead to the formation of layers of NOM which are more or less parallel to the surface, known as *soil horizons*. Each horizon can be considered to present a certain homogeneity in terms of chemical and physical properties (density, pH, ionic conduction, colour, etc.) and the state of transformation of the NOM (figure 2.1b). NOM can be mobilised by the water circulating in drainage basins, in particulate form (alone or in organo-metallic complexes), in colloidal form, or as solutes. NOM from different sources (various precursors, different horizons) circulate in drainage basins where they are mixed together (figure 2.1c). In some conditions, NOM can accumulate in datable "sedimentary archives" (lake sediments, stalagmites) (figure 2.1d).

Given the complexity of the structures and reactivity of NOM, it initially appears extremely difficult to reconstitute their transfer. However, we will see that EPR tracing efficiently contributes to overcoming this hurdle and provides information on a certain number of basic processes:

▷ the dynamics of NOM transformation in soils, which has significant consequences on the mechanisms through which organic carbon is stored,

▷ the dynamics of transfer into water. This would make it possible, for example, to determine the origin of NOM which can alter the water's quality,

▷ the formation of paleo-environments and paleo-ecosystems, by revealing the origin of NOM archived in sediments and/or stalagmites.

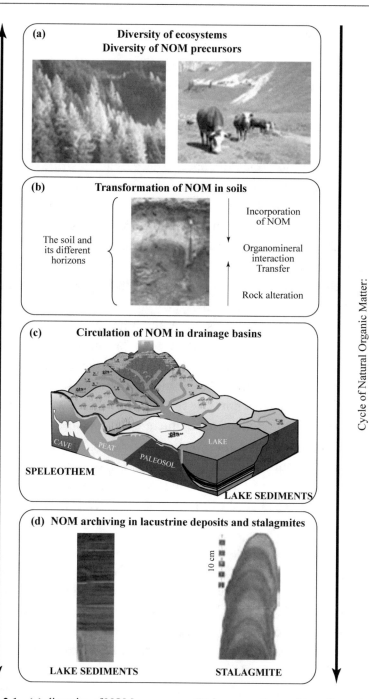

Figure 2.1 - **(a)** diversity of NOM precursors, **(b)** incorporation and transformation of NOM precursors in soil horizons, **(c)** circulation of NOM from soils to water, **(d)** archiving NOM in natural recorders, sediments, peat, ice, stalagmite (represented here).

2.3 - EPR signature of natural organic matter

The EPR spectrum recorded at room temperature for a sample of NOM results from the superposition of the signals produced by its semiquinone radicals. It depends on the nature, organisation and dynamics of the aromatic entities present in complex organic structures. When the sample contains only one type of these structures, its spectrum contains a single line which can be characterised by four parameters: its position, defined by the g factor; its shape, defined by the Lorentzian percentage (% L) of a linear combination of Lorentzian and Gaussian; its peak-to-peak width ΔB_{pp} expressed in mT; and its concentration [I] in number of spins per unit of organic carbon mass. The g_{iso} factor for semiquinone radicals is between 2.0028 and 2.0055 [Riffaldi and Schnitzer, 1972; Barančíková *et al.*, 1997; Yabuta *et al.*, 2005] and its value is sensitive to the chemical structure into which the electron is delocalised. For example, the g_{iso} factors for the radicals which have notable spin density on oxygen, nitrogen or sulfur atoms are higher than those of radicals where the spin density is mainly delocalised on carbon atoms [Volume 1, section 4.2.2]. The shape of the line, its width and the spin concentration are also sensitive to the environment of the unpaired electron, but not in a simple manner. Additional experiments, such as those described in chapter 7 for the investigation of primitive carbonaceous matter, would be necessary to obtain detailed information on the structure and dynamics of these organic structures. But, for the needs of the application described in this chapter, we can consider that the set of parameters (g, % L, ΔB_{pp}, [I]) constitutes a characteristic *signature* of the nature and dynamics of the complex organic matter in which the electrons are trapped. As the EPR lines of all the samples of NOM sampled in this study have a *Lorentzian* shape, the (% L) parameter will not be used here.

When the sample contains several types of complex organic matter, the spectrum is a superposition of the components corresponding to the different types. The characteristic parameters can be readily determined by simulating the experimental spectrum using a sum of "pure" components. It is possible to determine the absolute number of spins [I_i] corresponding to each component, but this calculation is only justified when it is useful to quantify the masses of the different types of NOM. This is not the case in the methodological study presented here, and we will therefore neglect this parameter. In contrast, the *relative intensities* of the different components of the spectrum, which are very easy to determine, are very important in the context of this study aiming to simply visualise *variations in composition* of the NOM during its transformation and transfer. We will see that it is convenient to use a "barcode" in which each component is represented by a coloured bar, the height of which is proportional to its relative intensity.

2.4 - Study areas and protocol applied

2.4.1 - Description of the drainage basins studied

In figure 2.2 we have indicated the sites of the three drainage basins which we used to test the EPR tracing method, as well as the Fontaine de Vaucluse basin which was the object of a specific application of this method.

▷ The study of the Mercube river drainage basin **(a)** was used to test tracing of NOM from soils (sources of NOM) down to superficial waters, within the water network and finally to the basin's spillway. This small river drains a 302-ha basin and flows into lake Geneva close to Yvoire in Haute-Savoie. The basin has gentle slopes. Part of the basin is characterised by the presence of clay-rich soils and a high-altitude layer where the water stagnates at the surface during the winter. This zone, which is difficult to exploit agriculturally is occupied by a forest (123 ha). The remainder of the basin is composed of a one-metre deep, well drained silt-rich soil. These soils are used to produce wheat and corn (120 ha) or as grassland (50 ha).

▷ The **(b)** Coulmes/Choranche (Isère) and **(c)** Anterne (Haute-Savoie) sites were studied to test the persistance of EPR signatures and their relevance for NOM tracing from soils to waters, and from waters to natural recorders (speleothems and lake sediments) (figures 2.1c and 2.1d). The site of Coulmes/Choranche corresponds to a karst system in the Vercors mountain range. The Coulmes plateau, at an altitude of between 800 and 1 400 m, is covered with a gradient of vegetation from a lower-altitude forest of oaks and box trees to a mountain forest of spruce and fir. This environment has developed on hard limestone rocks. The dominant soils are characterised by the organic matter accumulated on the exposed hard limestone. Brown forest soils develop in dolines where clay has accumulated. The underground karst rivers fed by this plateau flow into the Choranche caves. The site of Anterne, in Sixt Nature Reserve (Haute-Savoie), corresponds to a small subalpine drainage basin (2.5 km^2) which feeds into a lake located at an altitude of 2 063 m. This drainage basin is currently covered by alpine grasslands of graminaea which is mainly developed on brown soils. However, on very steep slopes, little-evolved soils are found on schist and dark earth. A few peaty zones are present around the lake.

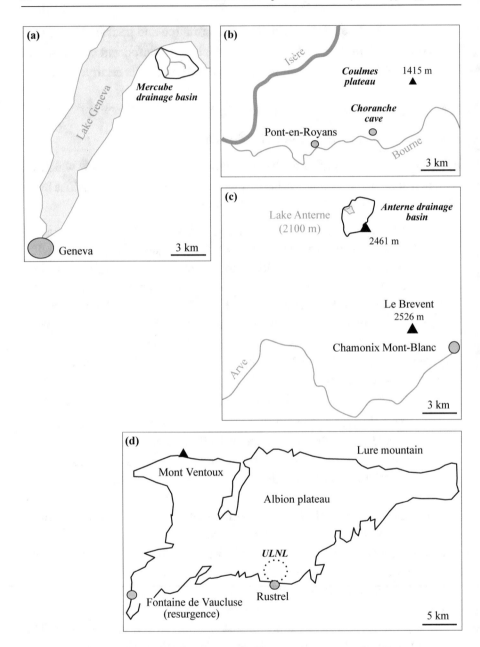

Figure 2.2 - Localisation of the different basins and sites studied:
(a) "Mercube" basin, (b) Coulmes/Choranche basin, (c) Anterne basin,
(d) Fontaine de Vaucluse basin. The localisation of the Underground Low-noise
laboratory (ULNL) where sampling was performed for this basin is shown.

▷ Fontaine de Vaucluse is one of the most extensive European karstic sources, with a mean flow-rate of 23 m^3 s^{-1}. It is the main egress for a 1500 m
thick limestone formation which extends over 1 115 km^2. We studied
the Rustrel site **(d)** which is part of its catchment area. It is located in a
Mediterranean karstic environment. This study aimed to test EPR tracing
of NOM from soils to subterranean water and to demonstrate its application
to determine the catchment area and the residence time for drinking water
in karst formations. The sampling site is shown in figure 2.6. The Rustrel-
Apt area underground low-noise laboratory (ULNL) provides privileged
access to several permanent flows from the "unsaturated zone" (where not
all the empty spaces in the karst are filled with water) for the Fontaine de
Vaucluse hydrosystem. The ULNL is composed of subterranean galleries
which randomly split up the karstic network. For this study, a permanent
flow located at 440 m under the surface (figure 2.6a) was monitored between December 2006 and December 2007. In this area, only two types of
organic matter sources are present at the surface [Emblanch *et al.*, 1998]:
sloping soil covered by disperse shrub-based (S1a) and grassy (S1b) vegetation, and soil that develops on the plateau (S2), which is covered by a
forest of holm oaks.

2.4.2 - Sample preparation

Soil samples from all of the areas studied were sampled by a soil survey performed on the four basins. Their preparation was limited to grinding and sieving
to 2 mm.

Sediments from the Anterne basin were collected using sediment traps installed
in the rivers and at the bottom of the lake at 13 m depth. Sediments were also
extracted from a core of lacustrine deposits. Here, we present the results for
deposits sampled at a core-depth of 28 – 29 cm, dating from around 1950.

In the Choranche caves, the sub-current (growth later than 1950) "stalagmitic
floor" was sampled in the subterranean river. This is a carbonaceous formation
similar to stalagmites but which constitute a surface layer which has formed
at the bottom of the cave.

Aqueous extracts from soils of the karstic Coulmes plateau were prepared
in the laboratory using a soil/solution ratio of 1/10 and orbital agitation at
300 revolutions per minute for 4 hours. They were filtered at 0.2 μm.

Waters from rivers were collected in the forest and agricultural sub-basins of the Mercube drainage basin during a winter low-flow period. Samples were also collected at the level of the spillway into lake Geneva.

Monthly sampling of the subterranean flow in the ULNL gallery was performed over the 2006–2007 hydrological cycle. All samples of natural water were gently evaporated at 40 °C to produce a dry residue (around 50 L filtered at 0.45 μm).

2.4.3 - Recording and simulation of EPR spectra

Solid samples (a few milligrammes) were placed in 3-mm diameter quartz tubes. Aqueous solutions were injected into 0.2-mm diameter quartz capillary tubes. EPR spectra were recorded at room temperature on an X-band EMX Bruker spectrometer, with a non-saturating power of 20 mW. The magnetic field and microwave frequency were measured independently to determine the g factor with a precision of $\pm$ 0.0002. The number of components present in the spectrum and the parameters characterising them (g, ΔB_{pp}, $[I]$, see section 2.3) were determined by simulating the part of the spectrum corresponding to semiquinone-type radicals thanks to a specifically-developed programme. We then determined the barcodes by attributing a height proportional to its relative intensity to each component.

2.5 - Assessing the EPR tracing method

In this section, we describe the studies which demonstrated the potential of the EPR tracing method. To better appreciate their field of application, they were performed on 3 very different drainage basins.

2.5.1 - EPR signatures can distinguish between different soil types and between horizons in a single soil

The EPR spectra for soil samples collected in agricultural and forest areas of the Mercube drainage basin are represented in figure 2.3. Figures 2.3a and 2.3b, respectively, show the spectra for the Bw horizon in agricultural soil and for the O horizon in forest soil recorded over around 60 mT. Figure 2.3c zooms in on the $g \sim 2$ region of the spectrum which corresponds to the semiquinone radicals of the Bw horizon and shows its simulation. Figure 2.3d shows how the spectrum for NOM in forest soil changes as a function of depth of sampling. The intensity of a broad line α specific to the surface horizon can be observed to decrease with depth, whereas several narrower lines β and γ appear. Similarly, for agricultural soil, the intensity of the broad line ε specific for the Ap horizon (figure 2.4a) is considerably decreased in the Bw horizon, but a narrower line φ appears (figure 2.3c).

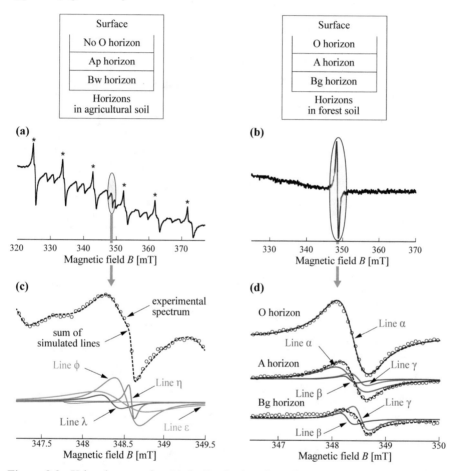

Figure 2.3 - X-band spectra for: **(a)** the Bw horizon in agricultural soil (asterisks indicate lines attributed to Mn^{2+} ions), **(b)** the O horizon for forest soil, recorded over 60 mT, the $g \sim 2$ region characteristic of NOM is presented in detail in **(c)** for the Bw horizon of agricultural soil and in **(d)** for the O, A and Bg horizons of forest soil. The experimental spectra (circles), their simulation (black dashes) and their various components are represented. The parameters used in the simulation and associated barcodes are indicated in table 2.1. Given the value of their g factor, some signals which appear during the simulation cannot be attributed to the NOM. They are therefore not listed.

This *narrowing* of the lines as we move away from the surface of the soil is attributed to *maturation* of the NOM [Barančíková *et al.*, 1997], in agreement with data published on the evolution of NOM in soils [Schulten and Schnitzer, 1997; Schulten and Leinweber, 2000; Leenheer, 2007]. The parameters used to simulate the spectra and associated barcodes can be found in table 2.1.

Table 2.1 - Simulation parameters and barcodes for soils and waters for the Mercube drainage basin.

	Line	g_{iso} factor	Linewidth ΔB_{pp} [mT]	Relative intensity [%]	Quantification of EPR signatures: «barcodes»
100% *Overall signature for MERCUBE drainage basin* α β γ ε φ η λ δ					
Soils and their EPR signature					
Forest «gleyic cambisol» α β γ ε φ η λ δ	O	α	2.0048	0.59	100	
	A	α	2.0048	0.59	82	
		β	2.0053	0.28	5	
		γ	2.0042	0.31	13	
	Bg	β	2.0053	0.28	27	
		γ	2.0042	0.31	73	
Agricultural «cambisol» α β γ ε φ η λ δ	Ap	ε	2.0036	0.64	98	
		φ	2.0032	0.31	1.7	
		η	2.0029	0.08	0.3	
	Bw	ε	2.0036	0.64	57.9	
		φ	2.0032	0.31	34.5	
		η	2.0029	0.08	1.5	
		λ	2.0043	0.25	6.1	
Waters and their EPR signature					
Forest sub-drainage basin α β γ ε φ η λ δ		α	2.0048	0.59	48	
		β	2.0053	0.28	7.5	
		δ	2.0036	0.51	44.5	
Agricultural sub-drainage basin α β γ ε φ η λ δ		ε	2.0036	0.64	98	
		φ	2.0032	0.31	2	
Spillway for the entire drainage basin α β γ ε φ η λ δ		α	2.0048	0.59	7.1	
		β	2.0053	0.28	44.1	
		φ	2.0032	0.31	19	
		δ	2.0036	0.51	29.8	

The capacity of the EPR signatures to distinguish between different soil types and the horizons in a single soil will be confirmed by studies performed on the three other sites. EPR can thus be used for a detailed study of NOM transfers within drainage basins.

2.5.2 - EPR signatures can be used to monitor the transfer of NOM of various origins to hydrographic networks

Analysis of the spectra in figure 2.4 demonstrates the capacity of EPR tracing to monitor the transfer of NOM from soils to the hydrographic network down to the spillway for the Mercube drainage basin. Figure 2.4a shows spectra for type A horizons for the soils from agricultural and forest sub-basins, and figure 2.4b shows those for the dry residues from water collected in these two sub-basins and at the level of the spillway into lake Geneva.

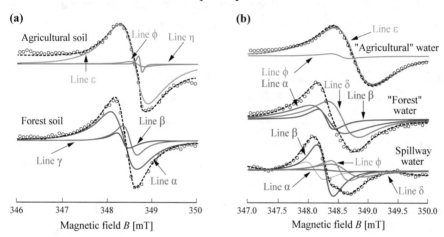

Figure 2.4 - EPR spectra recorded in the $g \sim 2$ region characteristic of NOM.
(a) Ap horizon for agricultural soil and A horizon for forest soil, **(b)** dry residue from water sampled in the agricultural and forest sub-basins and at the level of the spillway into lake Geneva. The experimental spectra (circles), their simulation (black dashes) and their various components are represented. The parameters used in the simulation and associated barcodes are indicated in table 2.1. Given the value of their g factor, some signals which appear during the simulation cannot be attributed to NOM. They are therefore not listed.

Comparison of these spectra leads to the following conclusions:

▷ Spectra for water samples collected in the forest and agricultural sub-basins include some of the lines detected in the corresponding soils: lines ε and ϕ characteristic of the agricultural soil are found in "agricultural" waters and lines α and β characteristic of the forest soil are detected in that for "forest" waters. All four lines are also found in water from the spillway into lake Geneva.

▷ Line γ characteristic of the forest soil is absent from "forest" water: it is said not to be *water available*.

▷ Conversely, on the spectrum for the water from the forest sub-basin and the spillway a line δ is observed which is not detected in any of the soils sampled. This may be a result of non-exhaustive sampling of the soils, the presence of a native source of NOM in the waters, or transformation of a type of NOM in the soil during its transfer.

The presence of characteristic lines specific to the forest and agricultural sub-basins in the spectrum of the waters collected from the spillway is coherent with the results of independent studies [Jordan-Meille *et al.*, 1998; Jordan-Meille and Dorioz, 2004]. It reveals the power of the EPR tracing method to simply and effectively study the transfer of the major types of NOM present in a drainage basin.

2.5.3 - EPR signatures can be used to trace transfer of NOM from soils to natural recorders

The studies carried out on the sites of lake Anterne and Coulmes/Choranche confirmed that EPR signatures can distinguish between different types of soil; it is thus possible to determine the origin of the NOM present in natural recorders: sediments and speleothems.

Anterne basin. We traced the NOM from the soils to the river and lake sediments. Figure 2.5a shows examples of spectra and their simulations, and all the results are presented in table 2.2. The main conclusions are as follows:

▷ A very narrow line σ is detected in all samples, whether soils or river or lake sediments. This line reveals the presence of an extensively matured NOM [Senesi and Steelink, 1989], probably derived from the mature organic matter trapped in sedimentary rocks.

▷ In the case of river deposits, in addition to line σ , a broader line, ν, is observed. This line is only found in the spectrum of little-evolved soils (lithosols), bare soil or under sparse vegetation, which are easily eroded. The presence of NOM from these soils in river sediments is therefore not unexpected.

▷ In the case of samples collected in sediment-traps placed at the bottom of the lake or in cores of lacustrine deposits, line ν is replaced by another line, κ. This line is detected in the spectrum for hydromorphic brown soils, which are water-logged for part of the year and are abundant on riverbanks. Riverbank sapping phenomena and regressive erosion of these formations are effectively observed in the basin.

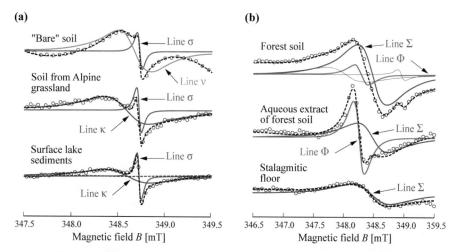

Figure 2.5 - Experimental spectra (circles), simulations (black dashes)
and different components for **(a)** Anterne basin: bare soil, brown soils of Alpine
grassland and surface sediments from lake Anterne, **(b)** brown forest soil from
the Coulmes plateau, its aqueous extract, stalagmitic floor from Choranche cave.
The simulation parameters and associated barcodes are indicated in table 2.2.

EPR tracing therefore reveals that the NOM for river and lake sediments are
derived from different soils: bare and little-evolved soils for river sediments,
more evolved soils from the banks of waterways for lake sediments. Sediments
from bare soils do not appear to accumulate in deltas, which explains why
they contribute little to the sediments in the centre of the lake. This result is
in agreement with the conclusion of a recent geochemical study (mineral and
organic) comparing current soils to lake sediments from the holocene period
(last 12,000 years) [Giguet-Covex *et al.*, 2011].

Choranche karst basin. The spectrum produced by the sample collected in the
stalagmitic floor in the Choranche caves consists of a single line Σ (figure 2.5b).
The same signature is found in the spectrum produced by forest soils from the
Coulmes plateau which covers these caves, as well as that of the aqueous extract
prepared in the laboratory. However, this line is completely absent from the
spectrum produced by organic soils developed on limestone and their aqueous
extracts. These results indicate that the main source of NOM in the stalagmitic
floor is brown soil. Without anticipating the significance of these results for
our understanding of how drainage basins function, a study of which will be
published later, we can already observe that EPR can be used to monitor NOM
from soils to sediments and from soils to speleothems, through water-based

transfer. This conclusion shows the potential of EPR tracing to monitor NOM in paleo-environments.

Table 2.2 - Simulation parameters and barcodes for waters and sediments from the Anterne and Choranche drainage basins.

Overall signature for main types of NOM detected at ANTERNE — 100% (κ σ φ ν)

Soils and their EPR signatures	Line	g_{iso} factor	Linewidth ΔB_{pp} [mT]	Relative intensity [%]	Quantification of EPR signatures: «barcodes»
«Alpine grassland sol» : Gleyic cambisol	κ	2.0034	0.47	81	
(κ σ φ ν)	σ	2.0029	0.05	19	(κ σ φ ν)
«Peat sol» : Histosol	φ	2.0041	0.58	99.8	
(κ σ φ ν)	σ	2.0029	0.05	0.2	(κ σ φ ν)
«Bare sol» : Regosol	ν	2.0032	0.35	94	
(κ σ φ ν)	σ	2.0029	0.05	6	(κ σ φ ν)
Sediments and their EPR signatures					
Core sediments	κ	2.0034	0.47	96	
(κ σ φ ν)	σ	2.0029	0.05	4	(κ σ φ ν)
Sediments trapped in the lake	κ	2.0034	0.47	96	
(κ σ φ ν)	σ	2.0029	0.05	4	(κ σ φ ν)
Sediments trapped in the flow	ν	2.0032	0.35	90	
(κ σ φ ν)	σ	2.0029	0.05	10	(κ σ φ ν)

Overall signature for main types of NOM detected at CHORANCHE — 100% (Φ Σ Ψ Θ)

Soils and their EPR signatures	Line	g_{iso} factor	Linewidth ΔB_{pp} [mT]	Relative intensity [%]	Quantification of EPR signatures: «barcode»
«Grassland cambisol»	Ψ	2.0030	0.45	93	
(Φ Σ Ψ Θ)	Θ	2.0049	0.29	7	(Φ Σ Ψ Θ)
Water extracted from the «grassland cambisol»	Ψ	2.0030	0.45	39	
(Φ Σ Ψ Θ)	Θ	2.0049	0.29	61	(Φ Σ Ψ Θ)
«Forest cambisol»	Φ	2.0048	0.18	7	
(Φ Σ Ψ Θ)	Σ	2.0033	0.5	93	(Φ Σ Ψ Θ)
Water extracted from the «forest cambisol»	Φ	2.0048	0.18	25	
(Φ Σ Ψ Θ)	Σ	2.0033	0.5	75	(Φ Σ Ψ Θ)
Stalagmitic floor	Σ	2.0033	0.5	100	
(Φ Σ Ψ Θ)					(Φ Σ Ψ Θ)

2.6 - Example of application of EPR tracing: identification of the inflow basin and the residence time for water in a karst formation

The aim of this study was to use EPR tracing to determine the inflow zone and the residence time for drinking water from the karstic site of Fontaine de Vaucluse, and to compare the results to those given by other methods which are much more difficult to implement.

As indicated in section 2.4.1, between December 2006 and December 2007 we collected two types of soil samples, S1 and S2, which exist on this site, and water which flows at 440 m underground in the ULNL laboratory. The EPR spectrum produced by a sample of water was compared to that obtained for soils S1 and S2 in figures 2.6b and 2.6c. The simulation parameters and associated barcodes are indicated in table 2.3. Analysis of all these results leads to the following conclusions:

▷ The EPR spectrum for water is very similar to that for soil S1 (the spectra for soils S1a and S1b are identical) and very different from that of soil S2. More precisely, the simulation shows that the spectrum for water only has a single line ω due to NOM; the spectrum for S1 includes the line ω and another line π, whereas the spectrum for S2 contains only line π. We can deduce that lithosol 1 belongs to the catchment basin for the flow studied. This result is entirely coherent with the external structure of the network of faults (figure 2.6a).

▷ Absence of the line π from the water spectrum shows that it is due to a type of NOM which has a very low potential for transfer to water. This low transferability may reflect low solubility, or strong adsorption or biodegradation.

▷ During 2007, the intensity of the line ω in the spectrum for soil S1 passes through a very marked maximum in the month of April (figure 2.6e). This peak of NOM is due to the heavy precipitation which occurred during that particular month (figure 2.6d), but it was only observed in the flow of subterranean water when significant infiltration occurred in December 2007, i.e., *8 months later* (figure 2.6f). This result allows us to propose a transit time for water within this karstic system of around eight months, which is in perfect agreement with the results of a fluorescence study shown in figure 2.6g [Blondel *et al.,* 2010]. It should be noted that the hypothesis of *instantaneous infiltration* of meteoritic waters towards the outlet in December 2007 is not compatible with the presence of magnesium at relatively high concentrations (4.14 mg L^{-1}) and very low concentrations of total organic carbon (0.8 mg L^{-1}) in samples collected at this time [Batiot *et al.,* 2003].

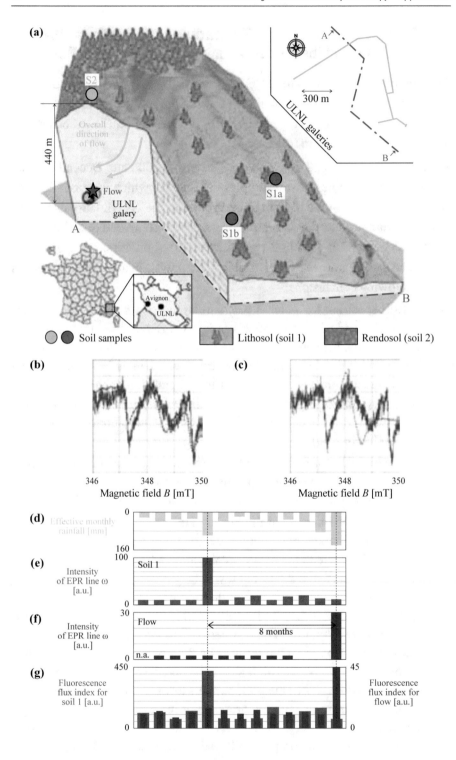

Figure 2.6 - (a) Localisation of the ULNL laboratory and its network of faults, **(b)** comparison of EPR spectra for water flow (continuous line) and S1 soil (dashed line), **(c)** comparison with spectra for S2 soil (dashed line). The following figures show the monthly variations for different parameters between December 2006 and December 2007, **(d)** rainfall, **(e)** intensity of line ω for soil S1, **(f)** intensity of line ω for water, **(g)** fluorescence index for soil S1 (red) and outflow of water (purple). Simulation parameters and associated barcodes can be found in table 2.3. [From Blondel *et al.*, 2010]

Table 2.3 - Simulation parameters and barcodes
for Fontaine de Vaucluse soils and waters.

100% ω π	Overall signature for main types of NOM at FONTAINE DE VAUCLUSE				
Soils and their EPR signatures	**Line**	**g_{iso} factor**	**Linewidth ΔB_{pp} [mT]**	**Relative intensity [%]**	**Qantification of EPR signatures: "barcodes"**
Soil 1a ω π	ω	2.0040	0.62	39	ω π
	π	2.0046	0.52	61	
Soil 1b ω π	ω	2.0040	0.62	22	ω π
	π	2.0046	0.52	78	
Soil 2 ω π	π	2.0046	0.53	100	ω π
Sampled water (see fig. 2.6) ω π	ω	2.0040	0.62	100	ω π

2.7 - Conclusion

In this chapter, we have described the first studies of EPR tracing of natural organic matter performed at the scale of drainage basins. The representation of the relative intensities of the different components of the spectrum by a "barcode" facilitates tracking of different types of NOM, from the soils of the drainage basins to natural recorders such as sediments and speleothems. This tracing method can be used to study the dynamics and reactivity of NOM at this scale, it can also be applied to paleo-environments. This novel method can also be used to obtain information on the impact of terrestrial organic matter on lacustrine ecosystems. For all these reasons, we consider that it provides a useful tool, well adapted to the management of drainage basins. In addition, comparison of variations in the EPR signatures due to the effect of various constraints (chemical, UV radiation, γ, etc.) to those obtained by other techniques could be used to develop a *bank of EPR signatures* for NOM associated with their main structural and reactive characteristics. More sophisticated EPR studies (ENDOR, pulsed EPR) could also be used to obtain more detailed information on the structure of the radicals present in the different types of NOM, but this would detract from the simplicity and rapidity of the method which are its main advantages and which make it applicable at large scales of time and space.

References

BARANČÍKOVÁ G., SENESI N. & BRUNETTI G. (1997) *Geoderma* **78**: 251-266.

BATIOT C., EMBLANCH E. & BLAVOUX B. (2003) *Comptes Rendus Geoscience* **335**: 205-214.

BATJES NH. (1996) *European Journal of Soil Science* **47**: 151-163.

BINET L. *et al.* (2002) *Geochimica Cosmochimica Acta* **66**: 4177-4186.

BLONDEL T. *et al.* (2010) *Isotopes in Environmental and Health Studies* **46**: 27-36.

BLYTH A.J. *et al.* (2008) *Quaternary Science Reviews* **27**: 905-921.

EMBLANCH C. *et al.* (1998) *Geophysical Research Letter* **25**: 1459-1462.

GIGUET-COVEX C. *et al.* (2011) *The Holocene* **29**: 651-665.

JACOB J. *et al.* (2008) *Journal of Archaeological Science* **35**: 814-820.

JERZYKIEWICZ M. (2004) *Geoderma* **122**: 305-309.

JORDAN-MEILLE L., DORIOZ J.M. & WANG D. (1998) *Agronomie* **18**: 5-26.

JORDAN-MEILLE L. & DORIOZ J.M. (2004) *Agronomie* **24**: 237-248.

KELLEHER B.P. & SIMPSON A.J. (2006) *Environmental Science and Technology* **40**: 4605-4611.

LAL R. & BRUCE J.P. (1999) *Environmental Science and Policy* **2**: 177-185.

LEENHEER J.A. (2007) *Annals of Environmental Science* **1**: 57-68.

LUGO A.E. & BROWN S. (1993) *Plant and Soil* **149**: 27-41.

NICKEL-PÉPIN-DONAT B., JEUNET A. & RASSAT A. (1990) "ESR study of the structure, texture and reactivity of the coals and coal macerals of the CERCHAR-GRECO minibank" in *Advanced Methodologies in Coal Characterization*, CHARCOSSET H., NICKEL-PÉPIN-DONAT B., eds, Elsevier, Amsterdam.

PÉPIN-DONAT B. & BRUNEL L.C. (1994) *Journal de Chimie Physique* **91**: 1896.

PETERS K.E., WALTERS C.C. & MOLDOWAN J.M. (1993) "Biomarkers and Isotopes in the Environment and Human History" in *The Biomarker Guide* Cambridge University Press.

POULENARD J. *et al.* (2004) *European Journal of Soil Science* **55**: 487-496.

RAICH J.W. & POTTER C.S. (1995) *Global Biogeochemical Cycles* **9**: 23-26.

RIFFALDI R. & SCHNITZER M. (1972) *Soil Science Society of America Proceeding* **36**: 301-305.

SCHULTEN H-R. & SCHNITZER M. (1997) *Soil Science* **162**: 115-130.

SAAB S.C. & MARTIN-NETO L. (2008) *Journal of the Brazilian Chemical Society* **19**: 413-417.

SCHNITZER M. & KHAN S.U. (1972) *Humic substances in the environment,* Marcel Dekker, New York.

SCHULTEN H.-R. & LEINWEBER P. (2000) *Biology and Fertility of Soils* **30**: 399-432.

SENESI N. & STEELINK C. (1989) "Application of ESR spectroscopy to the study of humic substances" in *Humic Substances II: In Search of Structure* Hayes, M.H.B., MacCarthy P., Malcolm R.L., Swift R.S., eds, Wiley, New York.

SENESI N. (1990) "Application of electron spin resonance (ESR) spectroscopy in soil chemistry" in *Advances in Soil Science*, Hayes, M.H.B., MacCarthy P., Malcolm R.L., Swift R.S., eds, Wiley, New York.

SIERRA M.M.D. *et al.* (2005) *Chemosphere* **58**: 715-733.

SIMONEIT B.R.T. (2002) *Applied Geochemistry* **17**: 129-162.

STEDMON C.A., MARKAGER S. & BRO R. (2003) *Marine Chemistry* **82**: 239-54.

STEELINK C. & TOLLIN G. (1967) "Free radicals in soil" in *Soil Biochemistry* McLaren A.D., Peterson G.M., eds, Marcel Dekker, New York.

SZAJDAK L., JEZIERSKI A. & CABRERA M.L. (2003) *Organic Geochemistry* **34**: 693-700.

Volume 1: BERTRAND P. (2020) *Electron Paramagnetic Resonance Spectroscopy - Fundamentals*, Springer, Heidelberg.

YABUTA H. *et al.* (2005) *Geochemistry* **36**: 981-990.

Chapter 3

Detection and characterisation of free radicals after spin trapping

Lauricella R. and Tuccio B.

Aix-Marseille University–CNRS, UMR 7273: Institute of Radical Chemistry, Team of Spectrometry applied to Structural Chemistry, St Jérôme Campus, Marseille.

3.1 - Introduction

EPR's capacity to detect low concentrations of paramagnetic species and the sensitivity of the spectrum to their environment and mobility, make it an ideal technique for the study of free radicals in solution. But as these species often have a very short lifetime, their concentration can be insufficient to allow their detection and/or it varies too quickly for their spectrum to be recordable with a conventional spectrometer. Two types of method can be used to detect paramagnetic species with very short lifetimes:

▷ time-resolved EPR techniques can produce a spectrum and monitor it over time. They rely on specific devices and/or spectrometers that are specially designed to accelerate spectrum acquisition. These techniques are described in chapter 10 of this volume,

▷ an indirect method can detect and identify free radicals using a conventional EPR spectrometer. This method, known as *spin trapping*, will be the subject of this chapter.

Although two experiments of this type were previously described [Mackor *et al.*, 1966; Iwamura and Inamoto, 1967], it was the practically simultaneous description of the method by the Janzen and Lagercrantz teams in 1968 which truly marked the birth of *spin trapping* [Janzen and Blackburn, 1968; Lagercrantz

© Springer Nature Switzerland AG 2020
P. Bertrand, *Electron Paramagnetic Resonance Spectroscopy*,
https://doi.org/10.1007/978-3-030-39668-8_3

and Forshult, 1968]. This method has since progressed considerably, with multiple applications in biology, biochemistry, polymerisation, electrochemistry, chemistry under irradiation and more generally in all the fields of in-solution chemistry which involve radical intermediates. Older review articles give a relatively general view of these applications [Evans, 1979; Janzen, 1971 and 1980; Perkins, 1980].

Spin trapping consists in causing a short-lived radical $R^\bullet$ to react with a diamagnetic molecule trap P, to create a paramagnetic spin adduct $(PR)^\bullet$ with a sufficiently long lifetime to allow its accumulation in the medium and its detection by EPR (figure 3.1). Analysis of the EPR spectrum for the $(PR)^\bullet$ adduct provides information on the $R^\bullet$ species. In favourable cases, $R^\bullet$ can be identified from the characteristics of the spectrum of the adduct (g_{iso} values, hyperfine coupling constant). However, the adducts formed by trapping of different radicals often produce very similar, or even identical, spectra. In this case, the method should be considered to identify a type of radical rather than the precise nature of the trapped species.

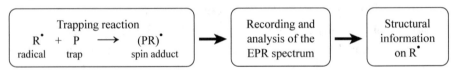

Figure 3.1 - General principle of a radical trapping experiment followed by EPR detection.

This method therefore has its limits and can cause artefacts which may lead to erroneous interpretations. Despite these limits, and because of its simplicity of implementation and its applicability in all types of media, over the last four decades *spin trapping* has allowed major advances in different fields related to biochemistry and radical chemistry. The aim of this chapter is to describe the main characteristics of the method and the expected results. From a practical point of view, several questions arise when performing a *spin trapping* experiment:

▷ Is the trap stable in the medium considered? Is it likely to produce reactions other than trapping of the target radical?

▷ What precautions should be taken during sample preparation and recording of the spectrum?

▷ Will trapping be quick enough and how stable will the adduct be?

▷ Has EPR data likely to identify the spin adduct been published?

▷ How should the EPR spectrum be interpreted and what information can be deduced from it?

By providing elements of response to these questions, we hope to guide those who, without being specialists, may have to perform a spin trapping experiment, and more generally help everyone to better understand this method and how it is used in the literature.

3.2 - Implementing the experiment

3.2.1 - Trap selection

The traps most commonly used are either nitroso compounds or aldonitrones, and the addition of a radical $R^\bullet$ to these molecules leads to the formation of an aminoxyl, or nitroxide[1] radical, for which the spectrum can be recorded (figure 3.2).

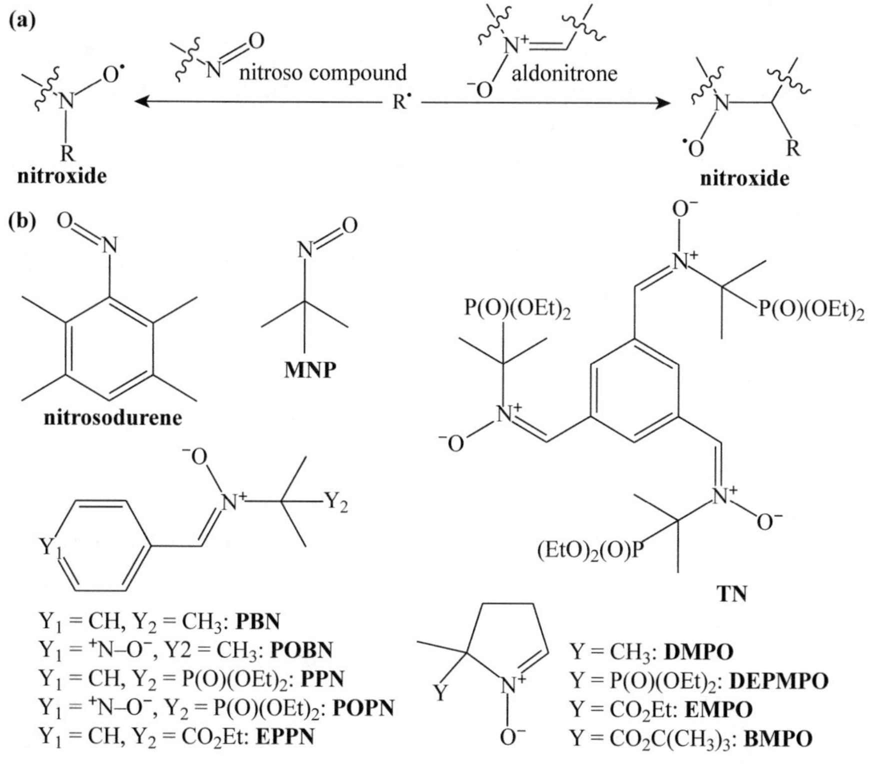

Figure 3.2 - **(a)** Trapping of a radical $R^\bullet$ by a nitroso compound and a nitrone to produce a nitroxyde observable by EPR, **(b)** main traps used.

---

[1]   Although the term "nitroxide" is incorrect to designate an aminoxyl radical, its use is so widespread that in the bibliography from November 2012, we found more than 17,500 references relating to the term "nitroxyde" compared to around 330 for "aminoxyl radical". To avoid any confusion, we will conform in this chapter to the common usage by designating an aminoxyl radical by "nitroxide".

The nature of the traps used explains why *spin trapping* is sometimes called the "nitroxide method" [2]. To simplify the notation, the spin adduct resulting from the addition of a radical R˙ to a trap P is noted P–R; for example, DMPO–OH designates the nitroxide formed by trapping of the hydroxyl radical on the DMPO nitrone.

◇ *Nitroso compounds*

Although a few uses of nitrosarene in *spin trapping* have been described [Terabe *et al.*, 1973; Konaka *et al.*, 1982; Wang *et al.*, 2000], the most commonly used nitroso trap is 1-methyl-1-nitrosopropane (MNP). The advantage of using a nitroso compound resides in the fact that the radical R˙ can be added directly to the nitrogen (figure 3.2). In the nitroxide that is formed, the R fragment can be subjected to additional hyperfine coupling which facilitates the EPR identification of the species trapped. For example, the spectrum for the MNP–Ph adduct reveals hyperfine coupling with the five protons of the aromatic nucleus and that of MNP–CH_2OH reveals coupling of the two equivalent protons in the methylene (figure 3.3a).

Nitroso compounds also have disadvantages which limit their field of application [Makino *et al.*, 1980; Davies, 2002]. Nitrosoalkanes are generally in equilibrium with a dimeric form which is inactive in *spin trapping* (figure 3.3b). This form dominates in polar solvents, making the use of these traps in aqueous medium somewhat complex. In addition, the adducts of oxygenated radicals on MNP are too unstable to be detected by EPR in classical conditions. MNP decomposes upon exposure to heat or daylight, producing the MNP–$C(CH_3)_3$ adduct (figure 3.3c), a very stable nitroxide which hinders the detection of any spin adducts. Nitroso compounds also react with carbon-carbon double bonds through an ene-reaction resulting in the formation of a nitroxide (see section 3.6). Finally, the toxicity of these compounds makes them impossible to use *in vivo*. In summary, the use of nitroso-type traps should be avoided when seeking to detect oxygenated radicals, in aqueous media, or if heating is required. In contrast, they can be very useful to identify radical intermediates formed in organic solvents, in particular when the reaction being studied does not involve any alkenes. All these characteristics are schematically represented in figure 3.3.

---

[2] The development of *spin trapping* is a direct consequence of the work on nitroxide physicochemistry, of which André Rassat was one of the forerunners, with 160 articles published on this subject alone between 1962 and 2006.

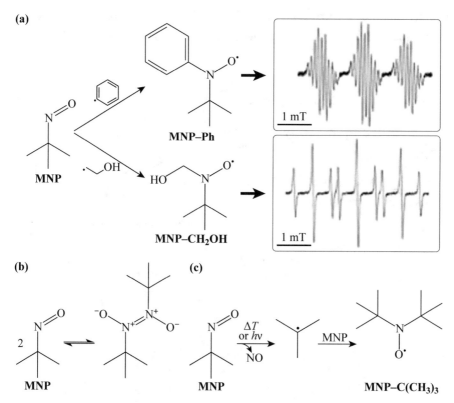

Figure 3.3 - **(a)** Trapping of phenyl and hydroxymethyl radicals by MNP and EPR spectra of adducts, **(b)** equilibrium between mono- and di-meric forms of MNP, **(c)** thermal or photochemical decomposition of MNP.

◇ Nitrones

Aldonitrones with a proton in the β position relative to the nitrogen (the β position is defined in figure 3.6a) are almost exclusively used in *spin trapping*[3]. More precisely, these traps are either pyrrolidines like DMPO, or β-arylnitrones like PBN (figure 3.4). These molecules are generally very stable in all types of media and many can be used under irradiation or at relatively high temperatures. Their low toxicity makes them compatible with the high concentrations required for various biological studies. Hydrophilic and more lipophilic compounds are also available, and an appropriate trap can often be found for almost any study medium. Finally, nitrones produce stable adducts will all types of radicals, and we

---

[3] Some *spin trapping* experiments with ketonitrones have been described, but the trapping reactions are slow and the EPR spectra for the adducts provide little information.

will see that they are prime choices when studying oxygen-based radicals such as HO$^{\bullet}$ (section 3.4.2), HO$_2^{\bullet}$/O$_2^{-\bullet}$ (section 3.4.3) and RO$^{\bullet}$, RO$_2^{\bullet}$ (section 3.4.4).

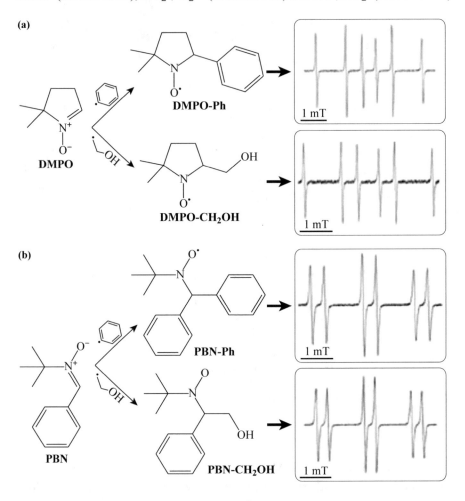

Figure 3.4 - Trapping of phenyl and hydroxymethyl radicals by **(a)** the nitrone DMPO and **(b)** the nitrone PBN; and EPR spectra of the adducts formed.

The main disadvantage of nitrones resides in the lack of information provided by the EPR spectra of their adducts. Indeed, an R$^{\bullet}$ radical adds to the carbon of the nitronyl group and not directly to the nitrogen. It follows that the nuclei with non-null spin for the R fragment are often too far from the radical centre of the spin adduct to produce observable hyperfine coupling: for most adducts, the EPR spectrum contains only six lines as a result of hyperfine coupling with the nuclei of nitrogen and hydrogen in β position with respect to the nitrogen. Consequently, after trapping by a nitrone, different radicals, like $^\bullet CH_3$ and $^\bullet CH_2OH$, can produce identical EPR spectra [Buettner, 1987]. In fact, not all nitrones present the same advantages and disadvantages, and it is important to choose the trap which is best adapted to the experiment that is to be performed. PBN-type nitrones, although very effective for the detection of carbon-based radicals, generally do not allow adducts of oxygen-centred radicals to be distinguished; in addition, the adducts of the hydroxyl radical decompose in aqueous medium, in particular producing nitroxides which hinder EPR observation. Adducts formed from cyclic nitrones such as DMPO produce more characteristic EPR spectra, but they are difficult to purify and their biological applications are generally limited to extracellular media because of their low lipophilicity. However, some cyclic nitrones such as BMPO can be purified by simple recrystallisation [Zhao et al., 2001]. With β-phosphorylated nitrones (DEPMPO, PPN, etc. figure 3.2b), the spectrum for the adducts displays strong hyperfine coupling with the ^{31}P nucleus, which generally depends on the structure of the radical trapped [Fréjaville et al., 1994 and 1995; Tuccio et al., 1996; Rizzi et al., 1997]. These signals are often not very intense as the lines are split, and sometimes show a complex pattern due to the possible formation of two diastereoisomer adducts and an equilibrium between two conformers. In the specific case of detection of superoxide [4], only a few nitrones produce adducts that are detectable by EPR in aqueous medium at neutral pH. Finally, some nitrones like the TN compound (figure 3.2b) present numerous advantages but are not yet commercially available [Roubaud et al., 2002; Allouch et al., 2007]. Table 3.1 summarises the main characteristics of ten of the most commonly used nitrones.

3.2.2 - Sample preparation

The operating conditions must be appropriate for the type of radical sought, for its mode of production, and for the trap selected. The solvent must solubilise all the reagents but remain inert with respect to the radicals, the trap and the spin adducts. The solvents to avoid include: toluene which can react with some

---

[4]   In line with the usage in the literature, "superoxide" refers to the hydroperoxyl radical/superoxide anion radical ($HO_2^\bullet / O_2^{-\bullet}$) pair, the trapping reaction for these two species by a nitrone N in an aqueous medium produces the same nitroxide $N–O_2H$.

radicals, basic solvents which promote nucleophilic additions, and any solvents which reduce the nitroxides, such as some alcohols.

Table 3.1 - Coefficients of octanol/water partition and main advantages and disadvantages of the different nitrones used as radical traps (the structures of the compounds are given in figure 3.2b).

Nitrone	K_p octanol/water	Main advantages	Main disadvantages
PBN	15 Average lipophilicity	Very good trap for carbon-based radicals in all media; commercially available; abundant bibliography.	Spectra do not allow precise identification of the trapped radical; rapid decomposition of PBN–OH and POBN–OH producing nitroxides; adducts PBN–O_2H and POBN–O_2H are not very stable in aqueous medium.
POBN	0.15 Hydrophilic	Very good trap for carbon-based radicals; can be used to detect some oxygen-based radicals; commercially available; abundant bibliography.	
PPN	10.2 Average lipophilicity	Very good traps for carbon- and oxygen-based radicals, including superoxide in aqueous medium; identification of the trapped radical facilitated by hyperfine coupling with the ^{31}P nucleus.	In aqueous medium, decomposition of the PPN–OH and POPN–OH adducts produces nitroxides; few bibliographic references; not commercially available.
POPN	0.21 Hydrophilic		
EPPN	29.8 Average lipophilicity	Good trap for carbon-based radicals; trap for superoxide radical in aqueous medium.	In aqueous medium, decomposition of the EPPN–OH adduct produces nitroxides; slow trapping of superoxide; not commercially available; few bibliographic references.
TN	> 300 Lipophilic	Excellent trap for all types of radicals, high trapping speeds; simple spectra which are characteristic of the type of radical trapped; good trap for superoxide; crystallised and water soluble nitrone despite is lipophilicity.	Not commercially available; few bibliographic references.

Nitrone	K_p octanol/water	Main advantages	Main disadvantages
DMPO	0.1 – 0.06 Hydrophilic	Very abundant bibliography; simple spectra that are characteristic for the type of radical trapped; commercially available.	Oily substance often containing paramagnetic impurities; sensitive to nucleophilic addition of water; low persistence of DMPO–O$_2$H at neutral pH.
DEPMPO	0.16 Hydrophilic	Abundant bibliography; characteristic spectra for the type of radical trapped; good trap for superoxide.	Oily substance sometimes containing paramagnetic impurities; complex spectra for some adducts; high cost; difficult to source commercially.
EMPO	0.33 Hydrophilic	Simple and characteristic spectra for the type of radical trapped; good trap for superoxide.	Oily substance sometimes containing paramagnetic impurities; high cost; difficult to source commercially.
BMPO	0.5 – 1 Average hydrophilicity	Simple spectra that are characteristic of the type of radical trapped; good trap for superoxide; crystallised nitrone; good solubility in non-polar medium despite its hydrophilicity.	Bibliography not yet very abundant; high cost; difficult to source commercially.

An appropriate amount of the trap must be present from the start of radical production. At too low a concentration, the adduct forms in insufficient amounts and its detection by EPR is complicated. An excess of trap not only increases costs, but can also increase the speed of disappearance of the adducts. If the process studied requires the use of specific conditions (irradiation, ultrasound, heating, etc.), it is important to verify that the chosen trap remains stable. Sometimes dioxygen must be eliminated from the medium to refine the spectral lines, in particular in organic solvents where O_2 is highly soluble. This can be done before starting the reaction, or just before recording the spectrum if O_2 contributes to the radical process being studied.

The cell used must be appropriate for the solvent and the mode of radical production. Some experiments may require the use of specific tubes, for example, compatible with degassing or direct mixing of reagents inside the cavity.

However, simple glass tubes [5] are appropriate for most experiments performed in solvent. Aqueous media require the use of capillary tubes, a flat cell, or the AquaX® cell [6], because of the high dielectric constant of water [Volume 1, chapter 9, complement 4]. Besides, production of radicals *in situ* by photolysis requires the use of quartz tubes.

3.2.3 - Recording the spectrum

As for any EPR spectrum, it is essential to correctly adjust the different acquisition parameters before recording the spectrum. But *spin trapping* experiments often involve nitroxides the concentrations of which evolve over time, which requires specific precautions to be taken.

▷ Sometimes there is only a short time available to record the spectrum after the start of radical production. It is therefore recommended that the EPR spectrum for a solution of traps be recorded before any radical production, so as to adjust the cavity parameters for later use.

▷ Generally, at X-band a modulation amplitude of 0.1 to 0.15 mT is used, but it can be necessary to reduce this amplitude to values less than 0.01 mT to observe hyperfine coupling with distant nuclei, or to increase it to 0.25 mT to reveal adducts present at very low concentrations. A microwave power of 10 to 20 mW provides maximal spectrum intensity while avoiding saturation [Buettner and Kiminyo, 1992].

▷ As kinetic evolution is always possible, it is unwise to record the signal with a low scan rate and a great time constant: variation of the nitroxide concentration during acquisition could produce a spectrum where the high-field and low-field lines have different intensities. For the same reason, rather than attempting to improve the signal-to-noise ratio by accumulating spectra [Volume 1, section 9.6.1], it is preferable to rapidly record (scan rate around 0.5 mT s^{-1}) a series of spectra at regular intervals: if changes occur over time, they can be preserved and the noise filtered through the application of classical signal treatment procedures such as SVD (singular-value decomposition) [Lauricella *et al.*, 2004]; otherwise, spectra can be added to produce a signal identical to the one that would be obtained by accumulation.

---

[5] Glass produces a signal in the $g \approx 2$ region, but it is generally too weak to affect the spectrum for the adducts produced in *spin trapping* experiments.

[6] Developed and commercialised by Bruker.

3.3 - Analysis of EPR spectra

3.3.1 - The information contained in the spectrum

In liquid solution, nitroxide molecules rapidly explore all possible orientations relative to the magnetic field. This is the *isotropic regime* where the EPR spectrum is determined by the average g_{iso} of the principal values of the $\tilde{g}$ matrix and by the average A_{iso} of the principal values of the hyperfine matrices, known as the hyperfine coupling constant (*hfcc*) [Volume 1, section 2.2.3]. In EPR, coupling constants are generally expressed in MHz. In *spin trapping* the $a = A_{iso}/g_{iso}\beta$ quantity which measures the separation between the hyperfine lines of the spectrum is preferred; it is expressed in gauss [G] or millitesla [mT] [Volume 1, section 2.3]. As the g_{iso} value is always close to $g_e = 2.0023$, the following correspondence can be used:

$$a = 1G = 0.1mT \rightarrow A \approx 2.802\,MHz \qquad [3.1]$$

◇ g_{iso} *value*

g_{iso} is determined by measuring the field at the centre of the spectrum. As the g_{iso} values for nitroxides are generally between 2.005 and 2.007, this number provides little information on the nature of the species trapped. However, it is still useful to measure it as it can reveal the existence of a *mixture* of several radicals.

◇ *Hyperfine interaction with the nitrogen nucleus*

The hyperfine structure of the EPR spectrum generated by a spin adduct is produced by interactions between the unpaired electron and paramagnetic nuclei, and its analysis provides structural information on the trapped species. Nitroxides can be considered as π radicals, where the unpaired electron is mainly localised on nitrogen and oxygen atoms in the NO bond. Their EPR spectrum therefore includes at least three lines due to coupling with the ^{14}N ($I = 1$) nucleus and the a_N *hfcc* is proportional to the spin density ρ_N on the nitrogen [Volume 1, section 2.3]. In the mesomeric form description, this delocalisation is represented by an equilibrium between two limit forms **A** and **B** (figure 3.5a): the more **B** present, the higher a_N.

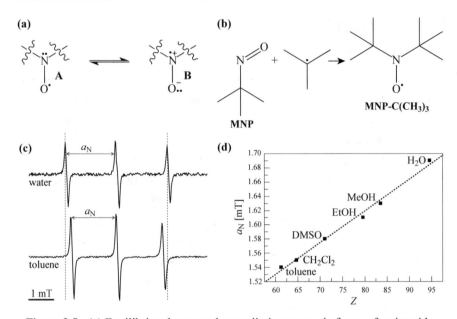

Figure 3.5 - **(a)** Equilibrium between the two limit mesomeric forms of a nitroxide, **A** and **B**, **(b)** formation of the MNP–C(CH$_3$)$_3$ adduct by trapping of the tert-butyl radical on MNP, **(c)** EPR spectrum of the MNP–C(CH$_3$)$_3$ adduct in water and toluene, **(d)** value of a_N measured for the MNP–C(CH$_3$)$_3$ adduct as a function of the polarity of the medium, as measuredwith Kosower's Z value.

The value of a_N can vary for different adducts with the same trap because of electron-withdrawing or electron-donating effects of groups in positions α or β relative to the nitrogen [Buettner, 1987]. This value is also sensitive to the polarity of the surrounding medium, since form **A** predominates in apolar media. Thus, for a given adduct, a_N will be larger in aqueous medium than in organic solvents, and its value is often linearly correlated with the polarity of the medium, for example as measured with Kosower's Z value [Brière *et al.*, 1964; Knauer and Napier, 1976; Deng *et al.*, 2002]. Figure 3.5b shows formation of the MNP–C(CH$_3$)$_3$ adduct and figure 3.5c shows its spectrum in water and in toluene. Figure 3.5d indicates the values of a_N measured in water (Z = 94.6), methanol (Z = 83.6), ethanol (Z = 79.6), dimethylsulfoxide (DMSO, Z = 71.1), dichloromethane (Z = 64.7) and toluene (Z = 61.3).

◇ *Other hyperfine interactions*

The EPR signal for an adduct is split by hyperfine interactions with the X paramagnetic nuclei in positions α, β, and more rarely γ, relative to the nitrogen of the aminoxyl function (figure 3.6a); the corresponding *hfcc* are respectively noted $a_{X\alpha}$, $a_{X\beta}$ and $a_{X\gamma}$. As interactions with nuclei in β or γ depend on the nitroxide conformation, the $a_{X\beta}$ and $a_{X\gamma}$ constants can provide important structural information. Coupling with nuclei in γ is very weak, and is only observed with a planar "W" arrangement of the axis for the nitrogen's p_z orbital and the three bonds separating the nitrogen from the Xγ nucleus (figure 3.6b). This coupling can be seen for example on the spectrum for adducts of superoxide on pyrrolidine nitrones.

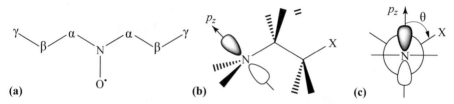

Figure 3.6 - (a) Definition of the positions α, β and γ in a nitroxide, **(b)** "W" geometry of a nitroxide favouring hyperfine coupling with a nucleus X in position γ, **(c)** Newman projection allowing visualisation of the dihedral angle θ involved in the Heller-McConnel relation for a nucleus X in position β (equation [3.2]).

The *hfcc* for an X nucleus in position β depends on the dihedral angle θ between the plane defined by the X–C_α–N nuclei, and the plane defined by the C_α–N bond and the axis of the nitrogen's $2p$ orbital (figure 3.6c). Variation of the coupling constant is described by the Heller-McConnell relation:

$$a_{X\beta} = (C_0 + C_1 \cos^2\theta)\rho_N \qquad [3.2]$$

In this expression, C_0 corresponds to the contribution of spin polarisation, C_1 accounts for the hyperconjugation mechanism, and ρ_N is the spin density on the nitrogen. For a given nitrone, $a_{X\beta}$ depends on the geometry of the molecule and therefore the nature of the radical trapped. For example, when X is a proton of a DMPO–R adduct, $C_0 \approx 0$ and $C_1 \approx 2.6$ mT and the constant $a_{H\beta}$ increases with the steric hindrance of the R fragment (figure 3.7).

In the spin adducts obtained with nitrones, positions α correspond to carbons, with 99 % ^{12}C nuclei ($I = 0$); therefore they do not generally produce line splitting. However, the concentration of adducts is sometimes large enough to observe hyperfine interactions due to ^{13}C nuclei at natural abundance (≈ 1 %). With nitroso compounds, hyperfine coupling in α appears on the spectrum every time a radical centred on a paramagnetic X nucleus is trapped, for example

$^{\bullet}$H, $^{\bullet}$D, $^{\bullet}$P(O)(OEt)$_2$, $^{13\bullet}$CH$_3$, $^{\bullet}$N$_3$. The value of $a_{X\alpha}$ in these cases is relatively characteristic of the nature of X.

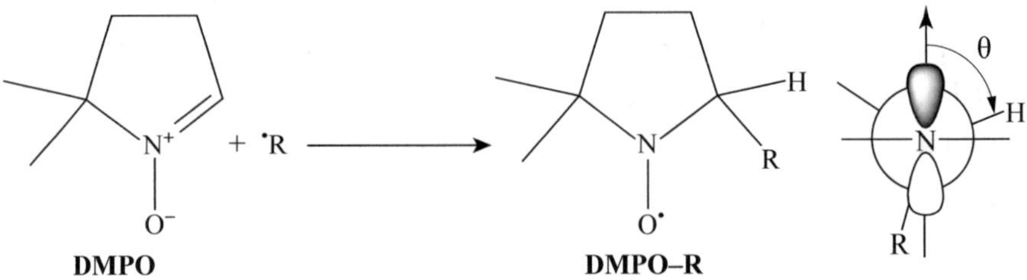

DMPO   **DMPO–R**

if R = OH, a_H = 1.49 mT ; if R = CO$_2$H, a_H = 1.87 mT
if R = CH$_2$OH, a_H = 2.27 mT ; if R = C(CH$_3$)$_2$OH, a_H = 2.40 mT

Figure 3.7 - Trapping of an $^{\bullet}$R radical by the DMPO nitrone in aqueous medium leading to the formation of the DMPO–R adduct and variation of a_H depending on the steric hindrance due to the trapped radical.

A final type of hyperfine interaction is observed when, in a nitroxide adduct, the unpaired electron is partly delocalised in a π system. This is the case with the MNP–Ph adduct, where spectra showing hyperfine coupling with the aromatic protons in ortho H$_o$, meta H$_m$ and para H$_p$ positions are observed: in benzene, for example, a_{Ho} = a_{Hp} = 0.2 mT and a_{Hm} = 0.09 mT. This coupling, although weak, complicates the spectrum but reveals that an aromatic radical is trapped (figure 3.3a).

3.3.2 - Analysis of spectra for spin adducts

Simple examination of the spectrum of a spin adduct already provides important information. In principle, the lines for a radical are symmetric relative to the centre of the spectrum which corresponds to g_{iso}; asymmetry generally indicates that the sample contains a mixture of several radicals. The total width L of the spectrum (distance between the first and last lines) is given by equation [3.3], where p is the number of groups of equivalent nuclei, n_i is the number of equivalent nuclei of spin I_i, and a_i is the corresponding hyperfine coupling constant [Volume 1, section 2.4.2]:

$$L = \sum_{i=1}^{p} 2n_i I_i a_i \qquad [3.3]$$

Since the spectrum of an adduct always has three lines split due to hyperfine coupling with paramagnetic nuclei in positions α, β or γ, it is natural to start by determining a_N. Separation between the first two lines gives the value of the smallest *hfcc*. To glean further information, a hypothesis must be made on the structure of the nitroxide, and the stick diagram predicted by this structure must be compared to the experimental spectrum. It is important to remember that a group of n equivalent nuclei of spin $I = \frac{1}{2}$ produces a pattern of $n + 1$ equidistant lines; the relative intensities of these lines are given by Pascal's triangle [Volume 1, chapter 2, complement 2]. When the spin of the equivalent nuclei is greater than $\frac{1}{2}$ or when non-equivalent nuclei are present, it is preferable to construct a diagram. Finally, it should be remembered that when several isotopes exist, their abundance determines the relative intensities of their corresponding lines.

In *spin trapping* experiments, spectra of mixtures of radicals are often obtained. To analyse these spectra, we generally start by determining approximate values of the parameters by following the principles set out above. In a second step, more precise values are obtained by adjusting a calculated spectrum to fit the experimental spectrum with the help of numerical simulation [7] software.

For a given spin adduct, the trapped radical can be identified by comparing the values for the parameters deduced from analysis of the spectrum to those published in the literature, considering that the values of *hfcc* vary depending on the nature of the medium. Much data can be found in [Buettner, 1987] and in the database available on the https://www.niehs.nih.gov/research/resources/epresr.

3.4 - Examples of spectra for adducts of classical radicals

3.4.1 - Carbon-based radicals

As illustrated in figures 3.3 and 3.8, nitroso compounds are ideal traps for the detection of carbon-based radicals: analysis of the EPR spectrum for the adduct can be used to determine whether it is the result of trapping of a radical centred on a tertiary carbon (no hyperfine coupling with a proton in β), a secondary carbon (hyperfine coupling with a proton in β), or a primary carbon (hyperfine coupling with two protons in β), trapping of $^\bullet CH_3$ (hyperfine hyperfine coupling with 3 equivalent protons in β), or of a radical carried by an aromatic nucleus (hyperfine coupling with the aromatic protons).

---

7 A number of freely accessible programmes are available, among which, the most used in *spin trapping* are *EasySpin* (http://www.easyspin.org) and especially *WinSim* (http://www.niehs.nih.gov/research/resources/software/tox-pharm/tools/index.cfm).

All nitrones are also excellent traps for carbon-based radicals. They generally produce particularly stable adducts, and can be used in numerous cases where a nitroso trap is unsuitable. PBN-type nitrones are often preferred, in particular in biological media. For a given nitrone, the value of a_H is generally higher in the case of a nitroxide formed by trapping a carbon-based radical than for other adducts.

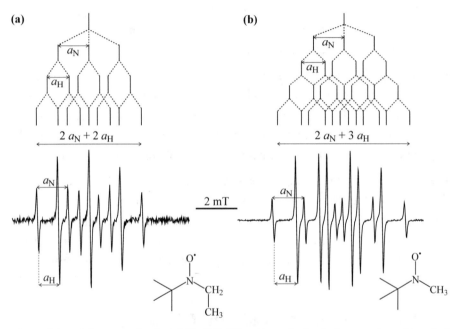

Figure 3.8 - (a) Spectrum for the MNP–CH$_2$CH$_3$ adduct obtained after trapping of the ethyl radical in toluene, and corresponding stick-diagram: a_N = 1.53 mT, a_H = 1.04 mT (2H), **(b)** spectrum for the MNP–CH$_3$ adduct obtained after trapping of the methyl radical in toluene, and corresponding stick-diagram: a_N = 1.53 mT, a_H = 1.13 mT (3H).

Thus, a_H is greater than 2 mT for DMPO–R adducts and greater than 0.3 mT for PBN–R adducts, where R is an alkyl radical [Buettner, 1987]; a_H is generally lower for other adducts. However, nitrones only rarely allow different carbon-based radicals to be distinguished (figure 3.4), as the coupling patterns observed on the spectrum are always very similar. This is all the more true that the *hfcc* can vary significantly with the medium, as shown in figure 3.9.

If the trapped radical is centred on a ^{13}C carbon, the spectrum shows additional hyperfine coupling due to this carbon which is present in the β position. This property can be used to elucidate a *reaction mechanism*, as shown by the example of the formation of the endoperoxide bridge in the phytohormone G3 (figure 3.10a).

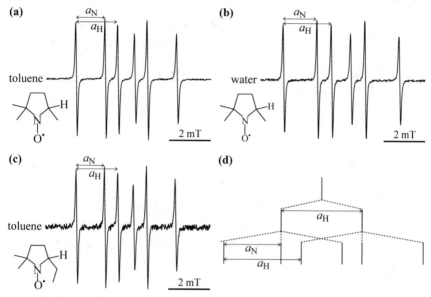

Figure 3.9 - **(a)** Spectrum for the DMPO–CH$_3$ adduct obtained by trapping of the methyl radical in toluene: a_N = 1.43 mT, a_H = 2.05 mT, **(b)** spectrum for the same adduct in water: a_N = 1.64 mT, a_H = 2.34 mT, **(c)** spectrum for the DMPO–CH$_2$CH$_3$ adduct obtained by trapping of the ethyl radical in toluene: a_N = 1.40 mT, a_H = 2.05 mT, **(d)** stick-diagram explaining the general appearance of spectra **(a) (b)** and **(c)**.

This reaction takes place in very mild conditions by spontaneous oxidation of the PG3 dienolic precursor by airborne oxygen, by a mechanism which had yet to be elucidated. Preliminary studies indicated that the reaction does not involve an excited state of the substrate, electron transfer, singlet oxygen, or acid-base catalysis. These observations led to the hypothesis that PG3 behaves like a radicaloid in a reaction with triplet oxygen crossing the spin barrier. To confirm this hypothesis, *spin trapping* experiments were undertaken with a range of nitrones (DEPMPO, POBN, TN) [Najjar *et al.*, 2005; Triquigneaux *et al.*, 2010a]. Whatever the trap used, oxygenation of a solution of PG3 containing the nitrone produced an intense EPR spectrum corresponding to the adduct of a carbon-based radical. As these first observations clearly favoured a self-oxidation radical mechanism for PG3, complementary *spin trapping* studies were performed using analogues of PG3 labelled with ^{13}C at various positions. Using the nitrone TN and the PG3* analogue represented in figure 3.10b, the

I_1 intermediate was shown to form and be trapped during the process. This result confirms the radical mechanism described in figure 3.10c. In figure 3.10d we have represented the EPR spectrum for the TN–^{13}C adduct which shows hyperfine coupling with the ^{13}C nucleus. The precise structure of the adduct, represented in figure 3.10e, was confirmed by tandem mass spectrometry.

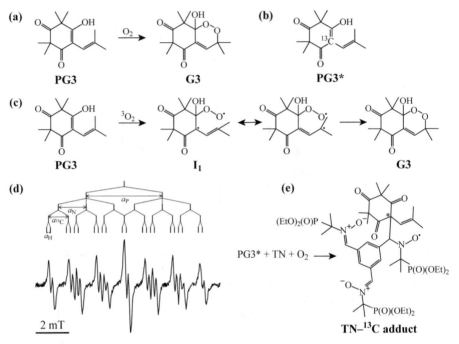

Figure 3.10 - **(a)** Formation of the G3 cyclic peroxide by self-oxidation of the dienolic precursor PG3, **(b)** formula for the ^{13}C-labelled PG3* precursor, **(c)** self-oxidation mechanism established by *spin trapping*, **(d)** spectrum for the TN–^{13}C adduct and stick-diagram explaining the hyperfine structure observed: a_P = 4.26 mT, a_N = 1.55 mT, a_{13C} = 1.14 mT and a_H = 0.19 mT, **(e)** structure of the TN–^{13}C adduct obtained after trapping by the TN nitrone of a radical intermediate formed during self-oxidation of PG3 (the position labelled with ^{13}C is marked by an asterisk *).

3.4.2 - The hydroxyl radical

The very reactive hydroxyl radical has a very short lifetime in all media. It can form during decomposition of peroxides in the atmosphere, during the decomposition of H_2O_2 in solution, during the Fenton reaction in aqueous or organic media, under the effect of ionising radiations, or *via* the Haber-Weiss reaction cycle in biological media. *In vivo*, the HO• radical has a lifetime of around 10^{-9} s and reacts with all types of molecules (amino acids, carbohydrates, nucleic acids, lipids, etc.) causing severe damage to biological cells. Its detection by EPR after *spin trapping* in aqueous medium therefore takes a particular importance in the biomedical context.

With cyclic nitrones such as DMPO, EMPO or BMPO (figure 3.2b), the spectrum for the adduct presents four lines with relative intensities 1:2:2:1 due to the almost equal a_N and a_H values (figure 3.11) [Buettner, 1987; Olive *et al.*, 2000; Zhao *et al.*, 2001; Stolze *et al.*, 2009]. In the case of β-phosphorylated cyclic nitrones such as DEPMPO, this pattern is duplicated due to the strong hyperfine coupling with the ^{31}P nucleus [Fréjaville *et al.*, 1994 and 1995]. It is nevertheless important to take care when interpreting these trapping experiments in aqueous medium, and to remember that detection of the nitroxide DMPO–OH (or EMPO–OH, or DEPMPO–OH, etc.) does not necessarily mean that HO[•] has been trapped: reactions other than *spin trapping* involving nucleophilic addition of water can produce the same EPR spectrum (see section 3.6).

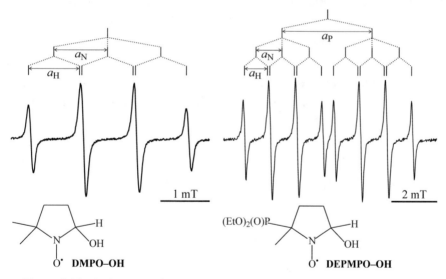

Figure 3.11 - (a) Spectrum for the DMPO–OH adduct in water and stick-diagram explaining the hyperfine structure observed: a_N = 1.40 mT, a_H = 1.34 mT, **(b)** spectrum for the DEPMPO–OH adduct in water and corresponding stick-diagram: a_N = 1.41 mT, a_H = 1.32 mT and a_P = 4.76 mT.

PBN, POBN, PPN, EPPN, etc. nitrone types generally cannot be used to detect the hydroxyl radical in aqueous medium as the adducts obtained are very unstable. Their decomposition produces an aldehyde and a nitroso compound which can lead to the formation of paramagnetic species either by reduction, or by trapping of a carbon-based radical which may be present in the medium [Janzen *et al.*, 1992; Rizzi *et al.*, 1997].

To circumvent this difficulty, an indirect method involving linear β-aryl-nitrones can be used to detect the hydroxyl radical. This method consists in adding DMSO, sodium formate or methanol, often termed *scavengers* of the hydroxyl radical, to the medium at the same time as a nitrone **N**; HO• reacts with these molecules to produce •CH$_3$, •CO$_2$Na or •CH$_2$OH radicals (figure 3.12a) which can be trapped by the nitrone. EPR detection of the **N–CH$_3$**, **N–CO$_2$H** or **N–CH$_2$OH** adducts proves that the hydroxyl radical was initially present in the medium (figure 3.12b). This type of study is easy to implement, but the *scavenger* must be added in sufficient quantities to avoid HO• reacting directly with the nitrone. This approach was successfully used in biological media, for example by R. Mason's team who have used the PBN trap in solution in DMSO to demonstrate the formation of the HO• radical in animal cells [Buettner and Mason, 1990; Burkitt and Mason, 1991; Kadiiska and Mason, 2002]. This technique of trapping a secondary radical was implemented in a study of the hepatic toxicity of copper. Using PBN as a trap in the presence of DMSO, the PBN–CH$_3$ adduct was detected by EPR, clearly demonstrating that copper salts caused the damage incurred by liver cells through a process involving the HO• radical [Kadiiska and Mason, 2002].

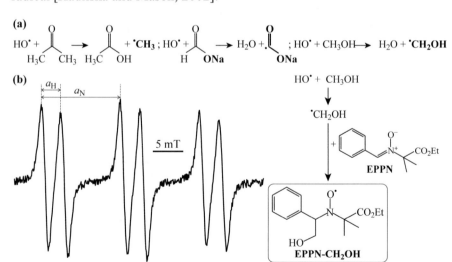

Figure 3.12 - (a) Reaction of the HO• radical with DMSO, sodium formate and methanol, **(b)** spectrum for EPPN–CH$_2$OH obtained in water when the HO• radical is produced in the presence of the EPPN nitrone and 10 % CH$_3$OH: a_N = 1.53 mT and a_H = 0.38 mT.

3.4.3 - The superoxide radical

The superoxide anion radical is generally produced by one-electron reduction of dioxygen, but it can also be obtained in the form of salts such as KO_2 by direct reaction of dioxygen with some metals. In aqueous medium, $O_2^{-\bullet}$ is in equilibrium with $HO_2^{\bullet}$, with a pKa of 4.88 [Behar *et al.*, 1970]. Although its reactivity is moderate, this radical naturally produced by aerobic organisms is a precursor of strong oxidants in biological systems, such as $HO^{\bullet}$, HOCl, ONO_2^- or $CO_3^{-\bullet}$. The toxicity of these compounds can be exploited by the immune system, but becomes damaging when an excess of superoxide leads to "oxidative stress". This occurs, for example, in the neurodegenerative diseases which are dealt with in chapter 4 of this volume. The implication of superoxide in various human diseases such as ischaemia-reperfusion syndrome, cancers, or even inflammatory phenomena has motivated particular interest in its detection by *spin trapping* in aqueous and biological media. Although the first superoxide trapping experiments were performed with the nitrone DMPO [Harbour and Bolton, 1975; Buettner and Oberley, 1978; Finkelstein *et al.*, 1979], only cyclic nitrones β-substituted with an electron-withdrawing group (ester, phosphonate, etc.) and trinitrone TN produce sufficiently stable adducts to allow effective detection of the superoxide in aqueous medium [Roubaud *et al*, 2002; Allouch *et al.*, 2007; Villamena *et al.*, 2007].

With pyrrolidinic nitrones (DMPO, EMPO, DEPMPO, see figure 3.2b), the EPR spectrum of the superoxide adduct can reveal long-distance hyperfine coupling. Thus, the spectrum for DMPO–O_2H shows 12 lines due to hyperfine coupling with the nitrogen nucleus, the proton in β and one of the protons in γ of the cycle (figure 3.13a). The DEPMPO–O_2H adduct produces a complex spectrum which results from several phenomena (figure 3.13b). Firstly, we expect twice as many lines due to the strong coupling with the ^{31}P nucleus. In addition, trapping of the superoxide on this nitrone produces two diastereoisomers, *cis* minor and *trans* major, with different EPR spectra. Finally, rapid equilibration occurs between two conformers of the *trans* adduct, which results in alternating broad and narrower lines [Fréjaville *et al.*, 1995]. The formation of two diastereoisomers and the conformational equilibrium should also be found with EMPO–O_2H or BMPO–O_2H adducts, but in the absence of coupling with the ^{31}P nucleus these phenomena do not significantly modify the spectrum (figure 3.13c).

β-aryl-nitrones are rarely used to detect superoxide in aqueous medium: PPN or EPPN-type traps showed some efficacy but are not as good as most cyclic nitrones [Tuccio *et al.*, 1996; Allouch *et al.*, 2005]. Trinitrone TN is a notable exception, at pH 7 the TN–O_2H adduct produces the intense spectrum shown in figure 3.13d [Roubaud *et al.*, 2002]. Although its synthesis and purification are simple, this trap is not yet commercially available, which considerably limits its use.

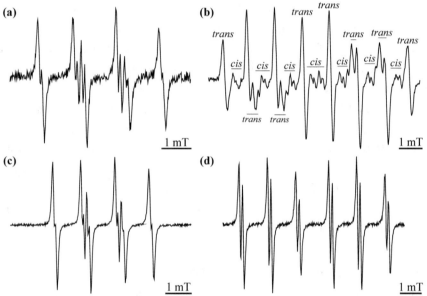

Figure 3.13 - Spectra obtained after trapping of superoxide with nitrones in an aqueous medium. **(a)** DMPO, producing DMPO–O_2H: a_N = 1.43 mT, $a_{Hβ}$ = 1.17 mT and $a_{Hγ}$ = 0.12 mT. **(b)** DEPMPO, producing DEPMPO–O_2H; first conformer of the *trans* diastereoisomer: a_N = 1.26 mT, $a_{Hβ}$ = 1.49 mT, $a_{Hγ}$ = 0.09 mT and a_P = 5.58 mT; second conformer of the *trans* diastereoisomer: a_N = 1.35 mT, $a_{Hβ}$ = 0.75 mT, $a_{Hγ}$ = 0.09 mT and a_P = 4.46 mT; *cis* diastereoisomer: a_N = 1.33 mT, $a_{Hβ}$ = 0.98 mT, $a_{Hγ}$ = 0.14 mT and a_P = 4.05 mT. **(c)** EMPO, producing EMPO–O_2H: a_N = 1.32 mT, $a_{Hβ}$ = 1.09 mT and $a_{Hγ}$ = 0.10 mT. **(d)** TN, producing TN–O_2H: a_N = 1.34 mT, $a_{Hβ}$ = 0.16 mT and a_P = 4.24 mT.

3.4.4 - Alcoxy and peroxy radicals

In organic chemistry, RO$^•$ and RO$_2$$^•$ radicals are produced in chain reactions or by homolytic cleavage of a peroxide bridge. It can therefore be interesting to detect them in organic solvent when seeking to determine a reaction mechanism. In the biomedical context, they are formed, in particular, from unsaturated fatty acids, and are considered reactive oxygen species (ROS). Although the β-aryl-nitrones generally produce persistent adducts with these radicals, cyclic nitrones are more commonly used to detect them. The spectra generally show hyperfine coupling with one of the protons in position γ on the cycle, but they can sometimes be confused with adducts of superoxide [Buettner, 1987; Bors *et al.*, 1992]. When the trap is a β-phosphorylated nitrone (*e.g.* DEPMPO), the value of a_P can be used to more readily distinguish between the adducts DEPMPO–OR ($a_P \approx 4.6$ mT) and DEPMPO–O_2R ($a_P \approx 5.0$ mT).

3.4.5 - Other radicals

Numerous other radicals bearing sulfur, halogens, phosphorous or nitrogen groups can be detected by EPR after trapping in organic, aqueous or biological media. The simplest case is that of H$^•$, formed for example by radiolysis of water. The spectra produced by adducts of aldonitrones are very characteristic as they show hyperfine coupling with two equivalent protons [Buettner, 1987]; that recorded after trapping on MNP is equally characteristic, because the *hfcc* for a_H and a_N just happen to be equal and produce four equidistant lines with relative intensities 1:2:2:1 [Makino *et al.*, 1980]. However, observation of these spectra is not absolute proof of trapping of H$^•$ as the same nitroxides can form by reduction of the nitroso or nitrone traps.

RS$^•$ radicals trapped by MNP produce adducts that can be readily identified by EPR thanks to the characteristic value $a_N \approx 1.8$ mT, but MNP–SR nitroxides can also be formed through nucleophilic addition of a thiol [Triquigneaux *et al.*, 2009]. Among the biological thiols, glutathione (GSH) is particularly important due to its anti-oxidant activity. The GS$^•$ radical is trapped by DMPO nitrone, but the DMPO–SG and DMPO–OH adducts produce similar EPR spectra. The ambiguity can be eliminated by using a β-phosphorylated trap such as DEPMPO, as the *hfcc* a_P allows the DEPMPO–OH ($a_P = 4.73$ mT, $a_N = 1.41$ mT, $a_{H\beta} = 1.38$ mT) to be distinguished from the DEPMPO–SG ($a_P = 4.58$ mT, $a_N = 1.42$ mT, $a_{H\beta} = 1.41$ mT) adduct [Karoui *et al.*, 1996].

Radicals centred on a paramagnetic nucleus are readily identifiable as the EPR spectra of their adducts show additional hyperfine coupling. In this field, it is important to mention the specific case of the $^•$NO radical. Although paramagnetic, this small molecule does not produce an EPR spectrum in solution at room temperature. Its detection by *spin trapping* is impossible with nitrones or nitroso compounds and requires specific traps. Cheletropic traps of $^{\bullet}$NO can be used, which produce a stable nitroxide after addition on a diene or dienolic compound [Korth *et al.*, 1992; Adam *et al.*, 1996; Lauricella *et al.*, 2010]. But there are currently only a few examples of this type of trap and the development of this approach requires significant synthesis work. In the vast majority of cases, the traps used are complexes of Iron(II)-dithiocarbamates, which may be liposoluble like diethyldithiocarbamate (DETC) or water-soluble like methyl glucamyl dithiocarbamate (MGD); the addition of $^{\bullet}$NO produces a very stable paramagnetic iron-mononitrosyl complex which is observable by EPR. This method has been extensively used in the biomedical field, and its applications have been the subject of several reviews [Vanin *et al.*, 2002; Berliner and Fujii, 2004; Yoshimura and Kotake, 2004].

3.5 - Kinetic aspects

The literature contains a limited number of kinetic studies of *spin trapping* relating almost exclusively to superoxide trapping, although some studies describe trapping of carbon-based radicals and the hydroxyl radical [*e.g*, Schmid and Ingold, 1978; Madden and Tanigushi, 1991; Pou *et al.,* 1994; Villamena *et al.*, 2003]. In addition, the authors of these studies are more often interested in the *persistence* of the adducts than the *trapping rates*. Finally, major discrepancies exist between published rate constants, often due to differing experimental approaches [Lauricella *et al.*, 2005]. Nevertheless, kinetic aspects are essential when implementing a *spin trapping* experiment: the trap is all the more effective when it reacts rapidly with a radical to produce the most durable spin adduct possible. It would not be possible to detail all the kinetic studies published in the field here as each experiment to trap a radical by a nitrone in a given medium constitutes a specific case. But some general considerations can be considered when choosing a trap for a *spin trapping* experiment.

A trap is effective if the kinetics of formation and decomposition of the adduct allow it to attain a sufficiently large steady-state concentration. In the case of radicals centred on a carbon atom, this condition is fulfilled whatever the trap. To detect HO$^{\bullet}$, it is preferable to use cyclic nitrones (DMPO, EMPO, etc.), as the adducts obtained with β-aryl nitrones decompose in protic media (section 3.4.2). Nitrones also generally produce persistent nitroxides after trapping of radicals such as H$^{\bullet}$, RO$^{\bullet}$, RS$^{\bullet}$, F$^{\bullet}$.

Kinetic studies of superoxide trapping in aqueous medium allow several conclusions to be drawn [Turner and Rosen, 1986; Goldstein *et al.*, 2004; Kezler *et al.*, 2004; Lauricella *et al.*, 2004; Allouch *et al.*, 2005 and 2007; Stolze *et al.*, 2007; Villamena *et al.*, 2007; Burgett *et al.*, 2008]. Formation of the adducts of superoxide is faster in acidic medium [8] and their degradation is slower. Cyclic nitrones trap superoxide more rapidly than β-aryl nitrones; the case of TN, which is one of the best traps for superoxide (section 3.4.3), constitutes an important exception to this rule. The presence of an electron-withdrawing group in position β relative to the nitrogen appears to accelerate trapping of the superoxide while also considerably slowing degradation of the adduct obtained; this effect was observed with nitrones EMPO, BMPO, DEPMPO, TN and PyOPN. Study of the kinetics of formation and decomposition of the adducts of superoxide with around ten nitrones showed that the EMPO, BMPO and TN traps were the best (see table 3.1). However, even with these molecules, trapping rates remain moderate.

In a given medium, trapping of a radical $R^\bullet$ is a bimolecular reaction which is always in competition with other pathways that $R^\bullet$ could engage in (duplication, oxidation or one-electron reduction, decomposition, etc.). To favour trapping, high trap concentrations, around 10 to 100 mmol L^{-1} must be used. However, it is recommended that the concentration remains below 150 mmol L^{-1} as the rate of degradation of adducts can increase with the concentration of trap [Tuccio *et al.*, 1995; Lauricella *et al.*, 2004]. In a *spin trapping* experiment, only some of the radicals formed will be trapped and the adducts can evolve towards diamagnetic species at a speed which varies depending on the medium. This is all the more true in biological medium where several molecules can eliminate the radicals, traps can be metabolised, and nitroxides rapidly reduced [Buettner and Mason, 1990 and 2003; Khan and Swartz, 2002; Mason and Kadiiska, 2005; Bezières *et al.*, 2010]. *Spin trapping* therefore cannot be used to *quantify* radical production, but it is a method of choice to *detect* and *identify* radicals in all sorts of media.

---

[8]   In fact, trapping of $HO_2^\bullet$ is much faster than trapping of $O_2^{-\bullet}$, to the point that trapping of the anionic radical is almost negligible in aqueous medium [Allouch *et al.*, 2007].

3.6 - Limitations of the method and precautions to avoid artefacts

We mentioned several times that reactions other than *spin trapping* could lead to the formation of nitroxides. These are mainly the ene-reaction, the Forrester-Hepburn reaction, and *Inverted Spin Trapping* (IST), which can lead to errors in interpretation [Ranguelova and Mason, 2011].

▷ The ene-reaction can take place when a nitroso compound is exposed to an alkene; as explained in figure 3.14a, a pericyclic reaction leads to the formation of a hydroxylamine which, in the presence of oxygen or any other oxidant, is rapidly converted into nitroxide. This reaction results in an EPR spectrum identical to that produced by trapping of a carbon-based radical. This is how MNP reacts with PG3, and this trap is therefore not suitable when studying the self-oxidation mechanism of the diene [Triquigneaux *et al.*, 2010b].

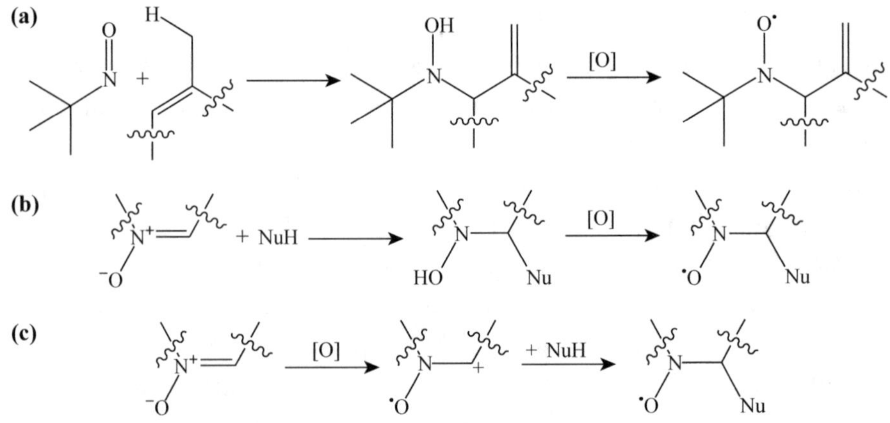

Figure 3.14 - Formation of nitroxides from nitrones or nitroso compounds without radical trapping: **(a)** ene-reaction between MNP and an alkene, **(b)** Forrester-Hepburn mechanism involving the addition of a nucleophilic compound NuH and oxidation of the intermediate hydroxylamine, **(c)** inverted spin trapping involving the oxidation of a nitrone as a cationic radical on which a nucleophilic NuH is added.

▷ The Forrester-Hepburn reaction consists in the addition of a nucleophile to the trap, followed by oxidation of the hydroxylamine formed, for example by reaction with the oxygen present in the medium (figure 3.14b) [Forrester and Hepburn, 1971]. This reaction is favoured in basic solvents such as pyridine or DMF, or in the presence of strong nucleophiles (hydride donors such as $NaBH_4$, phosphites, amines, etc.). It must always be considered in aqueous medium, in particular in the presence of traces of metal ions which can activate the nitrone function by complexation, making it more sensitive to the addition of water [Finkelstein *et al.*, 1980]; this reaction is particularly problematic as it produces an EPR spectrum identical to that observed by HO˙ trapping.

▷ The IST reaction involves oxidation of the trap producing a cationic radical onto which a nucleophile is added (figure 3.14c) [Eberson, 1992, 2000; El Hassan *et al.*, 2006]. [9] In the presence of strong oxidants (peroxides, H_2O_2, etc.) it must always be considered, and it can be promoted by UV or even visible light irradiation when using nitroso traps [Triquigneaux *et al.*, 2009]. If the nucleophile added is water, the same spectrum is observed as after trapping of the hydroxyl radical.

Finally, in a very general way, oxido-reduction reactions that the adducts can undergo produce some artefacts. For example, the **N–O₂H** adducts, obtained after trapping of the superoxide by a nitrone **N**, can evolve to produce **N–OH**, as shown in figure 3.15 (pathway d).

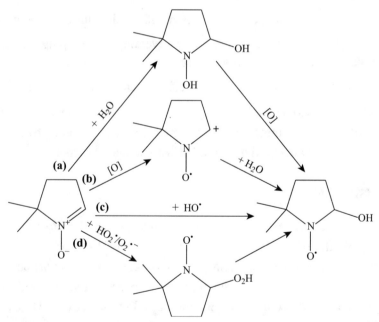

Figure 3.15 - Four ways to obtain the DMPO–OH nitroxide from the DMPO nitrone in aqueous medium: **(a)** Forrester-Hepburn reaction, **(b)** inverted *spin trapping*, **(c)** trapping of the HO˙ radical, **(d)** trapping of superoxide and decomposition of the DMPO–O₂H adduct.

---

[9] We can also observe the reduction of the trap to form an anionic radical followed by an electrophilic addition.

EPR spectroscopy can detect very small amounts of paramagnetic species in complex matrices which can contain several tens or hundreds of other compounds. However, it is important to always question whether what is observed by EPR corresponds to the main phenomenon or is the result of minor side-reactions. Before concluding on the trapping of a radical based on observation of a spectrum for an adduct, all the controls possible should be performed:

▷ verify the "window of potential" for the trap used, i.e., when it can be used without being reduced or oxidised; the data have been published for numerous compounds [McIntire *et al.,* 1980; Gronchi *et al.*, 1983; Tuccio *et al*, 1999]. If not already known, they must be determined by electrochemistry,

▷ avoid basic media (pH > 7.5 or solvents such as pyridine, DMF, etc.), or verify the chemical stability of the trap,

▷ for *in vitro* reactions, eliminate traces of metal ions using chelating agents if compatible with the phenomenon studied,

▷ always search for any possible reaction with nucleophilic compounds,

▷ examine how radical *scavengers* circumventing the trapping reaction affect the EPR spectrum for the adduct,

▷ always perform blank assays, in particular when an ene-reaction may occur.

This control can also be used to verify the absence of paramagnetic impureties in the traps, as some commercial batches may require purification by fractional distillation, recrystallisation or treatment with activated carbon.

3.7 - Conclusion

Through the examples presented in this chapter, we wished to demonstrate how the spin trapping method can be combined with EPR, and the extent of its potential. Recent developments in the biomedical field in particular have involved the synthesis of traps which can target particular sites of radical production, such as mitochondria or membranes [Hay *et al.*, 2005; Hardy *et al.*, 2007; Quin *et al.*, 2009], traps associated with cyclodextrins to stabilise adducts [Bardelang *et al.*, 2007; Han *et al.*, 2008; Stolze *et al.*, 2011], biotin-linked traps allowing antibody-based recognition [Mason, 2004; Lardinois *et al.*, 2010], or even traps labelled with fluorescent probes [Pou *et al.*, 1995; Hauck *et al.*, 2009]. Metabolisation of traps and adducts is a significant parameter which is too rarely taken into account [Liu *et al.*, 1996 and 1999; Bézières *et al.*, 2010; Ranguelova and Mason, 2011].

We cannot expect this method to provide more information than it can supply, and it is important to be aware of its limits. Although effective for the *detection* of free radicals, it should be avoided for their *quantification* because these trapping reactions are relatively slow, and adducts may be degraded. Even though the trapped *species* can sometimes be identified, it often only allows a *type of radical* (alkyl, peroxy, alkoxy, etc.) to be determined. Structural characterisation may require isotopic labelling of the trap or the substrate [Haseloff *et al.*, 1997; Timmins *et al.*, 1997; Conte *et al.*, 2009], or even the combined use of another analytical technique. Thus, recent studies have demonstrated the advantage of combining *spin trapping* , EPR and mass spectrometry (MS) [Tuccio *et al.*, 2006; El Hassan *et al.*, 2008; Reis *et al.*, 2009; Triquigneaux *et al.*, 2010a]: after the trapping reaction, the spin adducts are detected by EPR; the medium is then submitted to ESI-MS analysis without prior purification, and the precise structure of the adducts detected as well as that of their diamagnetic derivatives is established based on the results of tandem mass spectrometry analyses.

References

ADAM et al. (1996) *Tetrahedron Letters* **37**: 2113-2116.

ALLOUCH A. et al. (2005) *Organic and Biomolecular Chemistry* **23**: 2458-2462.

ALLOUCH A. et al. (2007) *Molecular Physics* **105**: 2017-2024.

BARDELANG D. et al. (2007) *Chemistry - a European Journal* **13**: 9344-9354.

BEHAR D. et al. (1970) *Journal of Physical Chemistry* **74**: 3209-3213.

BERLINER L. & FUJII H. (2004) *Antioxidants and Redox Signaling* **6**: 649-656.

BEZIERES N. et al. (2010) *Free Radicals in Biology and Medicine* **49**: 437-446.

BORS W. et al. (1992) *Journal of the Chemical Society Perkin Transactions* **2**: 1513-1517.

BRIERE R. et al. (1964) *Tetrahedron Letters* **27**: 1781-1785.

BUETTNER G. (1987) *Free Radicals in Biology and Medicine* **3**: 259-303.

BUETTNER G. & KIMINYO K. (1992) *Journal of Biochemical and Biophysical Methods* **24**: 147-151.

BUETTNER G. & MASON R. (1990) *Methods in Enzymology* **186**: 127-133.

BUETTNER G. & MASON R. (2003) *Critical Reviews of Oxidative Stresses and Aging* **1**: 27-38.

BUETTNER G. & OBERLEY L. (1978) *Biochemical and Biophysical Research Communications* **83**: 69-74

BUETTNER R. (2008) *Journal of Physical Chemistry A* **112**: 2447-2455.

BURKITT M. & MASON R. (1991) *Proceedings of the National Academy of Sciences of the USA* **88**: 8440-8444.

CONTE M. et al. (2009) *Journal of the American Chemical Society* **131**: 7189-7196.

DAVIES M. (2002) *Electron Paramagnetic Resonance* **18**: 47-73.

DENG C.H. et al. (2002) *Journal of Fluorine Chemistry* **116**: 109-115.

EBERSON L. (1992) *Journal of the Chemical Society Perkin Transactions* **2**: 1807-1813.

EBERSON L. (2000) *Toxicology of the Human Environment:* 25-47.

EL HASSAN I. et al. (2006) *Mendeleev Communications:* 149-151.

EL HASSAN I. et al. (2008) *New Journal of Chemistry* **32**: 680-688.

EVANS C. A. (1979) *Aldrichimica Acta* **12**: 23-29.

FINKELSTEIN et al. (1979) *Molecular Pharmacology* **16**: 676-685.

FINKELSTEIN et al. (1980) *Archives of Biochemistry and Biophysics* **200**: 1-16.

FREJAVILLE C. et al. (1994) *Chemical Communications:* 1793-1794.

FREJAVILLE C. et al. (1995) *Journal of Medicinal Chemistry* **38**: 258-265.

FORRESTER A. & HEPBURN S. (1971) *Journal of The Chemical Society Section C - Organic:* 701-703.

GOLDSTEIN S. et al. (2004) *Journal of Physical Chemistry A* **108**: 6679-6685.

GRONCHI G. et al. (1983) *Journal of Physical Chemistry* **87**: 1343-1349.

HAN Y. *et al.* (2008) *Journal of Organic Chemistry* **73**: 7108-7117.

HARBOUR J. & BOLTON J. (1975) *Biochemical and Biophysical Research Communications* **64**: 803-807.

HARDY M. *et al.* (2007) *Chemical Research in Toxicology* **20**: 1053-1060.

HASELHOFF R. *et al.* (1997) *Free Radical Research* **26**: 159-168.

HAUCK S. *et al.* (2009) *Applied Magnetic Resonance* **36**: 133-147.

HAY A. *et al.* (2005) *Archives of Biochemistry and Biophysics* **435**: 336-346.

IWAMURA M. & INAMOTO N. (1967) *Bulletin of the Chemical Society of Japan* **40**: 703.

JANZEN E. & BLACKBURN B. (1968) *Journal of the American Chemical Society* **90**: 5909-5910

JANZEN E. (1971) *Accounts of Chemical Research* **4**: 31-40.

JANZEN E. *et al.* (1992) *Tetrahedron Letters* **33**: 1257-1260.

JANZEN E. (1980) *Free Radicals in Biology* **4**: 115-154.

KADIISKA M. & MASON R. (2002) *Molecular and Biomolecular Spectroscopy* **58A**: 1227-1239.

KAROUI H. *et al.* (1996) *Journal of Biological Chemistry* **271**: 6000-6009.

KEZLER A. *et al.* (2003) *Free Radicals in Biology and Medicine* **35**: 1149-1157.

KHAN N. & SWARTZ H. (2002) *Molecular and Cellular Biochemistry* **234-235**: 341-357.

KNAUER B. & NAPIER J. (1976) *Journal of the American Chemical Society* **98**: 4395-4400.

KONAKA R. (1982) *Canadian Journal of Chemistry* **60**: 1532-1541.

KORTH H.-G. *et al.* (1992) *Angewandte Chemie International Edition* **31**: 891-893.

LAGERCRANTZ C. & FORSHULT S. (1968) *Nature* **218**: 1247-1248.

LARDINOIS O. *et al.* (2010) *Analytical Chemistry* **82**: 9155-9158.

LAURICELLA R. *et al.* (2004) *Organic and Biomolecular Chemistry* **2**: 1304-1309.

LAURICELLA R. *et al.* (2005) *Physical Chemistry Chemical Physics* **7**: 399-404.

LAURICELLA R. *et al.* (2010) *Chemical Communications* **46**: 3675-3677.

LIU K.J. *et al.* (1996) *Research on Chemical Intermediates* **22**: 499-509.

LIU K.J. *et al.* (1999) *Free Radicals in Biology and Medicine* **27**: 82-89.

MACINTIRE G. *et al.* (1980) *Journal of Physical Chemistry* **84**: 916-921.

MACKOR A. *et al.* (1966) *Tetrahedron Letters* **19**: 2115-2123.

MADDEN K. & TANIGUSHI H. (1991) *Journal of the American Chemical Society* **113**: 5541-5547.

MAKINO K. *et al.* (1980) *Analytical Letters* **50**: 311-317.

MASON R. (2004) *Free Radicals in Biology and Medicine* **36**: 1214-1223.

MASON R. & KADIISKA M. (2005) *Biological Magnetic Resonance* **13**: 93-109.

NAJJAR F. *et al.* (2005) *Tetrahedron Letters* **46**:2117-2119.

OLIVE G. *et al.* (2000) *Free Radicals in Biology and Medicine* **28**: 403-408.

PERKINS M.J. (1980) *Advances in Physical Organic Chemistry* **17**: 1-64.

POU S. *et al.* (1994) *Analytical Biochemistry* **21**: 76-83.

POU S. *et al.* (1995) *FASEB Journal* **9**: 1085-1090.

QUIN C. *et al.* (2009) *Tetrahedron* **65**: 8154-8160.

RANGUELOVA K. & MASON R. (2011) *Magnetic Resonance in Chemistry* **49**: 152-158

REIS A. *et al.* (2009) *European Journal of Mass Spectrometry* **15**: 689-703.

RIZZI C. *et al.* (1997) *Journal of the Chemical Society Perkin Transactions 2:* 2513-2518.

ROUBAUD V. *et al.* (2002) *Journal of the Chemical Society Perkin Transactions 2:* 958-964.

SCHMID P. & INGOLD K. (19878) *Journal of the American Chemical Society* **100**: 2493-2500.

STOLZE K. *et al.* (2007) *Bioorganic and Medicinal Chemistry* **15**: 2827-2836.

STOLZE K. *et al.* (2009) *Bioorganic and Medicinal Chemistry* **17**: 7575-7584.

STOLZE K. *et al.* (2011) *Bioorganic and Medicinal Chemistry* **19**: 985-993.

TERABE S. *et al* (1973) *Journal of the Chemical Society Perkin Transactions 2:* 1252-1258.

TIMMINS G. *et al.* (1997) *Redox Report* **3**: 125-133.

TRIQUIGNEAUX *et al.* (2009) *Journal of the American Society for Mass Spectrometry* **20**: 2013-2020.

TRIQUIGNEAUX M. *et al.* (2010) (a) *Organic and Biomolecular Chemistry* **2**: 1304-1309.

TRIQUIGNEAUX M. *et al.* (2010) (b) *Tetrahedron Letters* **51**: 6220-6223.

TUCCIO B. *et al.* (1995) *Journal of the Chemical Society Perkin Transactions 2:* 295-298.

TUCCIO B. *et al.* (1996) *Research on Chemical Intermediates* **22**: 393-404.

TUCCIO B. *et al.* (1999) *Electrochimica Acta* **44**: 4631-4634.

TUCCIO B. *et al.* (2006) *International Journal of Mass Spectrometry* **252**: 47-53.

TURNER M. & ROSEN G. (1986) *Journal of Medicinal Chemistry* **29**: 3439-3444.

VANIN A. (2002) *Methods in Enzymology* **359**: 27-42.

VILLAMENA F. *et al.* (2003) *Journal of Physical Chemistry* **107**: 4407-4414.

VILLAMENA F. *et al.* (2007), *Journal of the American Chemical Society* **129**: 8177-8191

Volume 1: BERTRAND P. (2020) *Electron Paramagnetic Resonance Spectroscopy - Fundamentals*, Springer, Heidelberg.

WANG Q. *et al* (2000) *Applied Magnetic Resonance* **18**: 419-424.

YOSHIMURA T. & KOTAKE Y. (2004) *Antioxidants and Redox Signaling* **6**: 639-647.

ZHAO H. *et al.* (2001) *Free Radicals in Biology and Medicine* **31**: 599-606.

Chapter 4

Copper complexation by peptides implicated in neurodegenerative diseases

Dorlet P.[a] and Hureau C.[b]

[a] *Laboratory of Oxidative Stress and Detoxification, UMR 8221, CNRS and CEA/iBiTec-S/SB²SM, Gif-sur-Yvette.*

[b] *Laboratory of Coordination Chemistry UPR 8241, CNRS, Toulouse.*

4.1 - Introduction

Neurodegenerative diseases (Alzheimer's disease, prion diseases such as Creutzfeldt-Jakob, Parkinson's disease, etc.) have raised serious medical and societal problems in the ageing populations in western countries. Their development in the brain is characterised by the following phenomena:

▷ Deposition in the extracellular medium of protein plaques, which perturb neuronal activity. The proteins found in these plaques are present in *soluble* form in the healthy brain, but in diseased brains, they adopt a conformation which favours their aggregation and the formation of plaques.

▷ Weakening of the cellular defence mechanisms combating "oxidative stress". During normal functioning of bioenergetic processes which occur in the cells of aerobic organisms, O_2 molecules are fully reduced to form H_2O molecules. But sometimes the O_2 is only partially reduced, which leads to the formation of *Reactive Oxygen Species* (ROS) such as peroxide H_2O_2, superoxide radicals $O_2^{-\bullet}$, and hydroxyl radicals $HO^\bullet$ (see sections 3.4.2 and 3.4.3 in chapter 3). These toxic species are normally eliminated by specialised enzymes, catalases, superoxide dismutases and

© Springer Nature Switzerland AG 2020
P. Bertrand, *Electron Paramagnetic Resonance Spectroscopy*,
https://doi.org/10.1007/978-3-030-39668-8_4

peroxidases. When this detoxification system fails to completely eliminate ROS, the cell is said to be subjected to oxidative stress. Because of their significant energy demands and their relatively low capacity for regeneration, brain cells are particularly exposed to oxidative stress and (post-mortem) brain tissues from patients with neurodegenerative diseases present evidence of excessive oxidation. For example, in the case of Alzheimer's disease, an abnormally high proportion of oxidised proteins, oxidised DNA and RNA, and peroxidised lipids are observed. However, it is impossible to deduce from these observations whether ROS are the *cause* or the *consequence* of neuronal death [Andersen, 2004].

Very rapidly it was hypothesised that transition ions could play an important role in neurodegenerative diseases. Indeed, pathogenic aggregates contain high concentrations of elements such as Fe, Cu and Zn (table 4.1) and most of the peptides and proteins involved in these diseases have sites where transition metal ions can bind.

Table 4.1 - Neurodegenerative diseases involving metal ions in Man.

Disease	Peptide/protein	Metal ions	Localisation in the brain
Alzheimer's	Amyloid-β peptide Tau protein	Cu(I/II), Zn(II), Fe(II/III)	Cortex, hippocampus, basal anterior brain, brainstem
Creutzfeldt-Jakob	Prion protein	Cu(II)	Cortex, thalamus, brainstem, cerebellum
Parkinson's	α-synuclein	Cu(I/II)	Substantia nigra, Cortex, locus coeruleus
Amyotrophic lateral sclerosis	Superoxide dismutase	Cu(I/II), Zn(II)	Cerebral cortex

In particular, Cu(II) ions can bind with high affinity; as a result, the Cu(II)-protein interaction may occur *in vivo* despite the low availability of Cu(II) ions [Hureau, 2012]. These ions could therefore participate in the aggregation process. In addition, transition ions which display redox activity, such as Fe(II)/Fe(III) and Cu(I)/Cu(II) couples can catalyse the production of ROS from dioxygen [Gaggelli *et al.*, 2006; Kozlowski *et al.*, 2009; Sigel *et al.*, 2006]. Figure 4.1 shows, for example, how a copper complex catalyses the reduction of O_2 by ascorbate to form $HO^{\bullet}$ and HO^{-}.

All these examples explain why numerous EPR studies have investigated the coordination of Cu(II) ions to peptides/proteins involved in neurodegenerative diseases. This is undoubtedly related to the development of high-resolution EPR techniques which are ideally suited to the study of the coordination sites of Cu(II) ions.

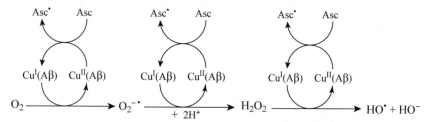

Figure 4.1 - O_2 reduction by ascorbate to form $HO^•$ and HO^- catalysed by a copper complex. The complex is Cu(Aβ), which is studied in section 4.4.

Below, we present the method used to determine the binding sites for a Cu(II) ion coordinated to a peptide (section 4.2), then we will apply it to the prion protein (section 4.3) and the amyloid-β peptide (Aβ) involved in Alzheimer's disease (section 4.4). Readers who are not very familiar with biological macromolecules will find a general presentation of peptides and proteins in appendix 4 to this volume. The amino acids making them up are designated by the one-letter and three-letter codes listed in table 1 of this appendix.

4.2 - EPR determination of the coordination of a Cu(II) ion by a peptide

4.2.1 - EPR of Cu(II) complexes

Complexes containing the Cu(II) ion in the $3d^9$ configuration have a spin $S = \frac{1}{2}$. When the ligands do not exert very strong geometric constraints, the complexes *spontaneously* adopt a stretched octahedral geometry (D_{4h} symmetry) or square-based pyramidal geometry (C_{4v} symmetry) which minimises their energy, and the unpaired electron occupies the $d_{x^2-y^2}$ orbital, where z is the symmetry axis [Volume 1, appendix 2]. This spontaneous symmetry-lowering takes place, for example, when the Cu(II) ions generated upon solvation of a copper salt in water are coordinated by 6 H_2O molecules. In this chapter we will see that it also occurs when these ions interact with highly flexible macromolecules, such as peptides and the "disordered" domains of some proteins.

The spectrum for a frozen solution of Cu(II) complexes reflects the axial symmetry of the $\tilde{\mathbf{g}}$ matrix (parameters $g_{//}$ and $g_\perp$) and of the hyperfine matrix due to interactions with ^{63}Cu and ^{65}Cu nuclei of spin $I = \frac{1}{2}$ (parameters $A_{//}$ and $A_\perp$). The values of $g_{//}$ and $A_{//}$ are very sensitive to the nature of the equatorial ligands, and empirical diagrams have been proposed to determine the coordination of the Cu(II) ion from the values of these parameters [Peisach and Blumberg, 1974] (figure 4.2).

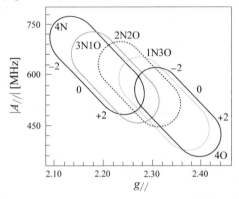

Figure 4.2 - Peisach-Blumberg diagram for a square planar Cu(II) complex. ($g_{//}$, $A_{//}$) domains are defined for different equatorial coordinations and several charges of the complex.

On the spectrum, the *superhyperfine pattern* due to interactions with the paramagnetic nuclei of ligands can sometimes be distinguished; this pattern is naturally very useful to determine the nature and number of ligands [Volume 1, section 4.4]. When this structure is not resolved or when it is too difficult to analyse, high-resolution techniques can be used, such as pulsed EPR (ESEEM and HYSCORE) or ENDOR. These techniques are presented in appendices 2 and 3 to this volume.

4.2.2 - Example of EPR spectra for Cu(II)(peptide) complexes

As examples, we have selected complexes formed with the small peptides Gly-His-Lys (GHK) and Asp-Ala-His-Lys (DAHK), high-affinity Cu(II) chelators present in human plasma. The GHK peptide, discovered by Pickart [Pickart and Thaler, 1973], was initially considered to be a cellular growth factor, but other functions have since been attributed to it (wound healing, tissue repair). As a result, it is now used in numerous "anti-ageing" products. The DAHK peptide is the N-terminal portion (see appendix 4) of human serum albumin, a protein involved in Cu(II) transport which is also

present at high concentrations in cerebrospinal fluid. Cu(II)(DAHK) and
Cu(II)(GHK) complexes are simple systems of known structure (figure 4.3),
they can be used to study the relations between the mode of coordination of
the Cu(II) ion and the EPR parameters [Hureau *et al.*, 2011].

▷ In the Cu(II)(DAHK) complex, the equatorial coordination of the Cu(II)
ion is ensured by 4 nitrogen atoms respectively supplied by the amine NH_2
of the N-terminal end, deprotonated amides from the Asp-Ala and Ala-His
peptide bonds, and the N_δ nitrogen atom from the imidazole group of the
histidine (see figure A4.1 in appendix 4). A molecule of water in apical
position completes the square-based pyramidal geometry.

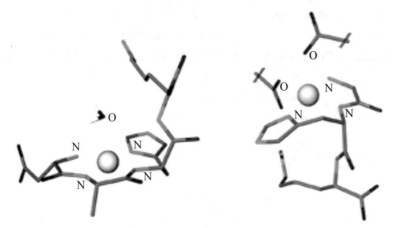

Figure 4.3 - Crystal structures of Cu(II)(DAHK) and Cu(II)(GHK) complexes.
Only the skeleton of the polypeptide chain is represented (see appendix 4).
The ligands of the Cu(II) ion are indicated.

▷ In the Cu(II)GHK complex, the Cu(II) ion is coordinated in the equatorial
plane by the N-terminal amine, the deprotonated amide of the Gly-His peptide bond and the N_δ nitrogen atom from the imidazole group of the histidine. The fourth equatorial position and an apical position are occupied by
an oxygen atom of carboxylate groups from another GHK molecule. These
two positions are *labile in solution* and we will see that other molecules
(amino acids) can substitute for carboxylates. As in the case of the Cu(II)
(DAHK) complex, the geometry is of the square-based pyramidal type.

X-band spectra of frozen solutions of Cu(II)(DAHK) and Cu(II)(GHK) complexes are represented in figures 4.4a and 4.4b.

The spectrum of the Cu(II)(DAHK) complex presents four hyperfine lines
centred at $g_{//} = 2.19$ spaced 20 mT apart ($A_{//}^{Cu} = 596$ MHz), and a line at

$g_\perp = 2.04$ without resolved hyperfine splitting (figure 4.4a). The spectrum for the Cu(II)(GHK) complex has four hyperfine lines centred at $g_{//} = 2.23$ spaced 18 mT apart ($A_{//}^{Cu} = 560$ MHz), and a line at $g_\perp = 2.05$ (figure 4.4b). Zooming on the $g \approx g_\perp$ region reveals seven distinguishable superhyperfine lines with relative intensities 1:3:6:7:6:3:1 produced by the interaction with three equivalent ^{14}N ($I = 1$) nuclei [Volume 1, complement 2 to chapter 2]. The fourth equatorial position is occupied by the oxygen atom from a water molecule. $g_{//}$ is found to be smaller and $A_{//}^{Cu}$ larger for the Cu(II)(DAHK) complex (4N equatorial coordination) than for the Cu(II)(GHK) complex (3N1O equatorial coordination), in agreement with the diagram in figure 4.2. This difference is apparent in the spectrum shown in figure 4.4a, where the fourth line in the "parallel" pattern appears clearly at a higher field than that of the "perpendicular" line. This is known as *overshot resonance*.

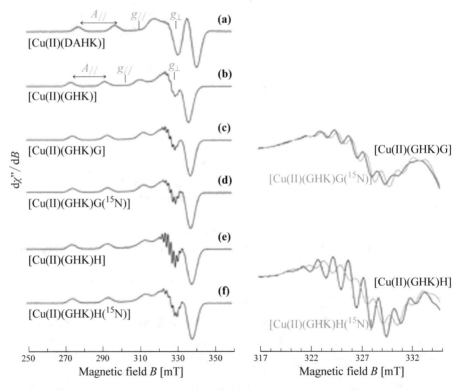

Figure 4.4 - X-band EPR spectra for Cu(II)(DAHK) and Cu(II)(GHK) complexes. Temperature 50 K, microwave frequency 9.42 GHz, power 0.13 mW. Modulation: frequency 100 kHz, peak-to-peak amplitude 0.5 mT.

As an equatorial position of the Cu(II)(GHK) complex is *labile* (figure 4.3), it can be modified by adding an amino acid to the solution. When glycine is added in excess, a slight change to the spectrum is observed, with $g_{//} = 2.22$; $A_{//}^{Cu} = 568$ MHz; $g_{\perp} = 2.05$ (figure 4.4c). The super-hyperfine pattern is also slightly modified. To determine the origin of these changes, the experiment can be repeated with ^{15}N-labelled glycine ($I = \frac{1}{2}$) (figure 4.4d). This produces variations in the superhyperfine pattern (right-hand part of the figure) suggesting that glycine binds to Cu(II) through the nitrogen from its amine function. If histidine is added instead of glycine, the modifications to the spectrum are more extensive (figure 4.4e). The parameters become $g_{//} = 2.21$; $A_{//}^{Cu} = 590$ MHz; $g_{\perp} = 2.05$, values close to those obtained with Cu(II)(DAHK). In the perpendicular region, we now observe a pattern of nine (rather than seven) superhyperfine lines with relative intensities 1:4:10:16:19:16:10:4:1, which indicates that the Cu(II) is now bound to four magnetically-equivalent nitrogen atoms in the equatorial plane. To determine whether the nitrogen atom is from the imidazole group or the amine function, an EPR spectrum was recorded with histidine labelled with ^{15}N on the imidazole nucleus (figure 4.4f). The significant change to the superhyperfine pattern (right-hand part of the figure) indicates that the histidine binds via the imidazole group.

This study shows that the Cu(II) ion can be coordinated in different ways by a peptide and that its coordination determines how the polypeptide chain folds (figure 4.3). It is a good illustration of the advantages of the diagrams in figure 4.2 and of the information that can be gleaned from the superhyperfine structure of the spectrum. These different elements will be exploited below.

4.3 - Coordination of Cu(II) by the prion protein

4.3.1 - Prion protein

Prion protein was the subject of much research and discussion in the 1990s at the time of the mad cow epidemic. This term, introduced by Stanley Prusiner (Nobel prize in physiology or medicine in 1997) in 1982, designates an infectious agent of purely protein nature (prion = *PRoteinaceous Infectious ONly particle*). In mammals, prion is responsible for communicable spongiform encephalopathies, such as Creutzfeldt-Jakob disease and Kuru in man, scrapie in sheep, and bovine spongiform encephalitis (mad cow disease). In healthy brains, this protein is present in a non-pathogenic form, termed PrP^C. It is found in large amounts in the central nervous system, in particular at the level of the synapses where it is tethered to the external surface of the neurones by its C-terminal part. It is necessary for this system to function correctly, and its capacity to bind copper ions led to hypotheses that it acted as a copper transporter, regulator or storage system [Millhauser, 2004]. In some circumstances, a conformational change causes proteins with the normal PrP^C form to adopt the pathogenic PrP^{Sc} form. This pathogenic form catalyses the change in conformation of other PrP^C proteins, which accelerates the process and rapidly leads to the formation of plaques. Even if the number of deaths due to prion diseases is extremely low in humans, it is nevertheless important to study plaque formation and the resulting cellular degeneration, not only to deal with animal health-scares such as the mad-cow epidemic, but also to understand the analogous process encountered in other diseases such as Alzheimer's and Parkinson's.

The PrP^C protein is *partially disordered*: its C-terminal domain is structured in three α helices, but its N-terminal domain is *unstructured in solution*. In the N-terminal domain of the PrP^C protein in mammals, an octapeptide PHGGGWGQ is repeated four times (figure 4.5).

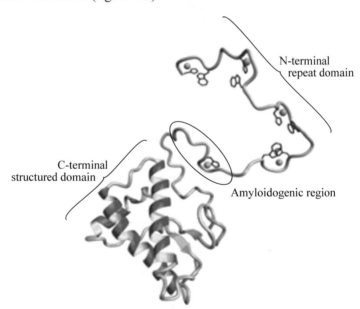

Figure 4.5 - Schematic representation of the peptide skeleton of the PrP^C protein. The imidazole groups of the histidines and the indole groups of the tryptophans mentioned in the text are indicated. [From Millhauser G., L. *et al.* (2004) *Accounts of Chem. Research* **37**: 79-85 © 2004 American Chemical Society, reproduced with permission]

When the C-terminal domain folds as β sheets, the PrP^{Sc} pathogenic form of the protein is produced. Hornshaw and collaborators were the first to mention PrPC's capacity to bind copper ions [Hornshaw *et al.*, 1995a; Hornshaw *et al.*, 1995b]. Binding takes place at the level of the repeat domain (fragment 51 to 89) and in the amyloidogenic region (fragment 90 to 126) which is involved in protein folding to produce the pathogenic form. It is therefore easy to understand the advantage of studying copper binding at this level. To determine how Cu(II) ions bind to the different domains of the PrP^{C} protein, synthetic peptides reproducing fragments of the protein's polypeptide chain are generally used. These peptides are of the type Ac–X$_1$X$_2$…X$_n$–NH$_2$, where X$_1$, X$_2$,…, X$_n$ are amino acids. Ac–X$_1$ is N-terminal acetylation and X$_n$–NH$_2$ is C-terminal amidation. These modifications aim to mimic the peptide bonds normally involving X$_1$ and X$_n$ in the protein.

4.3.2 - The sites of Cu(II) fixation on the N-terminal repeat domain

EPR study of synthetic peptides of variable length showed that the smallest fragment of the octapeptide which coordinates Cu(II) is HGGGW [Aronoff-Spencer *et al.*, 2000]. Indeed, the spectrum for the Cu(HGGGW) complex is identical to that of the repeat domain of the PrP^{C} protein saturated by 4 Cu(II) equivalents. It is characterised by $g_{//} = 2.23$; $g_\perp = 2.06$; $|A_{//}| = 496$ MHz, characteristic values for a 3N1O coordination in the equatorial plane (figure 4.2). This complex was crystallised and its structure revealed the Cu(II) ion to be coordinated in the equatorial plane by the nitrogen from the histidine, two deprotonated amides from His-Gly2 and Gly2-Gly3 peptide bonds and the carbonyl from the Gly3–Gly4 peptide bond (figure 4.6). A molecule of water in the apical position forms a hydrogen bond with the indole of the tryptophan residue. Pulsed EPR experiments showed that the structure *in solution* is identical to that observed in the solid state [Burns *et al.*, 2002].

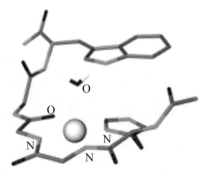

Figure 4.6 - Crystal structure of the Cu(HGGGW) complex. Only the skeleton of the polypeptide chain is represented. The ligands of the Cu(II) ion are indicated.

Since the HGGGW sequence is repeated in all four octapeptides, the four Cu(II) sites of the N-terminal domain produce identical spectra when this domain is saturated by four copper equivalents per protein. But if only one Cu(II) equivalent is provided per protein, it preferentially binds to the histidine nitrogens of the four octapeptides, and the 4N equatorial coordination produces a different spectrum [Valensin *et al.*, 2004]. Binding of Cu(II) ions is *anti-cooperative*, which means that binding of the first ion makes binding of subsequent ions more difficult [Walter *et al.*, 2006]. These experiments show how the *concentration* of Cu(II) ions determines folding of the N-terminal domain.

4.3.3 - The sites of Cu(II) fixation on the amyloidogenic domain

We studied the role of histidine His96 in the amyloidogenic domain (figure 4.5) in copper binding in detail [Hureau *et al.*, 2006]. Comparison of the EPR spectra of the different fragments of the polypeptide chain containing this histidine indicated that the shortest fragment supplying all the Cu(II) ligands is the pentapeptide GGGTH. In figure 4.7 we have represented the variations of the EPR spectrum for a frozen solution of Cu(II)(GGGTH) complexes as a function of pH.

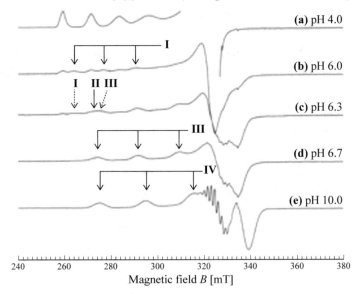

Figure 4.7 - X-band spectrum for a frozen solution of Cu(II)(GGGTH) complexes as a function of pH. Solvent: water, glycerol 10 %. Temperature 100 K, microwave frequency 9.38 GHz, power 0.03 mW. Modulation: frequency 100 kHz, peak-to-peak amplitude 0.5 mT. The $g_\perp$ region of spectrum **(a)** was deleted to improve readability. This spectrum is produced by Cu(II) ions not linked to the peptide.
[From Hureau C. *et al.* (2006) *J. Biol. Inorg. Chem.* **11**: 735-744]

An increase in pH results in a decrease of $g_{//}$, and to a lesser extent of $g_\perp$, and an increase of $A_{//}^{Cu}$. The UV-visible absorption spectrum for a solution of complexes also varies, with an increase in the d-d transition energy as the pH increases. The ligand field model allows this increase to be correlated with the decrease in $g_{//}$ and $g_\perp$ [Volume 1, annexe 2]. Detailed study shows that these spectral variations are due to the coexistence of *four forms of the complex* noted I to IV, with relative proportions varying with the pH as illustrated in figure 4.8.

This study was also used to determine the spectra of forms I, III and IV (figure 4.9), which were simulated with the parameters listed in table 4.2.

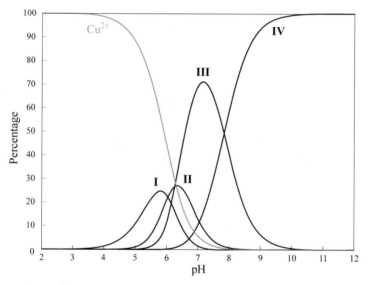

Figure 4.8 - Predominance diagram for the different forms of the Cu(II)(GGGTH) complex as a function of pH. [From Hureau C. *et al.* (2006) *J. Biol. Inorg. Chem.* **11**: 735-744]

A well-resolved superhyperfine pattern is visible in the $g_\perp$ region of the spectrum for form IV. It consists of nine lines with relative intensities corresponding to coupling with four equivalent nuclei of spin $I = 1$, thus the equatorial coordination is of type 4N (compare to figure 4.4e). The same pattern is observed on the EPR spectrum for a *liquid solution* at room temperature, confirming that it is a superhyperfine pattern and not hyperfine lines produced by the free Cu (figure 4.10).

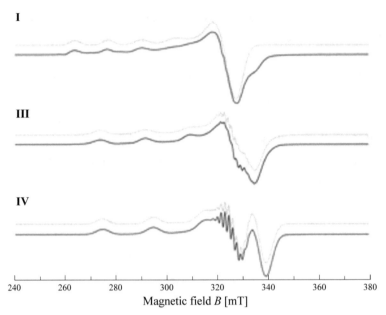

Figure 4.9 - Experimental spectra (black) for forms I, III and IV and their simulations (grey dashes). For form I, the contribution of unbound Cu(II) ions was subtracted. The parameters used in the simulation are listed in table 4.2.

[From Hureau C. *et al*. (2006) *J. Biol. Inorg. Chem.* **11**: 735-744]

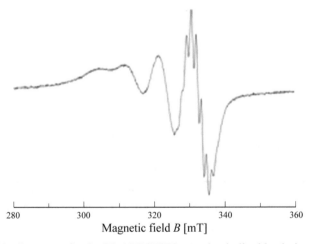

Figure 4.10 - Spectrum for the [Cu(GGGTH)] complex in liquid solution at pH 9.0. Temperature 298 K, microwave frequency 9.38 GHz, power 0.8 mW. Modulation: frequency 100 kHz, peak-to-peak amplitude 0.25 mT.

[From Hureau C. *et al*. (2006) *J. Biol. Inorg. Chem.* **11**: 735-744]

The diagram in figure 4.2 suggests that variations of the $g_{//}$ and $A_{//}$Cu parameters for the complex when the pH increases are due to the progressive replacement of oxygenated ligands (water molecules, oxygen from a carbonyl) by nitrogenated ligands.

Table 4.2 - EPR parameters of different forms of the Cu(II)(GGGTH) complex and of the (90–231) fragment of the Syrian hamster prion protein.

Complex	$g_{//}$ $(\sigma_{g_{//}}/g_{//})^b$	$g_\perp$	$A_{//}^{Cua}$ $(\sigma_{A_{//}})^b$	$A_\perp^{Cua}$	$A_{//}^{Na}$	$A_\perp^{Na}$	Width $l_{//}$; $l_\perp$ [mT]	Equatorial coordination
I	2.36 (0.005)	2.07	421 (−10)	45	35	35	Gaussian 2.0; 3.0	1N3O
III	2.23 (0.007)	2.05	540 (−9)	55	33	40	Lorentzian 0.9; 1.0	3N1O
IV	2.20 (0.006)	2.05	600 (−5)	50	35	42	Gaussian 0.70; 0.75	4N
rSHaPrP^c	2.22		541					

^{a} in MHz. ^{b} σ_g and σ_A: widths of the distributions of the g and A parameters. ^{c} fragment (90–231) for the Syrian hamster PrP^C protein, pH 7.4 [Burns *et al.*, 2003].

Pulsed EPR (ESEEM) experiments demonstrated that the imidazole from the histidine in the GGGTH peptide is coordinated to the Cu(II) ion at pH 5. Increasing the pH causes successive deprotonation of the amide functions of other amino acids, releasing N ligands and allowing the peptide to fold around the copper. In particular, the equatorial coordination of Cu(II) passes from 3N1O at pH 6.7 (form III) to 4N at high pH (form IV). The oxygenated ligand of form III is either a water molecule or a carbonyl from the peptide skeleton. ESEEM, HYSCORE and ENDOR experiments performed on the Cu(II) (GGGTH) complex indicate that interaction with exchangeable protons is not as strong as expected for protons from a water molecule in equatorial position. It is therefore the carbonyl group that supplies the oxygenated ligand [Hureau *et al.*, 2008]. We thus arrive at the models for forms III and IV represented in figure 4.11.

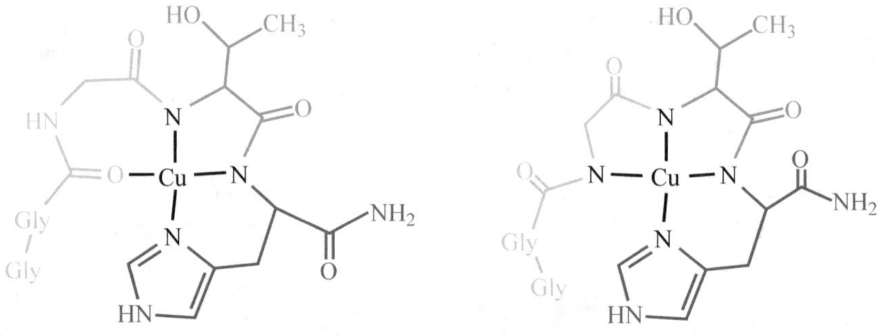

Figure 4.11 - Coordination models for forms III and IV of the Cu(II)(GGGTH) complex.

In table 4.2 we have also indicated the $g_{//}$ and $A_{//}^{Cu}$ parameters measured on the spectrum for fragment (90–231) of the Syrian hamster PrP^C protein, which lacks the repeat domain [Burns *et al.*, 2003]. Comparison with the parameters of the different forms of the Cu(II)(GGGTH) complex shows that the copper coordination is 3N1O in the protein at physiological pH. This coordination mode is important as it determines the redox properties and thus the reactivity of the protein for the production of ROS. For example, in cyclic voltammetry at pH 6.7, a quasi-reversible wave is observed at $E^{1/2} = 0.04$ V versus Ag/AgCl which corresponds to the *reduction* of Cu(II) to form Cu(I), and no wave is detected in oxidation. But at pH 9.0, *oxidation* of Cu(II) to produce Cu(III) occurs, with a quasi-reversible wave at $E^{1/2} = 0.66$ V versus Ag/AgCl. This oxidation is made possible by the stabilisation of Cu(III) due to the negative charges of the amides [Hureau *et al.*, 2006].

A similar study to that performed on the pentapeptide GGGTH was performed on the octapeptide GGGTHSQW, which contains a tryptophan (W), a potentially redox-active amino acid. The pH dependence of the spectrum and the different modes of coordination of the Cu(II)(GGGTHSQW) complex are identical to those of the Cu(II)(GGGTH) complex. Some differences in redox properties were, however, observed at pH 10. Fluorescence experiments revealed a strong influence of copper on the fluorescence of tryptophan at pH 10, suggesting that the folding of the GGGTHSQW polypeptide is such that this amino acid is close to the Cu [Hureau *et al.*, 2008].

The studies which we have just described relate to the binding of the copper to the His96 residue in the amyloidogenic region of the protein. It should be noted that His111, also present in this region, has been proposed as an alternative binding site [Jones *et al.*, 2004].

All these studies show the influence of the Cu(II) ions on the conformation of the PrP^C protein. The nitrogens from histidines are preferred ligands, but other amino acids can also contribute to folding of the peptide chain.

4.4 - Cu(II) coordination by the amyloid-β peptide

4.4.1 - Amyloid-β peptide

Alzheimer's disease severely affects the mental and physical capacities of those who develop it. In Europe, it currently affects one in twenty people over 65 years of age, which corresponds to over 8 million people. The health and societal challenges posed are therefore very extensive. Development of therapies relies on various approaches, but, in particular, it requires understanding of the molecular phenomena causing the disease. Two types of morphological lesions are observed in patients' brains: intracellular neurofibrillary degeneration and extracellular "senile plaques", principally composed of aggregates of peptides known as "amyloid-β peptides" (Aβ) [Holtzman *et al.*, 2011]. These peptides, mainly composed of between 40 and 42 amino acids, are produced from fragmentation of a membrane protein, APP (*Amyloid Precursor Protein*), and they are present in a *soluble monomeric* form within the synapses of healthy brains. Senile plaques contain an abnormally high concentration (of around mM) of Zn, Fe and Cu ions, which are very likely involved in the process of APP protein fragmentation and aggregation of Aβ peptides [Atwood *et al.,* 1998; Bush, 2000]. We have already mentioned that Cu and Fe ions can also play a significant role in the production of reactive oxygen species (ROS), which are particularly toxic to neurons [Hureau and Faller, 2009]. Aggregation of Aβ peptides and ROS production are the main consequences of the "amyloid cascade" (figure 4.12).

Although the availability of Cu(II) ions in the synaptic cleft is relatively low (< nM), their affinity for the Aβ peptide is quite high (K_d in the nM range), and their interaction is thus physiologically relevant [Hureau, 2012].

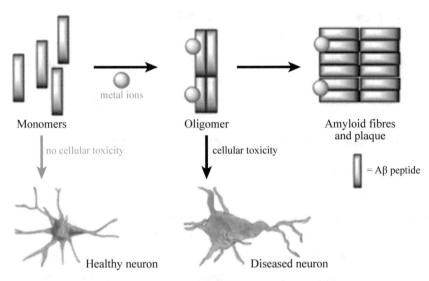

Monomers   Oligomer   Amyloid fibres
                                       and plaque

metal ions

no cellular toxicity   cellular toxicity

= Aβ peptide

Healthy neuron   Diseased neuron

Figure 4.12 - Schematic representation of the amyloid cascade. The Aβ peptides result from cleavage of the APP proteins. Their interaction with metal ions leads to the formation of toxic oligomeric forms.

4.4.2 - Determining the Cu(II) fixation sites on the human Aβ peptide

It is generally accepted that a single Cu(II) ion binds with a strong affinity to the Aβ peptide and that its ligands are supplied by the first 16 amino acids (sequence DAEFRHDSGYEVHHQK) of the peptide [Kowalik-Jankowska *et al.*, 2003]. Studies aiming to identify the ligands were therefore performed with the Aβ16 peptide represented in figure 4.13, which forms a soluble complex with Cu(II).

The results were a subject of considerable debate, which is not surprising given the complexity of the interactions of the Cu(II) ions with this peptide. A first source of complications was revealed by the pH-dependence of the EPR spectrum, which showed that two forms of the complex coexist at physiological pH, a low-pH form and a high-pH form [Karr and Szalai, 2007; Syme *et al.*, 2004]. In figure 4.14, we have represented the typical EPR spectra for these two forms and their simulations.

The EPR parameters of the two forms of the complex are similar, and the diagram in figure 4.2 suggests a 3N1O coordination. To go further it was necessary to perform high-resolution EPR studies on complexes of isotopically labelled peptides.

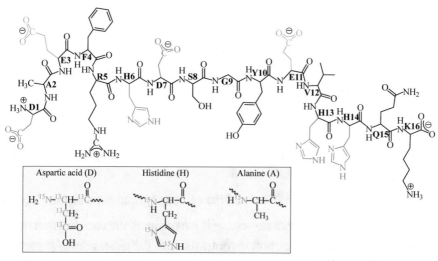

Figure 4.13 - The Aβ16 peptide. Isotopically labelled atoms (^{15}N and ^{13}C) used in high-resolution EPR studies are shown in the inset. The imidazoles of the three His residues and the carboxylates of the Asp and Glu residues are represented in grey.

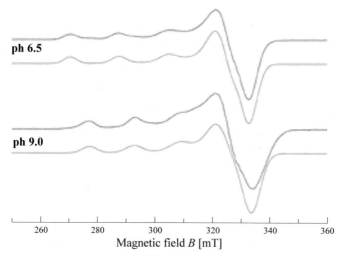

Figure 4.14 - EPR spectra (black) of the low-pH and high-pH forms of the Cu(II)(Aβ16) complex. Temperature 100 K, microwave frequency 9.38 GHz, power 0.5 mW. Modulation: frequency 100 kHz, peak-to-peak amplitude 0.5 mT. The simulations (grey) were obtained with the following parameters: pH 6.5: $g_{//} = 2.263$, $g_{\perp} = 2.057$, $A_{//} = 539$ MHz, $A_{\perp} = 50$ MHz, pH 9.0: $g_{//} = 2.228$, $g_{\perp} = 2.056$, $A_{//} = 494$ MHz, $A_{\perp} = 35$ MHz.

ESEEM and HYSCORE experiments (see appendix 2 to this volume) performed with Aβ16 peptides in which the three histidines were separately labelled with ^{15}N nitrogen ($I = \frac{1}{2}$) (inset in figure 4.13) showed these three residues to be involved in the equatorial coordination of the Cu(II) [Shin and Saxena, 2008]. This result was difficult to interpret as the experiments were performed at pH 7.4, and in these conditions both forms of the complex co-exist. Barnham's group therefore repeated the experiments at pH 6.3 and pH 8.0 on ^{15}N-nitrogen- and ^{13}C-carbon- labelled peptides ($I = \frac{1}{2}$) [Drew *et al.*, 2009a; Drew *et al.*, 2009b]. The results led the authors to propose the following equatorial coordinations (see figure 4.13):

▷ low-pH form: two histidines, the carboxylate from aspartate 1 (D1) and the terminal NH$_2$-amine,

▷ high-pH form: three histidines and the carbonyl group from alanine 2 (A2).

The latter suggestion appears surprising if we compare the coordination of the low-pH and high-pH forms, and it contrasts with the other results presented in the literature. We therefore performed high-resolution EPR experiments on Aβ16 peptides in which the amino acids proposed as ligands by Barnham's group were specifically labelled [Dorlet *et al.*, 2009; Hureau and Dorlet, 2012]: uniform labelling with carbon ^{13}C and nitrogen ^{15}N on D1, labelling with nitrogen ^{15}N on A2, and uniform labelling with nitrogen ^{15}N on each individual histidine (inset in figure 4.13). All complexes were prepared at pH 6.5 and at pH 9.0 to allow separate study of the low-pH and high-pH forms. Below, we summarise the main results deduced from ENDOR and pulsed EPR experiments. The ESEEM and HYSCORE spectra are analysed in detail in complement 1.

▷ **pH 6.5**. Comparison of the ENDOR spectra for the complex with ^{15}N-labelled D1 and the unlabelled complex confirms that the terminal amine –NH$_2$ of the peptide is an equatorial ligand. ESEEM experiments show that at least two histidines *simultaneously* coordinate the metal ion in the equatorial position (complement 1). HYSCORE spectra for complexes where the histidines are separately labelled with nitrogen ^{15}N indicate that histidine H6 is an equatorial ligand to the copper ion, and that a second equatorial position is occupied by H13 in some complexes, and by H14 in others. In addition, six-pulse HYSCORE experiments show that the 4th equatorial ligand is the *carbonyl* from D1 and that the *carboxylate* from D1 is an axial ligand. All these results contributed to the model presented in figure 4.15a.

▷ **pH 9.0**. ENDOR experiments show that the equatorial coordination by the terminal amine –NH$_2$ is conserved. ESEEM experiments indicate that a single histidine is coordinated to the Cu(II), but HYSCORE spectra for the complexes with ^{15}N-labelled histidines show that the three histidines contribute equally to the coordination. In addition, HYSCORE spectra indicate that the amide function of the D1–A2 peptide bond is deprotonated and that it is coordinated with the Cu(II) in the equatorial position. This interpretation is confirmed by the effect of labelling at D1 and A2 on the superhyperfine pattern on the EPR spectrum. On the HYSCORE spectrum, a signal typical for an amide nitrogen adjacent to a carbonyl bound to the Cu(II) is also observed. This carbonyl and amide can be attributed to the A2–E3 peptide bond. All these results contributed to the model presented in figure 4.15b.

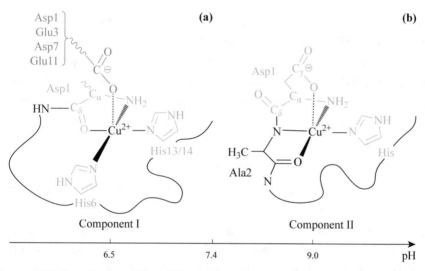

Figure 4.15 - Coordination of the Aβ16 peptide to the Cu(II) ion. **(a)** pH 6.5. Axial coordination by the carboxylates of Glu3, Asp7 and Glu11 was demonstrated by NMR, **(b)** pH 9.

The dynamic nature of the coordination of the Cu(II) ion by the Aβ peptide is clearly evident on the NMR spectrum recorded at room temperature [Hureau *et al.*, 2009]. Indeed, experiments performed at pH 6.5 indicate that the four carboxylate functions on the one hand, and the three imidazole functions of the histidines on the other, are similarly affected by the paramagnetism of the Cu(II) ion, demonstrating that there exist equilibria between equivalent functions for the same coordination position (figure 4.15a). Experiments performed at pH 9 show that, in alkaline conditions, these equilibria only concern the three imidazole functions. Although the EPR experiments were performed at cryogenic temperatures, their results are perfectly compatible with those obtained by NMR.

Determining the coordination of the Cu(II) ion by the Aβ peptide is complex. This problem could only be solved by using high-resolution EPR experiments on a series of isotopically labelled complexes [Dorlet *et al.*, 2009; Hureau and Dorlet, 2012]. The results reveal the specific role of the histidines and the first two amino acids in the peptide (Asp1 and Ala2) in coordinating the Cu(II) (figure 4.15). The presence of truncated Aβ(3 – 40/42) peptides in amyloid plaques could therefore be explained by oxidative cleavage due to the Cu(II) coordination in the high-pH form.

4.4.3 - Comparison of human and rat Aβ peptides

Although their brains contain Aβ peptides, *muridae* (mice and rats) do not develop an equivalent of Alzheimer's disease. The differences between "murine" and human peptides are found in fragment Aβ16, where arginine 5 (R5) is replaced by a glycine (G), tyrosine 10 (Y10) by a phenylalanine, and histidine 13 (H13) by an arginine (figure 4.13). These substitutions have several consequences:

▷ the pKa characterising the equilibrium between the low-pH and high-pH forms of complexes with Cu(II), which is 7.7 for human Aβ, is 6.2 for murine Aβ [Kowalik-Jankowska *et al.*, 2003],

▷ aggregation of the Aβ peptides is much slower in rats than in humans [Hong *et al.*, 2010].

▷ the deprotonated amide function which serves as a ligand in the high-pH forms of the complexes is derived from the D1–A2 peptide bond in the case of the human Aβ peptide (figure 4.15b) and the G5–H6 bond in the case of the murine peptide [Eury *et al.*, 2011].

Several studies showed that these differences were mainly due to the substitution of arginine 5 by a glycine [Hong *et al.*, 2010; Eury *et al.*, 2011].

To finish, we feel it is important to point out that interactions between Cu(I), Cu(II), Zn(II) and Fe(II) ions and the APP protein on the one hand, and the human Aβ peptide on the other, have also been studied in detail [Hureau, 2012; Faller and Hureau, 2009].

4.5 - Conclusion

The peptides and protein domains involved in neurodegenerative diseases have binding sites allowing them to form complexes with Cu(II) ions. The stability of these complexes is ensured by a 4-ligand equatorial coordination. We have shown that EPR spectroscopy can be used to determine these binding sites, even in situations where several coordination modes co-exist. The capacity of Cu(II) ions to induce folding of the polypeptide chains could promote or hinder the formation of pathogenic peptide aggregates characteristic of neurodegenerative diseases. This role is currently the topic of several discussions [Gagelli *et al.*, 2006; Thakur *et al.*, 2011].

If coordination of the Cu(II) ions largely determines the conformation of the peptides and protein domains studied in this chapter, it is because their polypeptide chains are particularly flexible. The situation is different in the case of structured proteins where folding of the polypeptide chain is mainly determined by inter-amino acid interactions and interactions between amino acids and the solvent (see appendix 4 to this volume). The metal centres present in these proteins are generally subjected to stringent constraints which impose low-symmetry structures. This is the case for copper-containing proteins like *laccases* which are dealt with in chapter 6, devoted to redox enzymes (see figures 6.5 and 6.6).

Complement 1 – Analysis of ESEEM and HYSCORE spectra for Cu(II)(Aβ16) complexes

Experiments performed at pH 6.5

3-pulse ESEEM experiments performed on unlabelled complexes show that at least two histidines simultaneously coordinate the metal ion in the equatorial position (figure 4.16. This spectrum can be compared to that in figure A2.4 in appendix 2.)

The 4-pulse HYSCORE spectra for complexes in which the histidines are separately labelled with ^{15}N nitrogen indicate that the three histidines are involved in copper coordination, but *not simultaneously*. Indeed, the signal intensity of complexes labelled on H13 and H14 is two-fold smaller than that of the complex where histidine H6 is labelled. From this result, we can deduce that H6 is an equatorial ligand and that a second coordination is supplied by H13 in half the complexes, and by H14 in the other half. The spectrum for the three complexes, in addition to the correlation peaks due to the non-binding nitrogen from the imidazole, includes another set of correlation peaks similar to that produced by a non-binding amide nitrogen when the adjacent carbonyl is bound to a metal ion. But unexpectedly, this signal does not appear with the unlabelled peptide. When the paramagnetic nuclei create deep modulations (which is the case here where two histidines are coordinated), the signals for the weakly coupled nuclei may not appear on the spectrum. In this case, it is better to record a 6-pulse HYSCORE spectrum. Indeed, the signal for the amide is observed when the spectrum for the unlabelled spectrum is recorded with this sequence. This signal disappears when the peptide is labelled with nitrogen ^{15}N on A2, which shows that the amide is that of the D1–A2 peptide bond. Comparison with the coordination of other Cu(II)(peptide) complexes strongly suggests that the carbonyl in the D1–A2 bond is an equatorial ligand to the copper.

HYSCORE spectra for complexes labelled with carbon ^{13}C on D1 show signals centred on the Larmor frequency ν_I of this nucleus (3.6 MHz at 334 mT and 3.3 MHz at 310 mT) (figure 4.17). The intensity of these signals is considerably weaker than that of the signals for the histidines. To get them to emerge more clearly, we once again used the 6-pulse sequence. The spectra recorded in the $g \approx g_{//}$ region reveal the presence of three carbon atoms. The fact that Cu(II) is coordinated by the terminal $-NH_2$ amine and by the carbonyl from the D1–A2 bond strongly suggests that the two strongly coupled nuclei are those of carbons C_α and C_δ from D1. The third, more weakly coupled nucleus, is attributed to the carboxylate from D1 in the apical position.

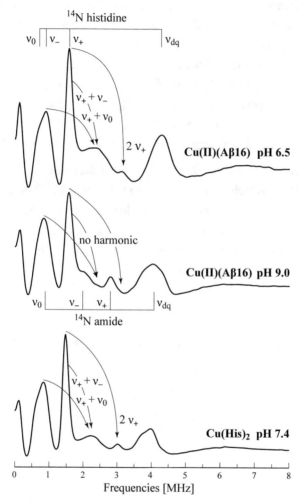

Figure 4.16 - Comparison of the 3-pulse ESEEM spectrum for Cu(II)(Aβ16) complexes at pH 6.5 (top) and pH 9.0 (middle) with that of the Cu(His)₂ complex at pH 7.4 (bottom). Temperature 4.2 K, microwave frequency 9.66 GHz, magnetic field 336 mT, τ = 140 ns.

Experiments performed at pH 9.0

The 3-pulse ESEEM spectra indicate that a single histidine is coordinated to Cu(II) in the equatorial position (figure 4.16), but the HYSCORE spectra for complexes where the histidines are labelled with nitrogen ^{15}N show that the three histidines contribute equally to this coordination. The HYSCORE spectrum for the complex where A2 is labelled with nitrogen ^{15}N is different to that for the unlabelled complex, and indicates that the amide for the D1–A2 peptide bond is deprotonated and coordinated to the Cu(II) in the equatorial position. Like at pH 6.5, a signal due to a non-binding amide nitrogen adjacent to the carbonyl coordinated to the Cu(II) is observed. Given that the terminal –NH$_2$ and the amide nitrogen in the D1–A2 bond are equatorial ligands located near to each other in the chain (figure 4.13), it could be hypothesised that the carbonyl ligand and the amide nitrogen detected on the ESEEM (figure 4.16) and HYSCORE spectra are those of the A2–E3 bond. This identification was recently confirmed by labelling the C=O in this bond with carbon ^{13}C [Drew *et al.*, 2009b].

The HYSCORE spectra for ^{13}C at pH 9 indicate the presence of three coupled ^{13}C nuclei, one of which is weakly coupled (figure 4.17). The carbon C_α and C_δ nuclei from D1 are more strongly coupled than at pH 6.5, which indicates that the bonds formed between the copper ion, the terminal amine and the deprotonated D1–A2 amide are stronger in these conditions.

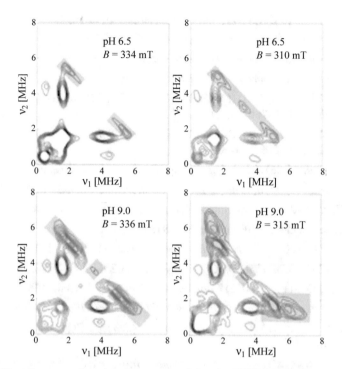

Figure 4.17 - Contour plots (quadrant ++) of 4-pulse HYSCORE spectra at pH 6.5 and pH 9 for the Cu(II)(Aβ16) complex in which D1 is labelled with ^{13}C carbon. The greyed-out zones indicate signals that are absent when unlabelled samples are analysed. Temperature 4.2 K, microwave frequency 9.66 GHz. The figure shows the sum of 6 spectra recorded by varying the duration τ between the first two pulses from 104 to 264 ns (see figure A2.5 in appendix 2). [Dorlet P. *et al.* (2009) *Angewandte Chemie International Edition* **48**: 9273-9276 © 2009 WILEY-VCH Verlag GmbH & Co. KGaA, Weinheim, reproduced with permission]

References

ANDERSEN J.K. (2004) *Nature Reviews Neuroscience* **5**: S18-S25.

ARONOFF-SPENCER E. *et al.* (2000) *Biochemistry* **39**: 13760-13771.

ATWOOD C.S. *et al.* (1998) *Journal of Biological Chemistry* **273**: 12817-12826.

BURNS C.S *et al.* (2002) *Biochemistry* **41**: 3991-4001.

BURNS C.S. *et al.* (2003) *Biochemistry* **42**: 6794-6803.

BUSH A.I. (2000) *Current Opinion in Chemical Biology* **4**: 184-191.

DORLET P. *et al.* (2009) *Angewandte Chemie International Edition* **48**: 9273-9276.

DREW S.C. *et al.* (2009) (a) *Journal of the American Chemical Society* **131**: 1195-1207.

DREW S.C. *et al.* (2009) (b) *Journal of the American Chemical Society* **131**: 8760-8761.

EURY *et al.* (2011) *Angewandte Chemie International Edition* **50**: 901-905.

FALLER P. & HUREAU C. (2009) *Dalton Transactions* 1080-1094.

GAGGELLI E. *et al.* (2006) *Chemical Reviews* **106**: 1995-2044.

HOLTZMAN *et al.* (2011) *Science Translational Medical* **3**: 77pc5.

HONG L. *et al.* (2010) *Journal of Physical Chemistry B* **114**: 11261-11271.

HORNSHAW M.P. *et al.* (1995) (a) *Biochemical and Biophysical Research Communications* **207**: 621-629.

HORNSHAW M.P. *et al.* (1995) (b) *Biochemical and Biophysical Research Communications* **214**: 993-999.

HUREAU C. *et al.* (2006) *Journal of Biological Inorganic Chemistry* **11**: 735-744.

HUREAU C. *et al.* (2008) *Journal of Biological Inorganic Chemistry* **13**: 1055-1064.

HUREAU C. *et al.* (2009) *Angewandte Chemie International Edition* **48**: 9522-9525.

HUREAU C. & FALLER P. (2009) *Biochimie* **91**: 1212-1217.

HUREAU C. *et al.* (2011) *Chemistry European Journal* **17**: 10151-10160.

HUREAU C. (2012) *Coordination Chemistry Reviews* **256**: 2164-2174.

HUREAU C. & DORLET P. (2012) *Coordination Chemistry Reviews* **256**: 2175-2187.

JONES C.E. *et al.* (2004) *Journal of Biological Chemistry* **279**: 32018-32027.

KARR J.W. & SZALAI V.A. (2007) *Journal of the American Chemical Soc*iety **129**: 3796-3797.

KOWALIK-JANKOWSKA T. *et al.* (2003) *Journal of Inorganic Biochemistry* **95**: 270-282.

KOZLOWSKI H. *et al.* (2009) *Coordination Chemistry Reviews* **253**: 2665-2685.

MILLHAUSER G.L. *et al.* (2004) *Accounts of Chemical Researchs* **37**: 79-85.

PEISACH J. & BLUMBERG W.E. (1974) *Archives of Biochemistry and Biophysics* **165**: 691-708.

PICKART L. & THALER M.M. (1973) *Nature New Biology* **243**: 85-87.

SHIN B.K. & SAXENA S. (2008) *Biochemistry* **47**: 9117-9123.

SIGEL A., SIGEL H. & SIGEL R.K.O. (2006) *Metal Ions in Life Sciences*, John Wiley & Sons, Ltd. Chichester.

SYME C.D. *et al.* (2004) *Journal of Biological Chemistry* **279**: 18169-18177.

THAKUR A.K. *et al.* (2011) *Journal of Biological Chemistry* **286**: 38533-38545.

VALENSIN D. *et al.* (2004) *Dalton Transactions* 1284-1293.

Volume 1: BERTRAND P. (2020) *Electron Paramagnetic Resonance Spectroscopy - Fundamentals*, Springer, Heidelberg.

WALTER E.D. *et al.* (2006) *Biochemistry* **45**: 13083-13092.

Chapter 5

Crystallochemistry of clay minerals, weathering processes and evolution of continental surfaces

Balan E., Allard T., Morin G., Fritsch E., Calas G.

Institute of Mineralogy and Physics of Condensed Media (IMPMC), Sorbonne Universities, Paris VI University, CNRS UMR 7590, IRD UMR 206, MNHN, Paris.

5.1 - Introduction

Precise analysis of the distribution and crystallochemistry of the minerals present in soils and rocks from the Earth's surface contributes to what we know about the history and mechanisms of the chemical evolution of our planet. This analysis provides information on the formation of soils and how they function as well as on the geological CO_2 cycle, and on the conditions for formation of deposits of industrial minerals (kaolins) and ores, such as bauxite for aluminium or garnierite for nickel. The study of the processes by which rocks are altered is also important with a view to long-term storage of radioactive or industrial waste in sub-surface conditions.

The clay minerals formed by rock weathering contain paramagnetic impurities (transition ions) and point defects (radicals) of which EPR can identify and quantify a large number. This approach is illustrated here through recent studies performed on kaolinite-group minerals, which are found in numerous environments on the Earth's surface. After describing the structure of these minerals, we will present the EPR spectra of the Fe^{3+} ions which are often present as substituted impurities, and we will discuss the information that can be obtained.

© Springer Nature Switzerland AG 2020
P. Bertrand, *Electron Paramagnetic Resonance Spectroscopy*,
https://doi.org/10.1007/978-3-030-39668-8_5

We will also show that the paramagnetic defects generated in these minerals by natural radioactivity provide unique constraints by which to determine the conditions in which the minerals were formed, and in some cases, to date the weathering processes.

5.2 - Kaolinite-group minerals

Kaolinite, for which the ideal formula is $Al_2Si_2O_5(OH)_4$, is a type of clay, layered silicate minerals made of very small crystallites (typical size: around one micrometer). Kaolinite can be found in a wide variety of environments, such as tropical soils, sedimentary basins and hydrothermal systems; it is also an important industrial mineral, used in three main sectors: paper and cardboard manufacturing (up to 55 % by mass); ceramics and refractory industries, and reinforcement of plastomeres and elastomeres (up to 80 % by volume). The need to handle the crystallochemical variability of kaolinites to optimise their use has triggered numerous studies of their optical, morphological and crystallochemical properties [Cases *et al.*, 1982] as well as the products of their thermal transformations [Djemai *et al.*, 2001].

The triclinic structure of kaolinite [Bish and Von Dreele, 1989; Bish, 1993] is based on a layer made of two sheets: a sheet (T) of tetrahedral SiO_4 silicates linked by their vertex corners, bound through oxygen atoms to a sheet (O) of $AlO_2(OH)_4$ octahedra linked by their edges. In the latter, only two of the three possible octahedral sites are occupied (dioctahedral nature of the layer) (figure 5.1). Adjustment of the (T) and (O) sheets and the dioctahedral character result in distortion of the coordination polyhedra [Bailey, 1988].

In particular, the two non-equivalent octahedral sites noted Al_1 and Al_2 are compressed in the direction perpendicular to the layer. The coherence of assembly of the (TO) layers is ensured by hydrogen bonds between the OH^- hydroxides in the upper plane of the $AlO_2(OH)_4$ octahedra and the oxygens from the basal plane of the SiO_4 tetrahedra. Kaolinite naturally exists in powdered form, with a crystallite size of around the µm.

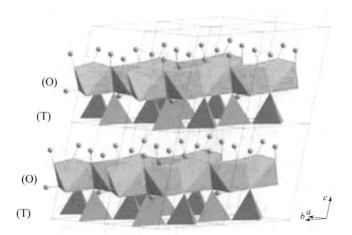

Figure 5.1 - Structure of kaolinite $Al_2Si_2O_5(OH)_4$. The structure results from the stacking of layers (2 layers are shown in the figure), each of which consists of a sheet of tetrahedral silicates and a sheet of octahedra centred on Al^{3+} ions. Three non-equivalent OH^- groups from octahedra, oriented sub-parallel to the c^* direction perpendicular to the layers, form hydrogen bonds with the oxygen of the tetrahedra from the adjacent layer.

Two polymorphous minerals of kaolinite exist with different stacking sequences for the TO layers:

> *Dickite* is formed in hydrothermal conditions or sedimentary basins [Beaufort *et al.*, 1998]. Its monoclinic cell it two-fold larger than kaolinite along the *c* axis, as one of the layers is the symmetric equivalent to the other due to the existence of a vertical glide plane [Bish and Johnston, 1993]. The hydroxides arrangement and distortion of the octahedral sites of dickite are slightly different from those reported for kaolinite.

> *nacrite* is a rare mineral formed in hydrothermal conditions [Buatier *et al.*, 1996]. It also has a double cell, but the stacking of the layers differs from that of kaolinite and dickite in that none of the SiO_4 tetrahedra from a layer face the $AlO_2(OH)_4$ octahedra of the adjacent layer [Giese, 1988].

Natural kaolinites present structural defects. Like in other layered silicates, the most frequent defects correspond to stacking defects, i.e., a succession of layers which, while maintaining inter-layer binding, does not reproduce the elements of symmetry (translations) of the ideal structure. The nature and proportions of stacking defects in kaolinites have been extensively debated, but no consensus has yet been reached [Bailey, 1988; Giese, 1988; Bookin *et al.*, 1989; Plançon *et al.*, 1989; Artioli *et al.*, 1995; Kogure and Inoue, 2005]. The most frequent chemical impurity detected in kaolinites is iron, which is incorporated during growth of crystallites from aqueous solutions [Herbillon *et al.*, 1976; Mestdagh *et al.*, 1982]. Its presence in natural samples is often associated with a reduction in the size of the crystallites and an increased proportion of stacking defects [Brindley *et al.*, 1986].

5.3 - Fe^{3+} centres in kaolinite-group minerals

EPR spectroscopy has shown that Fe^{3+} ions substitute for Al^{3+} in the dioctahedral layers of natural kaolinites [Meads and Malden, 1975; Bonnin *et al.*, 1982; Brindley *et al.*, 1986; Muller and Calas, 1993; Gaite *et al.*, 1993, 1997; Delineau *et al.*, 1994]. As kaolinite is found as a powder, it is impossible to determine the orientation of the principal axes of the zero-field splitting matrix for the Fe^{3+} sites [Volume 1, section 6.2] relative to the crystallographic axes. This limitation makes the EPR approach less discriminative than in the case of single crystal samples, which explains why the number and characteristics (distribution, symmetry) of the substitution sites have long been the subject of controversy. X-band EPR spectra of Fe^{3+} centres in kaolinite-group minerals vary as a function of the polymorph studied and the degree of order in the samples. In the case of kaolinite, three signals are observed, noted $Fe_{(II)A}$, $Fe_{(II)B}$ and $Fe_{(I)}$ (figure 5.2):

▷ the $Fe_{(II)A}$ and $Fe_{(II)B}$ signals are produced by Fe^{3+} ions located in the two octahedral sites normally occupied by Al^{3+}. They most frequently appear in the form of a single spectrum noted $Fe_{(II)}$ and can only be distinguished in particularly well-ordered kaolinite samples (see section 5.3.2),

▷ the $Fe_{(I)}$ signal is characterised by a quasi-symmetric structure at $g \approx 4.3$ (150 mT at X-band). Its origin has been extensively debated [Meads and Malden, 1975; Brindley *et al.*, 1986; Gaite *et al.*, 1993, 1997] and we will see below how it is currently interpreted.

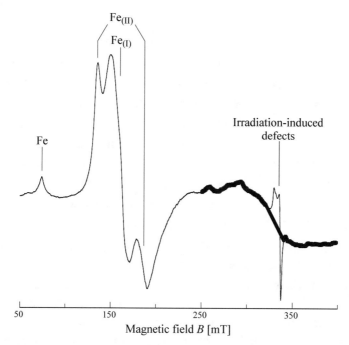

Figure 5.2 - X-band EPR spectrum (9.42 GHz) for a sample of natural kaolinite. The spectrum presents characteristic signals of structural Fe^{3+} ions substituted for Al^{3+} ions, of electronic defects induced by irradiation (A centres, see section 5.4), and a broad signal (highlighted part) centred at $g \approx 2$ produced by concentrated superparamagnetic iron phases (see section 5.3.2). The spectra were recorded on an ESP 300E spectrometer at room temperature. Power: 40 mW, modulation: frequency 100 kHz, peak-to-peak amplitude 0.5 mT.

5.3.1 - Interpretation of EPR signals for Fe^{3+} centres

◇ Sample preparation

Natural samples are subjected to granulometric sorting to eliminate the sandy fraction – mainly composed of quartz – and in some cases to chemical treatment (citrate-bicarbonate-dithionite) to selectively dissolve the iron oxides associated with kaolinite [Mehra and Jackson, 1960]. After heating, a constant sample volume is placed in a silica tube (Suprasil) and its weight (around 40 mg) is determined by double weighting for X-band analysis. For Q-band analysis, the sample volume is much smaller and the amount needs not be precisely determined.

◇ *Model of spectrum simulation*

To be able to apply EPR to the study of finely-divided minerals from soils required the development of a computational code to effectively simulate the spectra for polycrystalline samples. The ZFSFIT (*Zero-Field Splitting FITting*) code written by G. Morin and D. Bonnin [Morin and Bonnin, 1999] can calculate spectra from a spin Hamiltonian (spin $S \leq \frac{5}{2}$) including second order (parameters D, E) and fourth order zero-field splitting terms, and sometimes hyperfine interaction terms (nuclei of spin $I \leq \frac{7}{2}$). The theoretical bases and the calculation method are described in detail in [Morin and Bonnin, 1999]. This code presents numerous advantages compared to older algorithms [Gaite *et al.*, 1993; Chachaty and Soulié, 1995; Wang and Hanson, 1995]:

▷ Numerical diagonalisation of the spin Hamiltonian, already used by [Gaite *et al.*, 1993; Wang and Hanson, 1995], can be used to treat cases where perturbation theory cannot be applied due to comparable Zeeman and zero-field splitting terms. This situation is frequently encountered in the case of metal centres [Volume 1, section 9.5]. In addition, this method allows fourth-order zero-field splitting terms to be taken into account, which are very sensitive to the local symmetry of the ligand field. Hyperfine interactions are treated by second order perturbation theory when this approximation is valid. Otherwise, they are dealt with by complete diagonalisation.

▷ The calculation durations are reduced to a few seconds or minutes thanks to a triangular partition of the integration sphere, inspired by the very effective method developed to simulate NMR spectra [Aldermann *et al.*, 1986].

▷ Another novelty of the ZFSFIT code relates to the treatment of linewidth: inhomogeneous broadening due to fluctuations in the geometry of the paramagnetic centre in the polycrystalline material is simulated by introducing a distribution of the spin Hamiltonian parameters [Volume 1, section 9.5.3]. Indeed, studies carried out on the incorporation of Fe^{3+} and Cr^{3+} as substitutes for Al^{3+} in various minerals showed that it is essential to include a distribution of the zero-field splitting parameters in the calculation to correctly simulate the EPR spectra of natural minerals which generally present point defects or more extended defects [Morin and Bonnin, 1999; Balan *et al.*, 1999, 2000, 2002].

The ZFSFIT code was successfully used to identify the substitution sites for Fe^{3+} in aluminium (oxy)hydroxides such as gibbsite, α-$Al(OH)_3$ and boehmite, γ-$AlOOH$ [Morin and Bonnin, 1999]. Here, it allows analysis of the process by which Fe^{3+} has been incorporated into kaolinite-group minerals as well as of some irradiation-induced defects in phyllosilicates [Sorieul *et al.*, 2005].

◇ *Fe$_{(II)}$-type signals: a characteristic of well-ordered samples*

Simulation of the X-band and Q-band spectra for samples of kaolinite and dickite with exceptional crystallographic qualities made it possible to determine the zero-field splitting parameters for the sites producing the $Fe_{(II)A}$ and $Fe_{(II)B}$ signals (noted $Fe_{(D)A}$ and $Fe_{(D)B}$ for dickite). In figure 5.3 we show examples of kaolinite spectra and their simulations, and in table 5.1 we have listed the zero-field splitting parameters deduced from all the simulations [Balan *et al.*, 1999]. Spectra were simulated at several frequencies and several temperatures to determine these parameters with a high degree of precision. It should be noted that taking the fourth-order zero-field splitting parameters into account improves characterisation of the sites compared to results presented in a previous study [Gaite *et al.*, 1993].

The differences between the parameters obtained for kaolinite and dickite are due to relaxation of the internal structure of the layer linked to differences in stacking type (section 5.2). A study of iron-containing nacrite samples would complete this analysis of the relation between the stacking type and the relaxation of the dioctahedral layer. Unfortunately, none of the nacrite samples analysed up to now produce signals that could be attributed to Fe^{3+} centres.

Figure 5.3 - X- **(a)** and Q-band EPR spectra **(b)** for an exceptionally well crystallised sample of kaolinite from Decazeville. Spectra were recorded on an ESP 300E spectrometer **(a)** Power: 40 mW, modulation: frequency 100 kHz, peak-to-peak amplitude 0.5 mT. **(b)** Power: 40 mW, modulation: frequency 100 kHz, peak-to-peak amplitude 1 mT. Simulations allowed features produced by the $Fe_{(II)A}$ and $Fe_{(II)B}$ signals to be identified. The temperature-dependence of the spectrum indicates its sensitivity to minor structural variations.

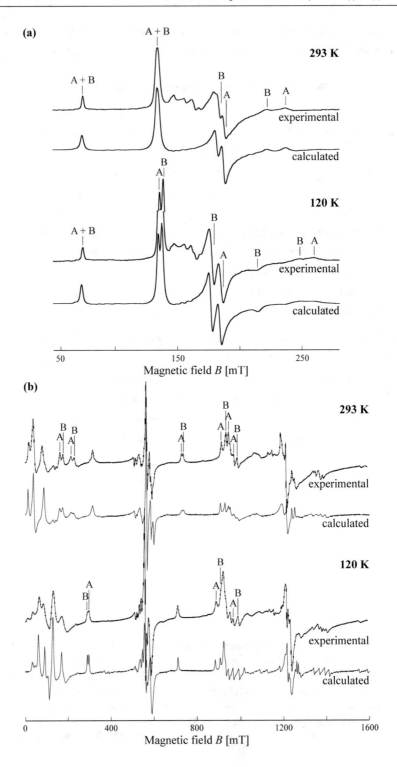

The EPR spectrum was also observed to be sensitive to the type of stacking in the case of Cr^{3+} ions. These ions are less commonly encountered than Fe^{3+}, but they have been observed to substitute for Al^{3+} ions in octahedral sites in samples of kaolinite and dickite from hydrothemal deposits [Balan *et al.*, 2002].

Table 5.1 - Zero-field splitting parameters for Fe^{3+} sites in kaolinite group minerals deduced from numerical simulations of $Fe_{(II)}$-type signals. $g = 2.00$ was assumed in the simulations.

Mineral	Temperature	Signal	D [cm^{-1}]	E [cm^{-1}]
kaolinite	293 K	$Fe_{(II)A}$	0.3312	0.0666
		$Fe_{(II)B}$	0.3171	0.0661
	120 K	$Fe_{(II)A}$	0.3468	0.0721
		$Fe_{(II)B}$	0.3360	0.0753
	4 K	$Fe_{(II)A}$	0.3594	0.0739
		$Fe_{(II)B}$	0.3450	0.0768
dickite	293 K	$Fe_{(D)A}$	0.3252	0.0851
		$Fe_{(D)B}$	0.3654	0.0993
	120 K	$Fe_{(D)A}$	0.3438	0.0901
		$Fe_{(D)B}$	0.3837	0.1031

◇ *EPR spectrum for a disordered sample: origin of the $Fe_{(I)}$-type spectrum*

When cations from transition elements are inserted into very disordered structures, their EPR spectrum cannot be simulated with a single set of zero-field splitting parameters: a broad, or even multimodal, distribution of these parameters is required [Kliava, 1986]. An inversion method with constraints can be used to deduce this distribution from the experimental spectrum, by discretisation of the second order zero field splitting parameters [Balan *et al.*, 1999]: the spectra corresponding to the nodes on a grid of zero-field splitting parameters can be calculated using the ZFSSIT code and the spectrum for the disordered material can be obtained by performing a weighted sum of the spectra. The weighting factors correspond to the distribution sought. Simultaneous inversion of experimental data acquired at X- and Q-band can be used to determine this distribution.

For some samples of disordered kaolinites, the distribution of the zero-field splitting parameters is multimodal (figure 5.4). One of the modes basically corresponds to the characteristic parameters of the octahedral sites of the ordered *kaolinite* ($Fe_{(II)}$ signals) and the two others to the parameters of the octahedral sites of dickite ($Fe_{(D)A}$ and $Fe_{(D)B}$ signals) (see table 5.1). Therefore, in these samples, local dickite-type stacking very probably exists. This result is also suggested by infrared spectroscopy and X-ray diffraction experiments [Balan *et al.*, 2002, 2011]. However, other samples of disordered kaolinite, in particular those collected in lateritic soils from Cameroon, do not display this multimodal distribution [Balan *et al.*, 2011]. Thus, several types of disorder are possible in kaolinite-group minerals. The relations between these types of disorder and the growth conditions for the crystallites remain to be determined.

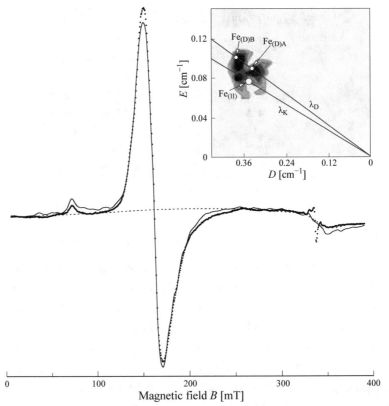

Figure 5.4 - X-band EPR spectrum (dotted lines) for a sample of disordered kaolinite sampled at the top of a lateritic profile in the region of Manaus (Brazil) (sample B4 in figure 5.7). Same experimental conditions as in figure 5.2. The spectrum is dominated by an $Fe_{(I)}$-type signal. The inset shows the distribution of the (D, E) parameters deduced from the simulation (continuous line). The grey scale reflects the values of the weighting factors (see text). The distribution includes three modes centred on values close to those characterising the $Fe_{(II)}$ signals for kaolinite and $Fe_{(D)A}$, $Fe_{(D)B}$ for dickite (table 5.1), indicated by white circles. The small shift is probably linked to the fact that the fourth-order zero-field splitting parameters were not taken into consideration in the simulation. The straight lines shown in the inset indicate that the shift occurs at constant rhombicity ($\lambda = E/D$).

5.3.2 - Determining the absolute concentration of structural iron

To interpret the EPR spectrum for Fe^{3+} ions substituted into kaolinites, we implicitly assumed them to reliably reflect the crystal order of the samples. This is only true if these ions are present at sufficiently high concentrations and if they are distributed homogeneously within the crystallites. This is a first reason to determine the concentration of the Fe^{3+} ions producing the EPR spectrum. Another reason is that comparison of this concentration to that of the total iron contained in the sample supplies information on the physicochemical conditions of the medium where the kaolinites grew [Muller et al., 1995].

To determine the concentration of Fe^{3+} ions, the simplest method consists in comparing the intensity of the experimental spectrum to that produced by a reference kaolinite with a known Fe^{3+} concentration (for intensity measurements, see Volume 1, complement 4 to chapter 9). Unfortunately, all natural or synthetic samples of iron-containing kaolinites contain concentrated iron phases, which can be detected by diffuse reflectance spectroscopy [Malengreau et al., 1994] or thanks to their super-paramagnetic resonance signal [Bonnin et al., 1982]. In contrast, the incorporation of Fe^{3+} at high temperature and low concentrations in corindon α-Al_2O_3 occurs homogeneously, generating a system where the Fe^{3+} ions are diluted [De Biasi and Rodrigues, 1983]. A sample of corindon containing Fe^{3+} ions can therefore be used as a standard since the intensity of its EPR spectrum is proportional to the concentration measured by chemical analysis. The EPR spectra for Fe^{3+} centres in kaolinite and corindon are different, but the parameters of the spin Hamiltonian are known in both cases. Calibration can thus be performed by *calculating* the EPR spectra produced by the same quantity of Fe^{3+} ions in kaolinite and the standard. The concentrations of Fe^{3+} ions measured in this way in samples collected at different sites (France, England, Cameroon, Brazil) vary between a few hundreds of ppm and around 3,000 ppm for kaolinites from the Amazonian basin [Balan et al., 2000].

This information can be used to determine how the iron is distributed between Fe^{3+} ions diluted in the structure of the kaolinite and other forms of iron which are not quantifiable by EPR and resist selective chemical dissolution treatments [Mehra and Jackson, 1960]. These other forms can be substituted Fe^{2+} ions, or more frequently, closely-spaced Fe^{3+} ions: aggregates of Fe^{3+} ions substituted into the dioctahedral kaolinite layer [Schroeder and Pruett, 1996] or super-paramagnetic nano-phases of iron oxides or oxyhydroxides [Malengreau et al.,

1994]. Magnetic resonance cannot distinguish between these phases, which both produce a broad signal at around $g = 2$ at X-band (figure 5.2) (see Volume 1, section 7.4 for the EPR spectrum of aggregates, and chapter 12 of this volume for the FMR of super-paramagnetic particles). The total iron concentration measured by chemical analysis varies between 2000 and 8000 ppm, depending on the origin of the kaolinites. Comparison with the concentrations measured by EPR reveals that in the samples studied, dilute structural Fe^{3+} represents less than half of the total iron, which demonstrates that iron preferentially forms concentrated phases during kaolinite growth. The concentration of the dilute structural Fe^{3+} and the disordered nature of the samples are apparently not correlated [Balan *et al.*, 2000].

EPR has also been used to study other transition elements present in minerals in the kaolinite group, such as vanadium, chromium and manganese. In the cases of manganese and vanadium, identification of specific species (Mn^{2+} and VO^{2+}) reflects the specific reduction-oxidation conditions which prevailed during formation of the kaolinites [Muller *et al.*, 1993, 1995; Allard *et al.*, 1997].

5.4 - Paramagnetic defects produced by irradiation

The first paramagnetic defects due to natural irradiation of kaolinites were identified in the 1970s [Angel *et al.*, 1974]. Three types are now known to exist; they are all electronic holes localised on oxygen atoms, but are distinguished by their atomic environment, the orientation of the orbital occupied by the unpaired electron, and their thermal stability. We can thus distinguish A centres, A' centres and B centres [Clozel *et al.*, 1994] (figure 5.5).

A centres are generally the most abundant and are thus used for geological applications as they are stable over the timescales involved. Indeed, the half-life of these centres is around 10^{12} years at room temperature [Clozel *et al.*, 1994]. A centres can be modelled as an electronic hole which compensates the charge imbalance appearing when divalent ions such as Mg^{2+} are substituted for Al^{3+}. The unpaired electron from the oxygen is located in the π orbital of an $Si–O_{apical}$ bond.

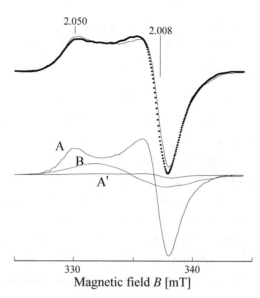

Figure 5.5 - Simulation of the X-band spectrum for the irradiation-induced defects in a kaolinite from the region of Manaus, showing the predominance of the signal due to A centres. Power: 40 mW, modulation: frequency 100 kHz, peak-to-peak amplitude 0.3 mT.

A centres are mainly produced by α, β, or γ ionising radiation. Their EPR signal has been used in several studies related to Earth sciences:

▷ to record past radioactivity in iron-containing nodules from lateritic profiles [Muller and Calas, 1989],

▷ to reveal past migrations of radio elements [Muller *et al.*, 1990, 1992; Ildefonse *et al.*, 1990, 1991] and quantify them [Allard and Muller, 1998; Allard *et al.*, 2007].

In section 5.5.3 we will see that the signal for an A centre can also be used to *date* kaolinites from soils and sediments from tropical areas [Balan *et al.*, 2005].

It should be noted that dickite, a polymorph of kaolinite (section 5.2), contains similar defects [Allard *et al.*, 2003]. The effects of irradiation are now known for various clays [Allard and Calas, 2009] and the recent discovery of irradiation-induced defects in clays such as *smectite, illite* and *sudoite* will make study of new geological environments possible [Sorieul *et al.*, 2005; Morichon *et al.*, 2008, 2010a, b].

5.5 - Application of kaolinite EPR to the study of weathering processes in tropical climates

5.5.1 - Alteration and formation of tropical soils

Tropical soils cover more than one third of the surface of emerged continents, and almost half of the world's continental water flow filters through these soils before entering the ocean [Tardy, 1993; Tardy and Roquin 1998]. These soils are produced by *geochemical alteration* of the continental crust, which consists in dissolving primary minerals and formation of less soluble minerals which accumulate within the soils. These minerals, formed at the Earth's surface, are mainly clay minerals and iron or aluminium oxides or hydroxides. For example, dissolution of the common feldspar, orthoclase, in acidic conditions can result in the formation of kaolinite and the leaching of soluble ions and molecules (in this case potassium and orthosilicic acid) into the hydrosphere:

$$2(Si_3AlO_8K)_{solid} + 9H_2O + 2H^+ \rightleftarrows (Si_2Al_2O_5(OH)_4)_{solid} + 2(K^+)_{aq} + 4(H_4SiO_4)_{aq}$$
$$[5.1]$$

The conditions of tropical alteration combine a high mean temperature, a high volume of precipitation and a high rate of biological production. These factors accelerate mineral dissolution, and tropical soils can be counted among the most altered media on the Earth's surface; their history may extend over several million years. It includes the evolution of the physicochemical conditions linked to climate variations and changes to the landscape [Muller and Calas, 1989; Girard *et al.*, 2000; Braun *et al.*, 2005] while also involving the link between chemical reactions and transport of reagents [Nahon and Merino, 1996]. The resulting structural and textural complexity must be deciphered to elucidate the role played by each mechanism.

In profiles of lateritic soils, clay minerals and iron oxides often present significant variations in size, chemical composition and structural order from the bottom to the top of the profile, indicating dissolution of ancient populations and crystallisation of more recent ones. At each level, minerals inherited from ancient populations and neoformed minerals are therefore likely to be observed in proportions which depend on the rate of dissolution and crystallisation. Depending on the hypotheses used to determine these rates, several evolutionary scenarios can be proposed for the systems:

▷ If we assume that dissolution and recrystallisation are rapid, the fraction of inherited materials will be low. In this case, the crystallochemical and isotopic characteristics observed are assumed to reflect the physicochemical conditions currently prevailing in the environments studied. Heterogeneity of the mineral populations is attributed to spatial or physicochemical fluctuations in these physicochemical conditions, linked for example to the coexistence of distinct micro-environments or to seasonal fluctuations in rainfall [*e.g.* Giral-Kacmarcik *et al.*, 1998].

▷ Conversely, if mineral transformation is slow, the fraction of inherited minerals will be high, and the minerals in the soil are likely to preserve the memory of ancient conditions, or even traces of continental palaeoclimates [Girard *et al.*, 2000].

The capacity of lateritic profiles to preserve an ancient message is attested by manganese oxides detected in lateritic crusts from West Africa and Brazil which were dated at more than 40 Ma [Henocque *et al.*, 1998]. Magnetostratigraphic experiments performed on profiles from French Guyana [Théveniaut and Freyssinet, 1999] also indicate ancient epochs. However, other approaches, such as dating based on uranium and thorium decay chains [Mathieu *et al.*, 1995; Dequincey *et al.*, 1999; Chabaux *et al.*, 2003a, b] and the radioactive isotopes produced by cosmic radiation such as ^{10}Be [Brown *et al.*, 1994], indicate more recent emergence, with ages of less than 2 Ma, or even 0.3 Ma in certain studies. This partial view of the history of lateritic profiles can also be explained by the fact that there is currently no general method to date successive generations of minerals neoformed during the different weathering processes involved. Interpretation of observations performed on mobile soil horizons is even more delicate. Indeed, the small size of the minerals observed and the circulation of fluids may lead to rapid evolution of these minerals.

In this context, EPR study of kaolinites can allow us to distinguish several populations using signals for Fe^{3+} centres to probe the local order of their structure, while also estimating their age by quantifying signals due to the paramagnetic defects induced by natural radioactivity.

5.5.2 - Tracing kaolinite generations in tropical soils using Fe^{3+} EPR

Kaolinites were sampled vertically throughout a section of lateritic soil from the centre of the Amazonian basin, in the region around Manaus (figure 5.6), and their X-band EPR spectra were recorded (figure 5.7). The characteristic spectral features of $Fe_{(II)}$ signals, typical of well-ordered kaolinites (figure 5.3a), and $Fe_{(I)}$, typical of disordered kaolinites were observed (figure 5.4). The relative proportions of the two signals varied throughout the profile, in a direction indicating an increase in crystal disorder from the bottom to the top. This increase in disorder correlates with a reduction in the size of the particles, as measured by granulometric sorting [Fritsch *et al.*, 2002].

Figure 5.6 - Vertical section showing the different horizons of a lateritic soil from the Manaus region. Samples were collected at the seven levels indicated by the circles. The structures observed on the section (discolouration, indurated iron-oxide-rich line) reflect recent re-mobilisation of iron due to fluctuations in the reduction-oxidation conditions which did not affect the kaolinites.

A more quantitative analysis reveals that the spectra are the result of a linear combination of two reference spectra (figures 5.7 and 5.8):

▷ the first, A1, is that of a well-ordered kaolinite typical of the sediment. It is very similar to the spectrum shown in figure 5.3a,

▷ The other, B4, is due to a very disordered kaolinite, representative of the upper horizons of lateritic soils from the Manaus region. This is the spectrum shown in figure 5.4.

In section 5.3.1, we saw that the spectrum for the disordered sample presents a strong "dickite"-type component (see inset to figure 5.4) reflecting the existence of a high proportion of stacking defects. This interpretation is consistent with the fact that the infrared spectrum for this sample presents extensive similarities to that of dickite in the range of frequencies characteristic of the OH-stretching modes (see figure 5.1) [Balan *et al.*, 2002, 2011].

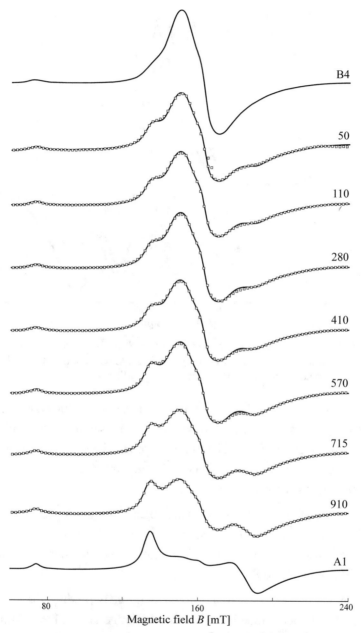

Figure 5.7 - X-band EPR spectra of structural Fe^{3+} ($Fe_{(I)}$ and $Fe_{(II)}$) for the seven samples of kaolinite collected from the profile shown in figure 5.6. The sampling depths are indicated in cm to the right of spectra. Experimental spectra are represented by dotted lines. The simulations obtained by linear combination of reference spectra A1 and B4, due to ordered and disordered samples, respectively, are represented by the continuous line. The experimental conditions are the same as those in figure 5.2.

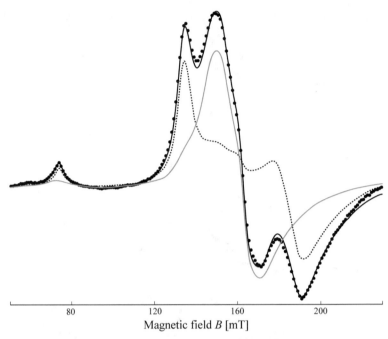

Magnetic field B [mT]

Figure 5.8 - EPR spectrum for the sample collected at 910 cm depth (black dots)
and its simulation (continuous black line) obtained by linear combination of the A1
(black dashes) and B4 (continuous grey line) components.

The spectra for samples collected on the profile show that they contain a mixture of two types of kaolinites, an ordered one from sedimentary levels, and a
disordered one typical of soil. Their relative proportions, deduced from simulations of the EPR spectra, are in very good agreement with those determined
by infrared spectroscopy (figure 5.9).

The persistence of ordered kaolinite in the upper levels of the profile suggests a
relatively slow rate of transformation of the kaolinites by dissolution/recrystal-
lisation, and thus the presence of ancient kaolinite populations within lateritic
profiles [Balan *et al.*, 2007].

In addition, the transformation of ordered kaolinite to produce disordered kaolinite appears to occur independently of other mineralogical transformations
observed in the same lateritic profile, for example, transformation of iron oxides
which was studied by Mössbauer spectroscopy, X-ray diffraction and diffuse
reflectance UV-visible spectroscopy [Fritsch *et al.,* 2005]. Haematite (Fe_2O_3)
and goethite (FeOOH) observed at the bottom of the profile are progressively
replaced by aluminium-substituted goethite ($Fe_{0.67}Al_{0.33}OOH$) towards the top,

through a dissolution and re-precipitation reaction. This transformation, limited to the upper part of the alteration profiles, is superposed on that of the kaolinites but does not modify it; it occurred after the appearance of the disordered kaolinite populations. This more recent dissolution/recrystallisation of iron oxides can also lead, due to the vertical migration of Fe and Al, to individualised discoloured pockets at meter or hectometer scale. These pockets are limited at their base by fine iron-oxide-enriched levels (figure 5.6) [Fritsch *et al.*, 2002].

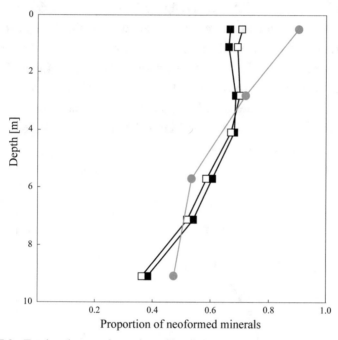

Figure 5.9 - Tracing the transformation of kaolinites throughout the profile shown in figure 5.6, as determined using EPR (white squares) and infrared (black squares) spectroscopies, superposed on the transformation of iron oxides as determined by Mössbauer spectroscopy (grey circles). Simulation of the spectra allows the replacement of older mineral populations by more recent ones to be quantified. The results show that the changes occurring in kaolinites and iron oxides are not synchronous.

5.5.3 - EPR-dating of kaolinites from lateritic soils

EPR can also be used to estimate the age of kaolinites. This approach was inspired by previous work in which the defects produced by irradiation of minerals from the kaolinite group were used to analyse the transfer of radionuclides in geological environments [Allard and Muller, 1998; Allard *et al.*, 2007]. The principle of EPR-dating is the same as that of the archaeological

dating described in detail in section 1.2 of chapter 1. We will simply indicate
the specific elements used to date kaolinites from three sites in the Manaus
region [Balan *et al.*, 2005].

◇ *Determining the palaeodose*

The irradiation dose recorded by kaolinite until now (palaeodose) was determined by using EPR to measure the concentration of irradiation-induced
defects, which are mainly A centres (figure 5.5). The calibration curve, which
links this concentration to the received dose, was obtained by performing artificial irradiation experiments. In these experiments, we used 1.5-MeV He$^+$
ions produced by the ARAMIS accelerator at the CSNSM from Université
Paris-Sud, the characteristics of which are described in [Bernas *et al.*, 1992].
Two representative samples of ordered and disordered kaolinite were subjected
to a large range of irradiation doses (figure 5.10).

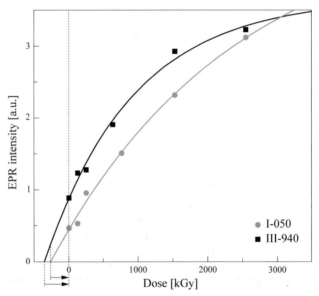

Figure 5.10 - Sensitivity curves obtained by irradiating two representative kaolinite
samples from the Manaus region with 1.5-MeV He$^+$ ions. The ordinate represents
the intensity of the signal for the A centre in arbitrary units. The arrows indicate
the palaeodoses determined.

It should be noted that these doses are much larger than those used for carbonates
or phosphates, which are around a kGy (figure 1.3 in chapter 1) [Ikeya, 1993] as
it is more difficult to create irradiation defects in kaolinite than in these minerals.

The results are well described by the following equation, which is identical to equation [1.3] in chapter 1 [Grün, 1991; Allard et al., 1994]:

$$I = I_0[1 - e^{-\mu(D+P)}] \qquad [5.2]$$

where I is the intensity of the EPR signal, which is proportional to the concentration of the irradiation-induced defects, I_0 is the saturation intensity, D is the applied dose in grays (1 Gy = 1 J kg^{-1}), P is the palaeodose, and μ is the efficiency coefficient, which depends on the mineral studied and the type of radiation used.

The values of μ deduced from these studies (7.8 × 10^{-4} kGy^{-1} for the ordered sample and 3.7 × 10^{-4} kGy^{-1} for the disordered sample) were used to extrapolate the palaeodose for the other samples from the intensity of the signal for their defects. The palaeodoses obtained by this method varied between 100 and 1000 kGy (figure 5.12).

◇ *Determining the dose rate*

For stable defects like the A centre, the palaeodose is the integral of the dose rate over time since the formation of the kaolinites (equation [1.1] in chapter 1). To convert the palaeodose determined by EPR into an age, we must therefore know the rate of the dose recorded by the kaolinite and any variations it has undergone over time. The effect of cosmic radiation is considered negligible due to the depth at which these samples were collected. The *current* annual dose can be calculated from the concentration of the radioactive elements, mainly U and Th, contained in the samples studied. Alpha radiation, for which the depth of penetration in matter is around 10 μm, is known to efficiently create irradiation-induced defects. To determine its contribution to the dose rate, we must therefore analyse the spatial distribution of the radioactive elements in the soil samples:

▷ The spatial distribution of uranium was determined at the μm scale on thin sections using cartography of the induced fission traces (figure 5.11). These maps reveal that uranium is concentrated in zircon grains and that it is also present, to a lesser extent, in titanium and iron oxides. Given its shallow depth of penetration, it is likely that the alpha radiation produced by uranium decay chains contributes little to the dose recorded by the kaolinites, and its contribution was therefore omitted when determining the dose rate.

▷ The spatial distribution of thorium could not be analysed directly as no appropriate imaging method was available. Analysis of series of isolated zircon grains [Balan *et al.*, 2001] indicated that thorium is not a major element in these grains, and that it is probably diluted throughout the clay matrix. The corresponding dose of alpha radiation was therefore taken into account when calculating the dose rate.

A significant difficulty in dating soils is linked to the open geochemistry of these media, which is likely to cause the dose rate to vary. In the case of EPR dating, this may be due to uranium leaching, as this element is soluble, and therefore mobile, in oxidising conditions. In formations from the Manaus region, the concentration of uranium within zircons showing low self-irradiation damage (low degree of amorphisation) limited migration of this element, as indicated by the relatively constant ratio between the uranium and thorium concentrations; thorium has relatively low geochemical mobility due to its very low solubility. These observations suggest that the dose rate varied little over time. It was therefore considered constant, and equal to the current annual dose calculated.

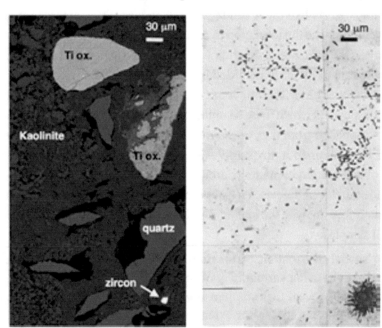

Figure 5.11 - Image obtained by scanning electron microscopy with backscattered electrons (left) and corresponding map of the traces of induced fission in uranium (right). Note the heterogeneous distribution of the uranium, which is mainly concentrated within the zircon grain (lower right) and within two larger-volume titanium oxides grains (top).

◇ Age of kaolinites

If we represent each sample of kaolinite by a point on a graph with the dose rate on the abscissa and the palaeodose as ordinate, the age of the sample will be given by the *slope* of the line joining this point to the origin (figure 5.12). The points representing kaolinite from sedimentary horizons practically align, suggesting that they are of similar age, slightly greater than 25 Ma. This result confirms that these kaolinites formed within the sediment by alteration of material initially deposited around the end of the Cretaceous period. Although more recent, kaolinite from the soil was found to have an age of around 6 to 10 Ma. This result appears to refute the rapid dissolution/recrystallisation model which would maintain them in equilibrium with current conditions in the medium (section 5.5.1).

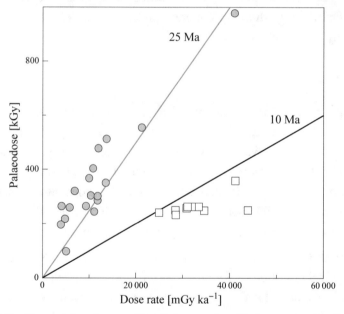

Figure 5.12 - Determining the age of kaolinite samples. Circles represent samples from sedimentary horizons and squares correspond to soil samples. If the dose rate is constant, the isochrone curves are straight lines passing through the origin. Although more recent than kaolinites from sediment, the significant concentration of irradiation defects observed in soil kaolinites reflects their old age.

The relative resistance of kaolinite to dissolution/precipitation at the Earth's surface is compatible with the ancient nature of the transformation of ordered kaolinite to produce disordered kaolinite, as observed vertically in soil profiles.

Through its composition in stable isotopes of oxygen and hydrogen, kaolinite is therefore likely to supply palaeoclimatic information. In contrast, a significant contribution of kaolinites to the rapid exchange of silica between the soil and plant coverage [*e.g.* Lucas *et al.*, 1993, 1996] is questionable, and the role of other silica-containing phases (*e.g.* biological silica) deserves to be investigated. It should be noted that the site of Manaus where this study was performed is exceptional in that it shows very limited migration of radioactive elements. Application of this method to profiles developed on endogenous rocks presenting several uranium-containing minerals, such as titanite ($CaTiSiO_5$) or monazite ($LaPO_4$), and potassium-containing minerals, such as micas, may therefore be much more difficult. In addition to the variation in dose rate, the presence of water in the system and the spatial distribution of radionuclides can lead to significant uncertainty in the age determined for kaolinites. This uncertainty can be minimised by studying samples collected in environments which may be dated by other methods.

5.6 - Conclusion

Despite the heterogeneity and variability of natural samples, their study by spectroscopic techniques sensitive to the local environment of the atoms and defects, such as EPR, reveals unexpected regularities at the atomic scale. Thus, thanks to its unique sensitivity, EPR can be used to obtain quantitative information on the structural order of the minerals, and on concentrations of defects and trace impurities. EPR can therefore provide very novel information, both on the formation media and on modifications to the environmental conditions.

The use of EPR to monitor ongoing environmental modifications due to chemical or radiological pollution constitutes a significantly underexploited field (nevertheless, see application described in chapter 2). EPR dating of minerals from soils constitutes another challenge with significant potential, and represents a unique possibility to quantify the historical evolution of continental surfaces, which is essential if we are to study historical climate change.

References

ALDERMAN D. W. *et al.* (1986) *Journal of Chemical Physics* **84**: 3717-3725.

ALLARD T. *et al.* (1994) *Physics and Chemistry of Minerals* **21**: 85-96.

ALLARD Th. *et al.* (1997) *Comptes Rendus de l'Académie des Sciences Paris*, Série II, Sciences de la Terre et des Planètes. **325**: 1-10.

ALLARD T. & MULLER J.-P. (1998) *Applied Geochemistry* **13**: 751-765.

ALLARD T. *et al.* (2003) *European Journal of Mineralogy* **15**: 629-640.

ALLARD Th., ILDEFONSE Ph. & CALAS G. (2007) *Chemical Geology* **239**: 50-63.

ALLARD T. & CALAS G. (2009) *Applied Clay Science* **43**: 143-149.

ANGEL B.R., JONES J.P.E. & HALL P.L. (1974) *Clay Minerals* **10**: 247-255.

ARTIOLI G. *et al.* (1995) *Clays and Clay Minerals* **43**: 438-445.

BAILEY S.W. (1988) "Introduction; polytypism of 1:1 layer silicates" in *Hydrous phyllosilicates (exclusive of micas)*, S.W. Bailey, ed., Reviews in Mineralogy, Mineralogical Society of America.

BALAN E. *et al.* (1999) *Clays and Clay Minerals* **47**: 605-616.

BALAN E. *et al.* (2000) *Clays and Clay Minerals* **48**: 439-445.

BALAN E. *et al.* (2001) *American Mineralogist* **86**: 1025-1033.

BALAN E. *et al.* (2002) *Physics and Chemistry of Minerals* **4**: 273-279.

BALAN E. *et al.* (2005) *Geochimica et Cosmochimica Acta* **69**: 2193-2204.

BALAN E. *et al.* (2007) *Clays and Clay Minerals* **55**: 253-259.

BALAN E. *et al.* (2011) *Comptes Rendus Géoscience* 343: 177-187.

BEAUFORT D. *et al.* (1998) *Clay minerals* **33**: 297-316.

BERNAS H. *et al.* (1992) *Nuclear Instruments and Methods in Physics Research* **B62**: 416-420.

BISH D.L. (1993) *Clays and Clay Minerals* **41**: 738-744.

BISH D.L. & JOHNSTON C.T. (1993) *Clays and Clay Minerals* **41**: 297-304.

BISH D.L. & VON DREELE R.B. (1989) *Clays and Clay Minerals* **37**: 289-296.

BONNIN D., MULLER S. & CALAS G. (1982) *Bulletin de Minéralogie* **105**: 467-475.

BOOKIN A.S. *et al.* (1989) *Clays and Clay Minerals* **37**: 297-307.

BRAUN J.-J. *et al.* (2005) *Geochimica et Cosmochimica Acta* **69**: 357-387.

BRINDLEY G.W. *et al.* (1986) *Clays and Clay Minerals* **34**: 239-249.

BROWN E.T. *et al.* (1994) *Earth Planet Sci. Lett.* **124**: 19-33.

BUATIER M.D. *et al.* (1996) *European Journal of Mineralogy* **8**: 847-852.

CALAS G. (1988) « Electron paramagnetic resonance » in *Spectroscopic methods in mineralogy and geology.*, F.C. Hawthorne. ed., Reviews in Mineralogy, Mineralogical Society of America **18**: 513-571.

CASES J-M. *et al.* (1982) *Bulletin de Minéralogie* **105**: 439-455.

CHABAUX F., RIOTTE J. & DEQUINCEY O. (2003) (a) *Reviews in Mineralogy and Geochemistry* **52**: 533-576.

CHABAUX F. *et al.* (2003) (b) *Comptes Rendus Géoscience* **335**: 1219-1231.

CHACHATY C. & SOULIÉ E. J. (1995) *Journal de Physique III France* **5**: 1927-1952.

CLOZEL B., ALLARD Th. & MULLER J-P. (1994) *Clays and Clay Minerals* **46**: 657-666.

DE BIASI R.S. & RODRIGUES D.C.S. (1983) *Journal of Material Science Letters* **2**: 210-212.

DELINEAU T. *et al.* (1994) *Clays and Clay Minerals* **42**: 308-320.

DEQUINCEY O. *et al.* (2002) *Geochim. Cosmochim. Acta* **66**: 1197-1210.

DJEMAI A. *et al.* (2001) *Journal of the American Ceramic Society* **84**: 1017-1024.

FRITSCH E. *et al.* (2002) *European Journal of Soil Science*, **53**: 203-218.

FRITSCH E. *et al.* (2005) *European Journal of Soil Science* **56**: 575-588.

GAITE J-M., ERMAKOFF P. & MULLER J-P. (1993) *Physics and Chemistry of Minerals* **20**: 242-247.

GAITE J-M. *et al.* (1997) *Clays and Clay Minerals* **45**: 496-505.

GIESE R.F., Jr (1988) « Kaolin Minerals: Structures and Stabilities » in *Hydrous phyllosilicates (exclusive of micas)*, S.W. Bailey, ed., Reviews in Mineralogy, Mineralogical Society of America **19**, 725 pp.

GIRAL-KACMARCIK S. *et al.* (1998) *Geochimica et Cosmochimica Acta* **62**: 1865-1879.

GIRARD J.-P., FREYSSINET Ph. & CHAZOT G. (2000) *Geochimica et Cosmochimica Acta* **64**: 409-426.

GRÜN R. (1991) *Nuclear Tracks and Radiation Measurements* **18** (1/2): 143-153.

HENOCQUE O. *et al.* (1998) *Geochimica and Cosmochimica Acta* **62**: 2739-2756.

HERBILLON A. J. *et al.* (1976) *Clay Minerals* **11**: 201-220.

IKEYA M. (1993) *New applications of electron spin resonance - Dating, Dosimetry and Microscopy*, World Scientific Publishing, Singapore.

ILDEFONSE P. *et al.* (1991) *Material Research Society Symposium Proceedings* vol 749-756.

ILDEFONSE P. *et al.* (1990) *Economic Geology* **29**: 413-439.

KLIAVA J. (1986) *Physica Status Solidi B* **134**: 411-455.

KOGURE T. & INOUE A. (2005) *European Journal of Mineralogy* **17**: 465-473.

LUCAS Y. *et al.* (1993) *Science* **260**: 521-523.

LUCAS Y. *et al.* (1996) *Comptes Rendus de l'Académie des Sciences de Paris* **322**: 1-16.

MALENGREAU N., MULLER J.-P. & CALAS G. (1994) *Clays and Clay Minerals* **42**: 137-147.

MATHIEU D., BERNAT M. & NAHON D. (1995) *Earth and Planetary Science Letters* **136**: 703-714.

MEADS R.E. & MALDEN P.J. (1975) *Clay Minerals* **10**: 313-345.

MEHRA O.P. & JACKSON M.L. (1960) *Clays and Clay Minerals* **7**: 317-327.

MESTDAGH M.M. *et al.* (1982) *Bulletin de Minéralogie* **105**: 457-466.

MORICHON E. *et al.* (2008) *Physics and Chemistry of Minerals* **35**: 339-346

MORICHON E. *et al.* (2010) (a) *Physics and Chemistry of Minerals* **37**: 145-152.

MORICHON E. *et al.* (2010) (b) *Geology* **38**: 983-986.

MORIN G. & BONNIN D. (1999) *Journal of Magnetic Resonance* **136**: 176-199.

MULLER J.P. & CALAS G. (1989) *Economic Geology* **84**: 694-707.

MULLER J.P., ILDEFONSE P. & CALAS G. (1990) *Clays and Clay Minerals* **38**: 600-608.

MULLER J.P. *et al.* (1992) *Applied Geochemistry* **1**: 205-216.

MULLER J.P. & CALAS G. (1993) « Genetic significance of paramagnetic centers in kaolinites » in *Kaolin genesis and utilization*, H. H. Murray, W. Bundy and C.Harvey, eds., The Clay Minerals Society, Boulder, 341 p.

MULLER J.P. & CALAS G. (1993) *Geochimica et Cosmochimica Acta* **57**: 1029-1037.

MULLER J.P. *et al.* (1995) *American Journal of Science* **295**: 115-155.

MURAD E. & WAGNER U. (1991) *Neues Jahrbuch Mineralogie Abhteilung* **162**: 281-309.

NAHON D.& MERINO E. (1996) *Journal of Geochemical Exploration* **57**: 217-225.

PLANÇON A. *et al.* (1989) *Clays and Clay Minerals* **37**: 203-210.

SCHROEDER P.A. & PRUETT R.J. (1996) *American Mineralogist* **81**: 26-38.

SORIEUL S. *et al.* (2005) *Physics and Chemistry of Minerals* **32**: 1-7.

TARDY Y. (1993) *Pétrologie des latérites et des sols tropicaux*, Masson, Paris.

TARDY Y. & ROQUIN C. (1998) *Dérive des continents. Paléoclimats et altérations tropicales*. Editions BRGM, Orléans.

THÉVENIAUT H. & FREYSSINET Ph. (1999) *Palaeo* **148**: 209-231.

Volume 1: BERTRAND P. (2020) *Electron Paramagnetic Resonance Spectroscopy - Fundamentals*, Springer, Heidelberg.

WANG D. & HANSON G.R. (1995) *Journal of Magnetic Resonance* A **117**: 1-8.

Chapter 6

Structure and catalytic mechanisms of redox enzymes

Bertrand P.

Laboratory of Bioenergetics and Protein Engineering, UMR 7281,
Institute of Microbiology of the Mediterranean,
CNRS & Aix-Marseille University, Marseille.

6.1 - Introduction

6.1.1 - Redox enzymes and their paramagnetic centres

In the cells of living organisms numerous chemical reactions characterised by $\Delta_r G < 0$ occur; these reactions should be spontaneous. However, if this were the case, the cells would be simple chemical reactors and complex living systems would never have evolved. In fact, these chemical reactions are blocked at room temperature for *kinetic* reasons and they are *specifically* catalysed by proteins with particular properties: enzymes (see appendix 4 for a general presentation of proteins). Interactions between enzymes and their substrates radically modify the profile for free energy variation for the reactions catalysed (figure 6.1):

▷ in the absence of an enzyme, the energy needed for the reactants to form the transition state is large (the energy barrier is high) and the reaction is very slow (figure 6.1a),

▷ in the presence of an enzyme, the reaction takes place in several steps char-acterised by low energy barriers, which considerably increases its speed (figure 6.1b). In general, enzymes also accelerate the reverse reaction.

During the reaction, the enzyme - which may be in a complex with the reactants or products - passes through forms know as "reaction intermediates". The se-quence of intermediates constitutes the catalytic cycle. The enzyme's efficacy depends on the speeds of passage from one intermediate to the next and thus on the concentration of the reactants involved in the bimolecular steps.

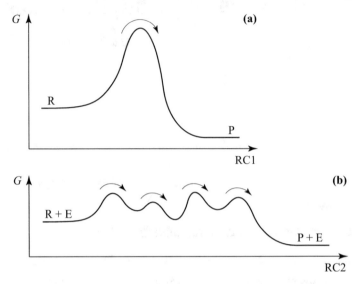

Figure 6.1 - Free energy profile for: **(a)** an uncatalysed reaction and **(b)** a reaction catalysed by an enzyme E. R represents the reactants, P the products of the reaction. The reaction coordinate is different in the two cases. The energy maxima correspond to transition states, the minima to reaction intermediates.

The initial rate of the reaction V, is proportional to the enzyme's concentration and is, by definition, its "activity". V is maximal at saturating reactant concentrations, i.e., high enough for the rate of the reaction not be limited by reactant availability. In these conditions, the activity is written

$$V_{max} = k_{cat}[E]$$

where [E] is the enzyme concentration. The rate constant k_{cat} is the maximum number of cycles per second that can be performed per molecule of enzyme. k_{cat} values are generally around 10^2 to 10^4 s^{-1}. Enzymes are sometimes purified in an inactive form and must be reactivated by appropriate chemical treatment, sometimes simply a reduction. To determine whether an enzyme is active, it is exposed to its reactants. If the reaction starts rapidly, the enzyme is in an active form, which is generally an intermediate of the catalytic cycle.

In cells in living organisms, numerous reactions involve electron exchange:

$$S_{red} + A_{ox} \rightarrow S_{ox} + A_{red} \qquad [6.1]$$

where S_{red} and A_{ox} are a substrate and an electron acceptor, respectively. These reactions are catalysed by *redox enzymes*, which contain (figure 6.2):

 ▷ an *active site* where the substrate S_{red} is activated. This process generates electrons and very often protons. The substrate often accesses the active site through a "channel",

 ▷ a *chain of redox centres*, which ensures rapid transfer of electrons towards the acceptor,

 ▷ in some cases, a chain of proton-acceptor groups for transfer of protons towards the solvent.

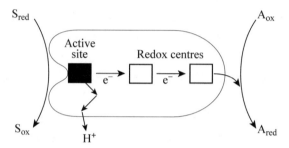

Figure 6.2 - Schematic representation of a redox enzyme. The arrows represent the steps in the reaction [6.1] which occur spontaneously under the influence of the driving force ($\Delta_r G < 0$), at high speeds thanks to low energy barriers. EPR provides information on the active site and the redox centres, and on the different steps of the catalytic cycle.

The importance of this enzymatic class can be understood when we consider that around 30 % of proteins present in living organisms are redox enzymes.

Some act in solution, others are membrane-bound. The latter are found in *bioenergetic systems* which convert chemical (respiratory systems) or light (photosynthetic systems) energy into a form that can be used by living organisms. The active sites and redox centres in redox enzymes are organic groups (flavins, chlorophylls, quinones) which are linked to the protein by weak or covalent bonds, or transition ions forming mono- or polynuclear complexes with O, S or N ligands supplied by the amino acids in the protein; in this case they are known as "metalloenzymes". These cations are derived from elements found in abundance in the Earth's crust, from the first (V, Mn, Fe, Co, Ni, Cu), second (Mo) or third (W) transition series. These centres are subjected to the constraints exerted by the protein and they generally adopt a very distorted geometry. Since the functions of the active site and the redox centres require them to readily change oxidation state, the enzyme can often be prepared in a *paramagnetic* state. This explains the advantage of EPR and other magnetic spectroscopies in the study of this enzymatic class.

6.1.2 - Redox enzymes and EPR spectroscopy

The first EPR experiments on redox enzymes were performed in the 1950s, at a time when very little was known about their structure and when any information was welcome. These experiments consisted in recording the spectrum for an EPR tube filled with membrane fragments or a solution of partially purified enzymes at room temperature. Signals for radicals and metal centres were observed, but low sensitivity and poor reproducibility hindered their interpretation. Decisive progress was made when improved purification and concentration techniques (typical values 10 μM to 1 mM) were developed and spectra were recorded at cryogenic temperatures. Spectroscopists then discovered a wide variety of spectral forms which differed markedly from those produced by high-symmetry chemical compounds. Their structural interpretation presented several challenges. Experimentally, the spectra for frozen solutions recorded at X- or Q-band were poorly resolved due to *g-strain* broadening [Volume 1, section 9.5.3]. Theoretically, the parameters for the spin Hamiltonian deduced from simulations of the spectra were difficult to analyse with "ligand field" type models, which are not well adapted to centres with low symmetry. Interesting results were nevertheless obtained, in particular for iron-containing centres for which Mössbauer spectroscopy provides complementary information. Polynuclear metal centres are frequently present in redox enzymes and coupling models were very useful to analyse their EPR spectra [Volume 1, section 7.4].

Since the 1980s, the 3D structures of many redox enzymes have been determined, and spectroscopic studies are now attempting to decipher their *reaction mechanisms*, i.e., to determine and characterise the steps in the catalytic cycle. *Site-directed mutagenesis* techniques, which can be used to modify the amino acids linking the active site and the redox centres (see appendix 4), are a major asset. The range of information accessible by EPR was considerably extended by the development of high-resolution techniques, pulsed EPR and ENDOR (see appendices 2 and 3 at the end of this volume). The emergence of high-field EPR, which allows the *g* values for the radicals to be precisely determined, and unambiguously identifies them also launched new avenues of study. The data provided by these techniques can be interpreted in detail thanks to new quantum chemistry models and to progress in numerical calculation tools. In particular, the hyperfine couplings measured by high-resolution techniques can be compared to the spin density distribution calculated by applying these models.

Below, we will illustrate various aspects of EPR applied to redox enzymes through studies performed on soluble enzymes, *laccases* and Ni−Fe *hydrogenases*, and on a membrane-bound bioenergetic complex, *photosystem II*.

6.2 - Laccases: enzymes oxidising substrates with a high potential

Laccases are enzymes with a molecular weight between 50 and around 100 kDa, which catalyse the oxygen-driven oxidation of organic substrates such as phenols, polyphenols and aromatic amines, according to the following type of reaction:

$$4 \text{ phenol} + O_2 \rightarrow 4 \text{ phenoxyl} + 2H_2O \qquad [6.2]$$

Laccases are found in higher plants and fungi, but also in insects and some bacteria. Their name is derived from the *lacquers* prepared from the sap of trees of the *Rhus* genus, which have long been used in furniture making in eastern Asia. At the end of the 19th century, it was realised that treatment of these lacquers was facilitated by "ferments" extracted from the same trees. These ferments were named "laccases" by the French chemist Gabriel Bertrand [Bertrand, 1895]. Fungi use laccases to degrade lignin, a complex polymer that is far too bulky to interact directly with the enzyme. In this case, oxidation of the substrate involves natural *mediators* (figure 6.3).

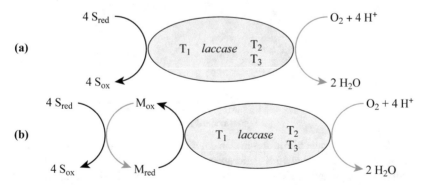

Figure 6.3 - Oxygen-based oxidation of a substrate catalysed by a laccase.**(a)** no mediator, **(b)** with a mediator M. T_1, T_2, T_3 are the Cu-containing centres of the laccase.

The use of *artificial* mediators considerably broadens the range of substrates that can be oxidised by laccases and opens up possibilities for a wide range of applications described in a number of patents: paper pulp and textile bleaching, improving food quality, bioremediation of soils and waters, polymer synthesis, development of biosensors and fuel cells, etc. [Kunammeni *et al.*, 2008].

6.2.1 - A puzzle for spectroscopists: the structure of T_1 and T_2 copper-containing centres

Analysis of the metal atom content of laccases indicated that each molecule contains four copper atoms, and EPR spectrometry rapidly revealed their organisation in the enzyme. At the end of the 1960s, the EPR spectrum for a frozen solution of oxidised laccases from the mushroom *Polyporus versicolor* was shown to contain *two components* of which the intensity indicated a stoichiometry of one centre per molecule: T_1 characterised by $g_{//} = 2.190$, $g_\perp = 2.042$, $|A_{//}| = 270$ MHz, $|A_\perp| = 29$ MHz, and T_2, characterised by $g_{//} = 2.262$, $g_\perp = 2.036$, $|A_{//}| = 530$ MHz, $|A_\perp| = 85$ MHz [Malström *et al.*, 1968]. All laccases produce this type of spectrum in the oxidised state. For example, in figure 6.4 we have represented the spectrum for the laccase from *Thielavia arenaria*, a mushroom from the same evolutionary branch as yeasts (single-celled fungi), morels and truffles.

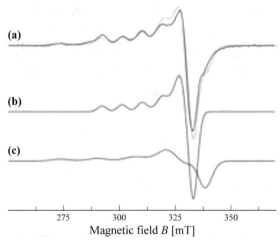

Figure 6.4 - X-band spectrum for a frozen solution of laccase from *Thielavia arenaria*. **(a)** spectrum recorded at 15 K. Microwave: frequency 9.434 GHz, power 1 mW. Modulation: frequency 100 kHz, peak-to-peak amplitude 1 mT. The simulation represented by the dashed line is the sum of the T_1 and T_2 components with the same intensity represented in **(b)** and **(c)**. These simulations were calculated by assuming that the $\tilde{g}$ and $\tilde{A}$ matrices had the same principal axes, with Gaussian lines with a full-width-at-half-maximum of σ. T_1: $g_{//} = 2.204$; $g_\perp = 2.040$; $A_{//} = 270$ MHz; $A_\perp = 30$ MHz; $\sigma_{//} = 140$ MHz; $\sigma_\perp = 135$ MHz. T_2: $g_{//} = 2.260$; $g_\perp = 2.040$; $A_{//} = 510$ MHz; $A_\perp = 105$ MHz; $\sigma_{//} = 300$ MHz; $\sigma_\perp = 220$ MHz.

To explain the "EPR silence" of the other two Cu^{2+} ions, it was proposed that they are coupled by a strong antiferromagnetic exchange interaction, resulting in a ground state with $S = 0$ (centre of type T_3) [Fee *et al.*, 1969; Malkin and Malmström, 1970]. This hypothesis was subsequently confirmed, and the classification of the T_1, T_2, T_3 centres based on EPR was applied to all copper-containing centres in proteins.

The structural interpretation of the EPR spectrum was much more difficult. The T_2 component resembles spectra produced by Cu^{2+} complexes with square planar coordination [Volume 1, figure 4.11], it was therefore attributed to a centre with this structure. Efforts were therefore concentrated on the T_1 centres, for which the spectroscopic properties are very atypical. Indeed, their hyperfine constants $(A_{//}, A_\perp)$ are abnormally weak, and the UV-visible spectrum displays an intense band at around 600 nm which causes the enzyme's blue colour [Malström *et al.*, 1968]. These properties are similar to those of the Cu^{2+} centre of some "blue" proteins such as azurin from the bacterium *Pseudomonas aeruginosa* ($g_{//} = 2.257$, $g_\perp = 2.055$, $|A_{//}| = 180$ MHz) [Broman *et al.*, 1963] and plastocyanin from the plant *Chenopodium album* ($g_{//} = 2.226$, $g_\perp = 2.053$, $|A_{//}| = 190$ MHz, $|A_\perp| = <$ 50 MHz) [Blumberg and Peisach, 1966]. The crystal structure of these proteins showed the Cu^{2+} ion to be coordinated by a methionine, a cysteine and two histidines (see appendix 4 to this book) in a distorted tetrahedral geometry, with two normal Cu–N(His) bonds, a relatively short Cu–S(Cys) bond and an abnormally long Cu–S(Met) bond (figure 6.5a) [Adman *et al.*, 1978; Colman *et al.*, 1978; Guss and Freeman, 1983].

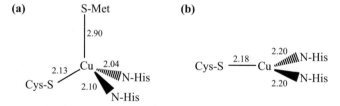

Figure 6.5 - Structure of the mononuclear Cu^{2+} centres from proteins **(a)** in plastocyanin, a blue protein [adapted from Guss *et al.*, 1983], **(b)** in the laccase from *Rigidoporus lignosus* [adapted from Garavaglia *et al.*, 2004]. The bond lengths are indicated in Angström.

This geometry is imposed by the protein. Indeed, it remains practically unchanged when the Cu^{2+} ion is reduced to Cu^+ or even when it is eliminated entirely [Guss *et al.*, 1986; Garret *et al.*, 1984].

Several crystal structures of laccases have been resolved over the last decade. They show that the geometry of the T_1 centre is quite different to that of centres contained in blue proteins: it is *planar*, with a deformed trigonal shape, has an S(Cys) ligand and two N(His) ligands (figures 6.5b and 6.6) [Ducros *et al.*, 1998; Piontek *et al.*, 2002; Hakulinen *et al.*, 2002; Garavaglia *et al.*, 2004]. The spectroscopic properties of Cu^{2+} centres in proteins have been interpreted based on these structures, and some indications are given in complement 1. Crystal structures also revealed that the T_2 centre, for which the EPR spectrum was considered "normal", in fact has a deformed trigonal geometry, with two N(His) ligands and one water ligand (figure 6.6). The dinuclear T_3 centre in turn is composed of two hydroxo-bridged copper atoms (Cu_a, Cu_b), each of which is also coordinated by three nitrogens N(His) (figure 6.6).

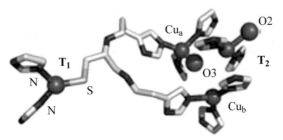

Figure 6.6 - Organisation of the copper centres in the laccase from *Rigidoporus lignosus*. O2 and O3 are oxygens from H_2O or OH^- ligands. Here, the O3 ligand, which bridges the (Cu_a, Cu_b) ions in the T_3 centre in oxidised laccases, is only linked to Cu_a, suggesting that the enzyme is partially reduced. The T_1 centre receives electrons from the substrate and transfers them to the active site (T_2, T_3) where oxygen reduction occurs. [File 1V10 in the *Protein Data Bank*]

The case of the T_1 and T_2 centres in laccases clearly illustrates the difficulties encountered when attempting to structurally interpret EPR spectra for protein centres with low symmetry. The unusual spectral characteristics of the T_1 centres reflect the specific geometry imposed by the protein, and were only interpretable once the geometry was known (complement 1). In addition, the spectrum for the T_2 centre, which is similar to those produced by Cu^{2+} complexes with square planar coordination, actually corresponds to a very different structure. The spectral features are easier to interpret when the Cu^{2+} ions form complexes with peptides or disordered domains from some proteins. In this case, the flexibility of the polypeptide chain ensures the 4-ligand equatorial coordination which stabilises the complex (see chapter 4).

6.2.2 - A reaction intermediate in oxygen reduction: a trinuclear Cu^{2+} complex

The spectral characteristics described above are those of laccases purified under air. These enzymes are *inactive*, but can be activated by reducing them with ascorbate. If a solution of activated laccases is placed in contact with its O_2 substrate, the catalytic cycle starts and stopped flow and freeze quench techniques (see section 10.2 in chapter 10) can be used to trap a *reaction intermediate* where the four copper ions are in the Cu^{2+} state. This form of the enzyme produces an EPR spectrum where the T_1 component, but not the T_2 component, is visible. At very low temperatures ($T < 20$ K) and high power, the signal of a centre of spin $S = \frac{1}{2}$ is also observed, with $g_x = 1.65$, $g_y = 1.86$, $g_z = 2.15$ [Lee *et al.*, 2002]. This signal is characterised by a strong temperature-dependence of the spin-lattice relaxation time T_1 due to an Orbach process involving an excited state of energy $\Delta \approx 140$ cm^{-1} [Volume 1, section 5.4.1]. All these results strongly suggest that, in this form of the enzyme, the three Cu^{2+} ions in the active site are coupled by strong antiferromagnetic exchange interactions. Other spectroscopic techniques confirmed that the three Cu^{2+} ions effectively constitute a *trinuclear complex* bonded by an μ_3-oxo from the substrate. This result made it possible to model the catalytic cycle of laccases [Yoon *et al.*, 2005; Solomon *et al.*, 2008]. A review of the current state of our knowledge relating to the mechanism of laccases can be found in [Augustine *et al.*, 2010].

6.2.3 - Oxidation of phenolic substrates or mediators

Up to now we have been interested in the T_1 centre and the (T_2, T_3) site, where oxygen reduction takes place. Electron transfer from T_1 towards (T_2, T_3) and oxygen reduction are rapid processes, and the oxidation of phenolic substrates (or mediators, figure 6.3) by the T_1 centre represents the limiting step for laccase activity. Research therefore seeks to optimise this step with a view to creating industrial applications, by developing powerful laccase-mediator systems, using cheap, non-toxic mediators. EPR spectroscopy has also been used in this context, for example to identify the radicals produced by oxidation of the different mediators [Brogioni *et al.*, 2008] or to count the phenoxyl radicals produced by reaction [6.2] during bleaching of paper pulp [Suurnäkki *et al.*, 2010].

6.3 - Hydrogenases: enzymes to oxidise and produce dihydrogen

Hydrogenases catalyse H_2 oxidation using a variety of electron-acceptors:

$$H_2 + 2A_{ox} \rightarrow 2H^+ + 2A_{red} \qquad [6.3]$$

The catalytic cycle necessarily includes the following steps: diffusion of H_2 to the active site, activation of H_2 within the active site, electron transfer towards A_{ox}, proton transfer towards the solvent. Experiments have shown that H_2 activation within the active site involves *heterolytic* cleavage:

$$H_2 \rightarrow H^- + H^+ \qquad [6.4]$$

The hydrogenases appeared very early during biological evolution, and they are currently found in bacteria using H_2 as a substrate or which produce H_2 to eliminate excess electrons. Because of their potential biotechnological applications, their catalytic properties have been extensively studied. Indeed, reaction [6.3] is one of the half-reactions exploited in O_2/H_2 fuel cells, and coupling the reverse reaction to a photosynthetic system to reduce A_{ox} would allow dihydrogen to be produced from solar energy (see figure 6.15b). However, industrial use of hydrogenases is complicated as they are inactivated by oxygen from air. This complex phenomenon is currently the topic of numerous studies. Due to the nature of their active sites, hydrogenases can be divided into two classes: Ni–Fe hydrogenases and Fe–Fe hydrogenases. Here, we are only interested in the Ni–Fe hydrogenases. The EPR study of Fe–Fe hydrogenases is described elsewhere [Lubitz *et al.*, 2007].

In hydrogenases, like in most redox enzymes, the active site differs from the redox centres and the redox centres are non-identical. In addition, some centres are paramagnetic in the oxidised state, while others are paramagnetic in the reduced state. To detect the EPR signals for all the paramagnetic centres in the enzyme, it is therefore necessary to perform a potentiometric titration monitored by EPR: a series of redox equilibria are generated by adding small amounts of an oxidant or a reductant to an enzymatic solution, and the potential at a platinum electrode immersed in the solution is measured relative to a reference electrode. By analysing the spectra produced by samples collected at each equilibrium, the variation in the concentration of paramagnetic centres can be monitored over a broad range of potentials, and their redox potential can be deduced thanks to the Nernst equation. This method was first applied to Ni–Fe hydrogenases in the 1980s. Combined with isotopic substitution experiments, it provided precious information on the active site and the number and nature of the redox centres well before their crystal structure was solved. The resolution of this structure [Volbeda *et al.*, 1995] led to new EPR experiments being developed, the results from which improved our understanding of the enzyme's catalytic mechanism. We will present these studies in two sections, one devoted to the active site, the other to the electron transfer chain.

6.3.1 - The active site of Ni–Fe hydrogenases

◇ The Ni–A and Ni–B signals

The X-band spectrum for a frozen solution of a hydrogenase purified in air has two components characteristic of centres of spin ½ (figure 6.7a).

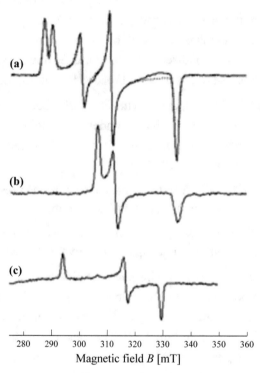

Figure 6.7 - X-band spectra produced by various states of the active site of the hydrogenase from the bacterium *Desulfovibrio desulfuricans*.
(a) After purification under air. Temperature 100 K. Microwave frequency 9.406 GHz. The simulation indicated by the dotted line was obtained by adding the following components: Ni–A (34 %) g = 2.315, 2.233, 2.008 and Ni–B (66 %) g = 2.338, 2.158, 2.006. **(b)** Ni–C signal. Temperature 70 K. Microwave frequency 9.407 GHz. **(c)** Ni–L signal. Temperature 70 K. Microwave frequency 9.420 GHz. For all spectra: power 10 mW, modulation: frequency 100 kHz, peak-to-peak amplitude 1 mT.

The total intensity of the spectrum is compatible with 0.6 paramagnetic centres per molecule of enzyme. Most Ni–Fe hydrogenases purified in air produce this type of spectrum, but the relative proportions of the two components depend on the nature of the enzyme and how it was purified. Analysis of the metal atom content indicates that each molecule contains one atom of nickel. ^{61}Ni is the only natural isotope of nickel to have a paramagnetic nucleus ($I = \frac{3}{2}$, abundance 1.1 %). If the culture medium for the bacteria is enriched in ^{61}Ni, the two components of the spectrum display a hyperfine structure corresponding to a nucleus with spin $\frac{3}{2}$, thus proving that the nickel atom contributes to these signals [Albracht *et al.*, 1982; Moura *et al.*, 1982]. These signals were therefore attributed to two states, named Ni−A and Ni−B, of a nickel-containing centre which form when the enzyme is purified in the presence of oxygen. One of the ligands of the nickel is probably an oxygenated group. Indeed, when the H$_2$-reduced enzyme is re-oxidised with ^{17}O-enriched oxygen ($I = \frac{5}{2}$), the Ni−A and Ni−B components are observed to broaden [van der Zwaan *et al.*, 1990].

In the Ni−A and Ni−B states, the enzyme is *inactive* but it can be *re-activated* by chemical reduction or by incubation under H$_2$. Detailed study of the inactivation and re-activation processes monitored by EPR led to the following conclusions [Fernandez *et al.*, 1985, 1986]:

▷ the formation of the Ni−A state appears to require the presence of oxygen. In this state re-activation is very slow (several hours),

▷ the Ni−B state can be produced by some oxidants in the absence of oxygen. In this state, the enzyme can be rapidly re-activated (a few tens of seconds).

From the crystal structure of a hydrogenase [Volbeda *et al.*, 1995], the active site was found to be a dinuclear Ni−Fe centre (figure 6.8). This hydrogenase was purified under air, and a polycrystalline powder produced an EPR spectrum similar to that shown in figure 6.7. The nickel, coordinated to 5 ligands with a very distorted geometry, is bridged to an iron atom by the thiolates from two cysteines and an unidentified ligand L. The iron remains in the Fe^{2+} state ($S = 0$) in all forms of the enzyme, which explains why it was never detected by EPR [Dole *et al.*, 1997; Huyett *et al.*, 1997].

The experiment mentioned above involving the ^{17}O isotope indicates that in the Ni−A and Ni−B states, L is an *oxygenated* species, probably H$_2$O or OH$^-$. This explains why the enzyme is *inactive* in these states, as we will see that this position is used during the catalytic cycle.

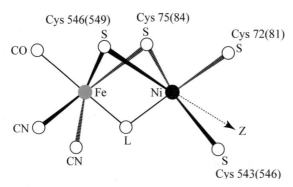

Figure 6.8 - Schematic representation of the active site of Ni–Fe hydrogenases. The nature of the bridging ligand L depends on the state of the active site. The numbering of the cysteines (see appendix 4) is that for the enzyme from *D. fructosovorans*, numbering for *D. vulgaris* is indicated in brackets.

The electronic structure of the nickel centre in the Ni–A and Ni–B states was then investigated. The first point to elucidate was the valence state of the nickel, which is on principle Ni^{3+} ($3d^7$) in the low-spin situation, or Ni^+ ($3d^9$). The g values for the Ni–A and Ni–B signals were analysed with a ligand field model adapted to an idealised (square-based pyramid type) geometry of the nickel centre, where the base is defined by the ligand L and three cysteine sulfurs; the Z axis is indicated in figure 6.8. According to this model, the orbital occupied by the unpaired electron is d_{z^2} for Ni^{3+} and $d_{x^2-y^2}$ for Ni^+; the Z axis corresponds to the smallest principal value ($g \approx 2.01$) in the first case, and to the largest ($g \approx 2.30$) in the second [Lubitz *et al.*, 2007]. The EPR study of a *single crystal* of the enzyme from the bacterium *Desulfovibrio vulgaris* showed that the magnetic axis corresponding to $g = 2.01$ is very close to the Z direction in figure 6.8 in the Ni–A and Ni–B states. This finding is a strong argument in favour of the nickel being present as Ni^{3+} [Trofanchuk *et al.*, 2000]. Data on the distribution of the spin density in the Ni–B state were obtained by ENDOR spectroscopy. ENDOR experiments with ^{61}Ni ($I = \frac{1}{2}$) were used to assess the Ni^{3+} spin population at 0.44 [Flores *et al.*, 2008] and proton ENDOR experiments on a *single crystal* revealed two strongly coupled non-exchangeable protons, and a weakly coupled exchangeable proton on ligand L. This ligand was therefore identified as OH^- [van Gastel *et al.*, 2006]. DFT-type calculations based on the *real geometry* of the active site reproduce the principal axes of the $\tilde{g}$ matrices relatively well, but not quite as well the principal values [Stadler *et al.*, 2002; Stein and Lubitz, 2004; van Gastel *et al.*, 2006]. They give a spin population of 0.5 for the d_{z^2} orbital of Ni, and a population of 0.34 on the bridging sulfur from cysteine 549 (*D. vulgaris* numbering), which identifies the two strongly coupled non-exchangeable protons detected by ENDOR as the β protons for this cysteine (see figure A4.1 in appendix 4). The spin population calculated for the OH$^-$ bridging ligand is weak, but sufficient to explain the broadening of the Ni–B signal due to interaction with ^{17}O and interaction with the exchangeable proton observed by ENDOR.

However, these nice EPR studies still do not explain the difference between the Ni–A and Ni–B states. New approaches based on modification of specific amino acids and using other techniques such as direct electrochemistry, are currently being developed to better understand the activation/inactivation processes of Ni–Fe hydrogenases [Lamle *et al.*, 2004; Abou Hamdan *et al.*, 2013].

◇ *The Ni–C and Ni–L signals*

A series of EPR experiments identified the sites where the H$^-$ and H$^+$ ions bind during heterolytic cleavage of H$_2$ (reaction [6.4]). When a Ni–Fe hydrogenase purified in air is incubated under hydrogen, the Ni–A and Ni–B signals disappear and are replaced by a different signal, characterised by $g = 2.01, 2.16, 2.19$ (figure 6.7b). This signal, which broadens when the enzyme is enriched in ^{61}Ni [Cammack *et al.*, 1987], was named Ni–C and it characterises an *active form* of the active site. The structure of the active site in the Ni–C state is similar to that in the oxidised state (figure 6.8), but no electron density appears at the site of the bridging ligand L [Higuchi *et al.*, 1999; Garcin *et al.*, 1999]. The disappearance of the oxygenated ligand was confirmed by ^{17}O ENDOR experiments [Carepo *et al.*, 2002]. However, proton ENDOR experiments and HYSCORE spectroscopy indicated that the L site is occupied by a *strongly-coupled exchangeable proton*, which was attributed to a *H$^-$ hydride* bridging the Ni^{3+} and Fe^{2+} ions [Fan *et al.*, 1991; Whitehead *et al.*, 1993; Foerster *et al.*, 2005]. These experiments, which explain the role played by the Fe^{2+} ion in catalysis, show that the Ni–C signal is characteristic of the state of the active site produced by the heterolytic cleavage of H$_2$ (equation [6.4]). The binding site for the H$^+$ ion remained to be determined. We will see that this site was discovered almost by chance through the study of a remarkable property of the Ni–C signal, its *photosensitivity*.

If a solution of enzymes in the Ni–C state is illuminated at low temperature ($T < 100$ K), the Ni–C signal progressively disappears and is replaced by the so-called Ni–L signal characterised by g = 2.01, 2.16, 2.19 (figure 6.7c). This shift is *reversible*, as the Ni–L signal spontaneously reconverts to Ni–C when the temperature exceeds around 130 K. The reconversion kinetics present a marked isotopic effect (D versus H) [Burlat, 2004] which suggests that H^- and/or H^+ transfers triggered by light cause the transformation of Ni–C into Ni–L. Transfer of H^- was confirmed by ENDOR and HYSCORE experiments which showed that the hydride is no longer bound to the nickel in the Ni–L state [Whitehead *et al.*, 1993; Brecht *et al.*, 2003]. Another result indicates that illumination also triggers *proton transfer*. Indeed, when the redox state of the enzyme is such that the active site is in the Ni–C state, magnetic interactions with a neighbouring iron-sulfur centre produce at low temperature a "split Ni–C" spectrum (section 6.3.2 and figure 6.10b). Simulation of this spectrum shows that the *exchange component* of the interactions, which is significant in the Ni–C state, disappears in the Ni–L state [Dole *et al.*, 1996]. According to the crystal structure, the "exchange pathway" connecting the two paramagnetic centres starts with a hydrogen bond between the sulfur from cysteine 543 and the oxygen from Glu25 (figure 6.9).

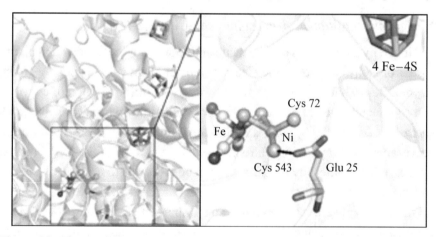

Figure 6.9 - The metal centres in the *D. fructosovorans hydrogenase*. The zoom shows the hydrogen bond between the sulfur from Cys 543 and the oxygen from Glu 25 which starts the exchange pathway between the Ni centre and the "proximal" 4Fe–4S centre visible in the figure. This bond is broken in the Ni–L state. [File 1YQW in the *Protein Data Bank*]

The rupture of this pathway in the Ni–L state can be explained if exposure to light results in *deprotonation* of this cysteine. Although the study was performed at cryogenic temperature, it strongly suggests that cysteine 543 could bind a proton at room temperature during heterolytic cleavage of H_2.

To test this hypothesis, we investigated whether residue Glu25 is involved in proton transfer towards the solvent generated by reaction [6.4]. Indeed, in proteins, protons are transferred along chains of proton acceptor groups (water molecules, carboxylic groups from glutamic and aspartic acid) separated by short distances, often connected by hydrogen bonds [Bertrand, 2014]. By substituting other amino acids for Glu25 and studying the effects on enzymatic activity, it was possible to show that this residue constitutes the first link in the proton transfer chain, thus confirming the relay role played by cysteine 543 [Dementin *et al.*, 2004].

6.3.2 - The electron transfer chain

In redox enzymes, electron transfers are often ensured by a chain of iron-sulfur centres. These centres and their magnetic properties are briefly presented in complement 2. In this section, we will first show how EPR was used to determine the composition of the electron transfer chain in Ni–Fe hydrogenases and its redox properties, then we will examine how this chain works.

◇ Potentiometric titration of iron-sulfur centres in an Ni-Fe hydrogenase

All Ni–Fe hydrogenases have the same active site, but the number and nature of the iron-sulfur centres in the electron transfer chain vary depending on the enzyme. Below, we describe the spectra observed during titration of enzymes with a molecular weight of around 90 kDa, such as those from the bacteria *D. fructosovorans* and *D. gigas*. At high potential and very low temperature, a weakly anisotropic spectrum centred at $g = 2.02$, typical of a $[3Fe–4S]^{1+}$ centre, is observed. The intensity of this spectrum corresponds to a single centre per molecule (figure 6.10a). As the temperature is increased, the spectrum broadens and finally disappears at around 40 K due to spin-lattice relaxation [Volume 1, section 5.4]. If a reductant such as sodium dithionite is added, the potential at the electrode drops, the amplitude of the spectrum decreases and a dissymetric signal due to the $[3Fe–4S]^0$ centre of spin $S = 2$ appears at very low-field. Things get more complicated at very low potential. Indeed, the electrons provided by the reductant are then used by the enzyme to reduce the protons in the aqueous solvent to form H_2 (reverse of reaction [6.3]). As a result, the equilibrium is displaced towards that of the H_2/H^+ couple in the experimental conditions.

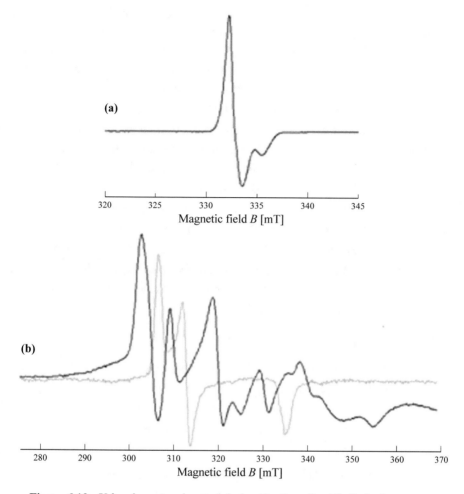

Figure 6.10 - X-band spectra observed during titration of an Ni−Fe hydrogenase.
(a) Spectrum for the [3Fe−4S]$^{1+}$ centre. Temperature 15 K. Microwave: frequency
9.406 GHz, power 0.4 mW. Modulation: amplitude 0.5 mT. The spectrum (Ni−A, Ni−B)
for the active site (figure 6.7a) is completely saturated in these conditions.
(b) "Split Ni−C" spectrum. Temperature 6 K. Microwave: frequency 9.4069 GHz,
power 10 mW. Modulation: amplitude 1 mT. To show the splitting of the lines, we have
reproduced the Ni−C spectrum from figure 6.7b. The structures observed beyond the
300−345 mT region are produced by the iron-sulfur centres.

To continue the titration, this equilibrium must be controlled by setting the partial
pressure of H$_2$ [Cammack *et al.*, 1987]. Samples prepared in these conditions
at low potential produce spectra which are very temperature-dependent:

▷ At very low temperature ($T \leq 10$ K), the "split Ni−C" spectrum due to interactions between the Ni−C centre ($S = \frac{1}{2}$) and the iron-sulfur centres is ob-

served. A broad, relatively unstructured signal due to the $[3Fe-4S]^0$ ($S = 2$) and $[4Fe-4S]^{1+}$ ($S = \frac{1}{2}$) centres is superimposed on this spectrum (figure 6.10b). Simulation of the "split Ni–C" spectrum at X-, Q- and S-bands shows that the Ni–C centre interacts with a single $[4Fe-4S]^{1+}$ centre, which is therefore the centre closest to the active site [Guigliarelli *et al.*, 1995].

▷ At temperatures exceeding around 30 K, the spin-lattice relaxation time T_1 for the iron-sulfur centres becomes short enough for their spectra to broaden and disappear, and the effects of the interaction with the $[4Fe-4S]^{1+}$ centre average to zero [Volume 1, section 7.5.2]. We then observe the Ni–C signal represented in grey in figure 6.10b.

The variations in intensity of all these signals with the electrode potential are well described by the one-electron Nernst equations, as a result it is possible to determine the redox potential of the different centres. In particular, the potential of the 3Fe–4S centre is found to be considerably more positive than that of the 4Fe–4S centres, for example + 65 mV versus – 340 mV in the enzyme from *D. fructosovorans* [Rousset *et al.*, 1998].

◇ *Is the electron transfer chain efficient?*

Unexpected information related to the *arrangement* of the centres in the electron transfer chain was provided by the crystal structure [Volbeda *et al.*, 1995] (figure 6.9). These centres are spaced at around 12 Å from each other, which favours rapid electron transfer [Bertrand, 1991]. However, they are arranged in the following order:

$$[Fe\ Ni]\quad [4Fe-4S]_{prox}\quad [3Fe-4S]_{med}\quad [4Fe-4S]_{dist}$$

The centre closest to the active site is a 4Fe–4S centre, as indicated by the study of the "split Ni–C" spectrum mentioned above, but the position of the 3Fe–4S centre was unexpected. Indeed, the presence of a median centre with high potential between the two low-potential 4Fe–4S centres appeared not to be very favourable for rapid electron transfer along the chain. To test this point, the 3Fe–4S centre was converted into a 4Fe–4S centre in the hydrogenase from *D. desulfuricans*. To do so, it is sufficient to replace an appropriately placed proline residue by a cysteine which provides the fourth binding site for a 4Fe–4S centre. The potentiometric titration of the modified enzyme monitored by EPR showed that the conversion of the 3Fe–4S centre into a 4Fe–4S centre decreased the potential of the median centre by around 300 mV. Despite this difference, the enzymatic activity remained essentially unchanged [Rousset *et al.*, 1998]. Thus, in contrast to what might be supposed, the presence of a high-potential 3Fe−4S centre in the middle of the electron transfer chain does not hamper the enzyme's efficacy. These experiments provide new information on the factors determining the efficacy of electron transfer chains in redox enzymes [Dementin *et al.*, 2011].

6.4 - Photosystem II: a solar energy-driven enzymatic complex

In photosynthetic organisms, membrane-bound "antenna" systems collect photons and transfer them to "reaction centres" where very rapid *charge separation* takes place (figure 6.11):

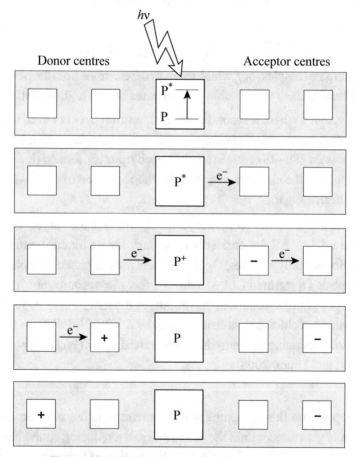

Figure 6.11 - Charge separation in a reaction centre.
- Pigment P absorbs a photon and becomes electronically excited: state P*.
- P* rapidly transfers an electron towards a chain of acceptor centres and becomes the cation P^+ of the ground state.
- P^+ is reduced by a chain of electron donor centres and returns to its initial state, P.

Absorption of a photon generates a reducing equivalent at the end of the acceptor chain and an oxidising equivalent at the start of the donor chain. In chloroplasts from plants and green algae and in cyanobacteria, charge separation is exploited to produce ATP and to reduce $NADP^+$ to form NADPH, which is then combined with electrons extracted from water to synthesise sugars. This "oxygenic" photosynthetic system plays an essential role on Earth, since it is responsible for synthesising biomass, fixing CO_2 and producing the oxygen that we breathe. Its study as a function of the radiation frequency shows that it involves three membrane-bound complexes functioning *in series* [Hill and Bendall, 1960]:

1- Photosystem II where a first charge separation causes oxidation of water by plastoquinones:

$$2\,H_2O + 2\,PQ \xrightarrow{\quad 4\,h\nu \quad} O_2 + 2\,PQH_2 \qquad [6.5]$$

2- A cytochrome complex, $b_6 f$, which receives the electrons from the plastoquinols and transfers them to a soluble protein containing a high-potential centre.

3- Photosystem I where a second charge separation event is used to transfer electrons from the *high-potential* centre in the soluble protein towards the *low-potential* 2Fe–2S centre in another soluble protein, ferredoxin. Ferredoxin then supplies the electrons required for the formation of the NADPH used to synthesise sugars.

We will add a few details on photosystem II, which is the subject of this section (figure 6.12). All the elements necessary for its function are linked to two homologous polypeptides, D_1 and D_2. The pigment is a chlorophyll linked to polypeptide D_1, termed Chl_{D1}. Within a few picoseconds, absorption of a photon by this pigment produces the $(P680^+, Phe^-)$ state, where $P680^+$ is a very oxidising dimer of chlorophylls characterised by $E^{\circ\prime}(P680^+/P680) = +1.26\,V$ and Phe^- is a very reducing pheophytin characterised by $E^{\circ\prime}(Ph/Ph^-) \approx -0.500\ V$ [Rappaport and Diner, 2008].

Charge separation then propagates by electron transfer, until an oxidising equivalent is generated within the active site and a reducing equivalent is generated on a plastoquinone (figure 6.12). Absorption of four photons therefore allows oxidation of two H_2O molecules and reduction of two plastoquinones (reaction [6.5]). The possibility of using water as a source of electrons thanks to solar energy opens numerous perspectives which have motivated, and continue to motivate, very active research into the functioning of this system.

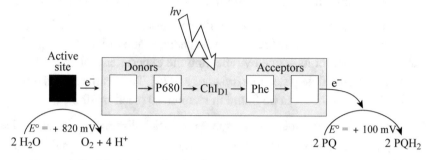

Figure 6.12 - Mechanism of action for photosystem II. The charge separation created by light energy causes electrons to circulate in the opposite direction to that imposed by the potentials.

In this section, we present the EPR experiments which identified the donor centres and the active site of photosystem II.

6.4.1 - The enigmatic signal 2

Many years ago, it was discovered that continuous illumination of chloroplasts from plants produces an EPR spectrum with two radical components. The first (signal 1) is centred at $g = 2.0025$ and has no apparent hyperfine structure, the second (signal 2) is centred at $g = 2.0045$ and displays a poorly resolved hyperfine structure [Commoner *et al.*, 1957]. Several experiments performed in the 1960s allowed signal 1 to be attributed to the $P700^+$ radical created by photo-oxidation of photosystem I, but the centre or centres producing signal 2 were much more difficult to identify.

◇ From signal 2 to Tyr_Z and Tyr_D centres

Signal 2 is produced by centres located on the *electron donor* side of photosystem II [Babcock and Sauer, 1973], therefore it must appear transiently when these centres are photo-oxidised, and disappear rapidly when they are reduced by the active site (figure 6.12). However, when a preparation of chloroplasts is exposed to a flash of light, signal 2 appears very rapidly but only part of it disappears rapidly, the remainder is stable for many hours in the dark. To understand this phenomenon, researchers monitored the evolution of the spectrum during the transient regime by using two methods:

▷ slow down the kinetics by *partially inhibiting* the active site, which makes it possible to monitor the changes over time to the whole of the spectrum with a conventional EPR spectrometer [Babcock and Sauer, 1975].

▷ reduce the spectrometer's response time to around 100 μs, which makes it possible to monitor how the amplitude changes at a point of the spectrum, and to reconstruct its shape by repeating the experiment for several values of the magnetic field [Blankenship *et al.*, 1975]. The details of this time-resolved EPR method can be found in chapter 10 (section 10.3.2).

In both cases, the component of the spectrum which disappears rapidly, called signal 2_f (*fast*), and that which disappears very slowly, called signal 2_s (*slow*), were observed to have slightly different shapes, and their intensity was found to correspond to a single centre adjacent to PSII. The 2_f and 2_s signals therefore derive from *two similar donor centres* which were named Z^+ and D^+. The Z centre, which is rapidly reduced by the active site is the physiological donor, whereas D only plays a secondary role. The redox potential of the (D^+, D) pair can be measured by potentiometric titration. Given the equilibrium constant between D^+Z and $D Z^+$, it can be deduced that the potential of the (Z^+, Z) pair is around 1 V. This value is compatible with the Z centre playing a redox relay role between the active site and P680 (figure 6.12) [Boussac and Etienne, 1984]. It was not until 1987 and 1988, i.e., 30 years after the first observation of signal 2, that the identity of the Z and D centres was finally revealed by site-directed mutagenesis experiments (see appendix 4) performed on photosystem II from *cyanobacteria*. It was first observed that deuteration of the tyrosine residues in the enzymes caused signal 2_s to narrow, indicating that D^+ is very probably a tyrosyl radical [Barry and Babcock, 1987]. Tyrosine 160 from the D_2 polypeptide appeared to be a good candidate. Its replacement by a phenylalanine causes signal 2_s to disappear, proving that this tyrosine corresponds to the D centre. Tyrosine 161 from the D_1 polypeptide, homologue of tyrosine 160 from the D_2 polypeptide, was then identified as Z [Debus *et al.*, 1988; Vermaas *et al.*, 1988]. The Z and D centres were subsequently denoted Tyr_Z and Tyr_D.

◇ *Subsequent studies*

The very specific nature of the Tyr_Z and Tyr_D centres suggest that their role is not limited to that of redox centre. The electronic structure of the $Tyr_D^•$ and $Tyr_Z^•$ radicals and their arrangement in the membrane were therefore studied.

▷ Proton ENDOR and ESEEM experiments were performed to determine the distribution of the spin density in these radicals. This did not pose any particular problems for $Tyr_D{}^\bullet$, which is stable [Hoganson and Babcock, 1992; Rigby et al., 1994; Warnke et al., 1994; Dole et al., 1997]. But to be able to work on the signal from $Tyr_Z{}^\bullet$, the signal from Tyr_D must be eliminated by replacing it by a phenylalanine, and inhibiting the active site to be able to generate $Tyr_Z{}^\bullet$ by continuous illumination and freezing of the sample. These experiments showed that the distribution of the spin density was practically the same for the two radicals, and similar to that of other tyrosyl radicals [Tommos et al., 1995].

▷ The orientation of Tyr_D in the membrane was determined by EPR experiments performed on membrane fragments deposited on a Mylar support. A first study performed at X-band suggested that the plane of the radical was almost perpendicular to the membrane [Rutherford, 1985]. High-field EPR provided more precise information. The principal values of the $\tilde{g}$ matrix for the $Tyr_D{}^\bullet$ radical were first determined precisely by simulating the spectrum for a frozen solution, recorded at 190 and 285 GHz (see figure 9.6 in Volume 1). These simulations were used to reproduce the variations of the spectrum recorded at 285 GHz for membrane fragments when the orientation of the support is varied relative to the magnetic field **B**. The resulting data indicated that Tyr_D forms a 64° angle with the membrane [Dorlet et al., 2000]. Study of single crystals by EPR produced the same result [Hofbauer et al., 2001].

▷ As some authors have suggested that the role of Tyr_Z might be to receive the hydrogen atoms (protons + electrons) produced by water oxidation (figure 6.12) [Tommos et al., 1995], efforts were made to determine the distance between Tyr_Z and the active site. The active site can be prepared in an inactive but paramagnetic state in which its magnetic interaction with the $Tyr_Z{}^\bullet$ radical produces splitting of the lines [Boussac et al., 1989]. Several studies of the interaction spectrum based on the point dipole approximation [Volume 1, section 7.3] have been published, the most convincing of which consisted in simulating, with the same set of parameters, the spectra recorded at frequencies between 9.4 GHz and 285 GHz [Dorlet et al., 1999]. These studies indicated distances of between 6 and 9 Å, which are too great for direct transfer of a hydrogen atom [Peloquin and Britt, 2001]. The first crystal structures of photosystem II confirmed that the distance between the four Mn atoms in the active site (see following section) and the phenolic oxygen from Tyr_Z is large, around 8 Å [Ferreira et al., 2004;

Loll *et al.*, 2005]. However, a high-resolution structure recently revealed that the active site is in fact an Mn_4CaO_5 complex in which the calcium ion is coordinated to two H_2O molecules which form hydrogen bonds with the phenolic oxygen from Tyr_Z. This arrangement is favourable for rapid transfer of protons or hydrogen atoms between the active site and Tyr_Z [Umena *et al.*, 2011].

Over the last 50 years, progress in the interpretation of signal 2 has closely followed progress in EPR spectroscopy: improved sensitivity of spectrometers, development of high-resolution spectroscopic methods (ENDOR, pulsed EPR, high-field EPR), development of time-resolved techniques. Retrospectively, it appears surprising that it took so long to identify the centres producing signal 2, but it must be remembered that the idea that an amino acid could play a redox role in an enzyme had yet to be raised, even though a precedent existed with ribonucleotide reductase [Reichard and Ehrenberg, 1983]. The existence of Tyr_D at first complicated the interpretation of signal 2, but the stability of the $Tyr_D^{\bullet}$ radical subsequently made it possible to identify $Tyr_Z^{\bullet}$ simultaneously. Indeed, it is stimulating to note that we still do not know what role Tyr_D plays in this system.

6.4.2 - The mysteries of the oxygen-evolving centre

In photosystem II, the oxidising equivalents produced by charge separation are used to extract the electrons from water in the active site, also known as the oxygen evolving centre (figure 6.12). Since absorption of a photon by the pigment Chl_{D1} creates one oxidising equivalent, oxidation of two H_2O molecules to produce one O_2 molecule requires the absorption of four photons (equation [6.5]). Since Pierre Joliot and collaborators' elegant experiments [Joliot *et al.*, 1969] and their interpretation by Kok [Kok *et al.*, 1970], we know that a sequence of light pulses causes the active site to complete a cycle of oxidation states which have been named S_1 to S_4, the index indicates the number of oxidising equivalents accumulated. The spontaneous passage from S_4 to S_0 leads to synthesis of one O_2 molecule (figure 6.13). Most of these oxidation states are paramagnetic, and EPR has played, and continues to play, an essential role in their characterisation.

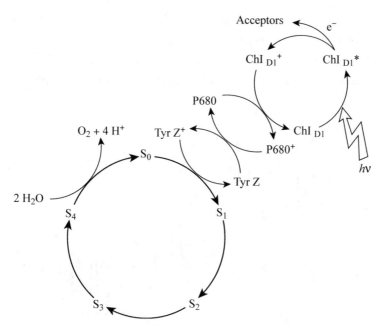

Figure 6.13 - Cycle of oxidation states for the active site of photosystem II.
The charge separation on the side of the electron donor produced by absorption
of a photon by the ChI_{D1} pigment, is detailed for the $S_0 \rightarrow S_1$ transition. The same
sequence is reproduced for the three subsequent transitions. The $S_4 \rightarrow S_0$ transition
is spontaneous and leads to oxidation of two H_2O molecules to form one O_2 molecule.

◇ The Mn$_4$ complex

Based on the above, the active site of photosystem II is able not only to activate
H_2O molecules, it can also *store four oxidising equivalents* by releasing an
electron upon each charge separation. Only polynuclear transition ion complexes are known to have such properties. Indeed, it has long been known that
this photosystem contains four manganese atoms which are essential to its
functioning. Manganese was first detected by EPR in the form of Mn^{2+} ions in
photosystem II complexes where the active site was inhibited [Blankenship and
Sauer, 1974]. But we had to wait several years before the spectrum for the active
site in a *functional state* was observed, by illuminating the sample with very
brief (25 ns) light pulses, freezing it rapidly and recording its EPR spectrum at
very low temperature [Dismukes and Siderer, 1981]. This spectrum, composed
of a pattern of at least 19 hyperfine lines separated by 8 to 9 mT and centred
at $g \approx 2$, was qualified as a *multiline* spectrum (figure 6.14). If we start with a
"dark adapted" sample, where the active site is in the S_1 state, the intensity of

the spectrum passes through a maximum after one or five light pulses, which indicates that it corresponds to the S_2 state (figure 6.13). The shape of the spectrum is typical of polynuclear complexes of spin $S = \frac{1}{2}$ in which the unpaired electron interacts with several ^{55}Mn nuclei ($I = \frac{5}{2}$) [Volume 1, section 7.4.3].

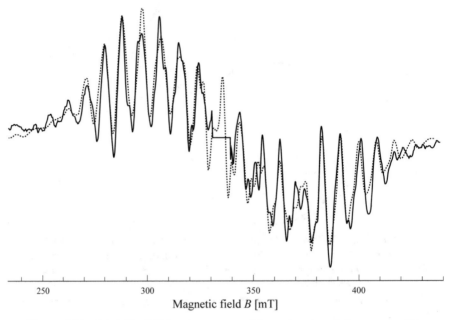

Figure 6.14 - "Multiline" X-band spectrum of the active site of photosystem II in the S_2 state. Temperature: 9 K. Microwave frequency: 9.421 GHz, power 20 mW. Modulation: frequency 100 kHz, peak-to-peak amplitude 2.5 mT. The simulation in dashed lines was obtained as indicated in the text and in [Charlot *et al.*, 2005].

Since the active site contains four manganese atoms, this spectrum can be produced by a tetranuclear complex or be the result of superposition of the spectra produced by two dinuclear complexes, or by a trinuclear complex and a mononuclear complex. A simplified simulation is sufficient to distinguish between the different possibilities. The spectrum of a complex with N strongly coupled Mn sites is described by the following spin Hamiltonian [Volume 1, section 7.4]:

$$\hat{H}_{S=1/2} = \beta\, \mathbf{B} \cdot \tilde{\mathbf{g}} \cdot \mathbf{S} + \sum_{n=1}^{N} \mathbf{S} \cdot \tilde{\mathbf{A}}_n \cdot \mathbf{I}_n \qquad [6.6]$$

The hyperfine matrix $\tilde{\mathbf{A}}_n$ is linked to the "local" matrix $\tilde{\mathbf{a}}_n$ of site n by

$$\tilde{\mathbf{A}}_n = K_n \tilde{\mathbf{a}}_n \qquad [6.7]$$

Where K_n are the "spin coupling coefficients" which verify

$$\sum_{n=1}^{N} K_n = 1 \qquad [6.8]$$

Using *isotropic* $\tilde{g}$ and $\tilde{a}_n$ matrices with values similar to those generally observed in manganese complexes, it can be shown that it is impossible to reproduce the structures from figure 6.14 by superimposing the spectra for two dinuclear complexes or one trinuclear complex and a mononuclear complex. In contrast, good simulations are obtained with different sets of parameters if the spectrum is assumed to be produced by a single *tetranuclear* complex, which we will hereafter term Mn$_4$ [Bonvoisin *et al.*, 1992; Hasegawa *et al.*, 1999; Peloquin and Britt, 2001; Carrell *et al.*, 2002].

◇ *Valence state and organisation of the Mn ions in the complex*

The following step consists in simulating the spectrum using the spin Hamiltonian [6.6] with four Mn sites and *anisotropic* $\tilde{g}$ and $\tilde{a}_n$ matrices. If, for simplicity, we assume that the principal axes of all the matrices are parallel, a good simulation is obtained with a quasi isotropic $\tilde{g}$ matrix ($g_x = 1.988$, $g_y = 1.985$, $g_z = 1.975$), three weakly anisotropic $\tilde{A}_1$, $\tilde{A}_3$, $\tilde{A}_4$ hyperfine matrices with values of $|A_{iso}|$ equal to 330, 245 and 190 MHz, and a more anisotropic $\tilde{A}_2$ matrix with $|A_{2iso}| \approx 260$ MHz (figure 6.14) [Charlot *et al.*, 2005]. The quasi isotropic nature of the $\tilde{g}$ matrix and the moderate anisotropy of the hyperfine matrices justify the hypothesis relating to the principal axes. The relevance of this simulation is confirmed by the fact that the values of g are practically identical to those *measured* on a spectrum recorded at 94 MHz (1.988, 1.981, 1.965) [Matsuoka *et al.*, 2006] and that the values of $|A_{iso}|$ are close to those determined by ENDOR experiments with ^{55}Mn (298, 238, 190 and 238 MHz) [Peloquin *et al.*, 2000]. As the hyperfine matrices for Mn^{3+} sites are generally more anisotropic than those for Mn^{4+} sites, these results indicate that in the S$_2$ state, the Mn$_4$ complex is composed of three Mn^{4+} ions of spin $S = \frac{3}{2}$ (sites 1, 3, 4) and one Mn^{3+} ion of spin $S = 2$ (site 2).

The spin coupling coefficients K_n in equation [6.7] represent the spin populations of the different Mn sites, which are determined by the valence state of the Mn ions and the relative values of the exchange parameters [Volume 1, section 7.4 and appendix 5]. Their values can be deduced from equation [6.7], using the hyperfine constants $(A_n)_{iso}$ provided by the simulation and local hyperfine constants $(a_n)_{iso}$ for Mn^{3+} and Mn^{4+} sites present in inorganic complexes or at the active site of enzymes. The set $a_{iso}(Mn^{3+}) = -192$ MHz and $a_{iso}(Mn^{4+}) = -237$ MHz obtained for the active site of catalase from *Thermus thermophilus* [Zheng *et al.*, 1994] produces values of K_n which effectively verify equation [6.8]. These values correspond to a spin coupling scheme in which one of the Mn ions in an Mn_3 triad interacts with a fourth Mn [Charlot *et al.*, 2005]. This pattern is perfectly consistent with the recently-obtained high-resolution crystal structure of photosystem II [Umena *et al.*, 2011]. Indeed, this structure effectively shows that the active site is an Mn_4CaO_5 complex composed of a cubane Mn_3CaO_4 structure bridged to the fourth Mn, and the values for the Mn−O and Mn−Mn distances indicate that the fourth Mn mainly interacts with a single Mn in the Mn_3 triad.

Complementary studies were used to localise the Mn^{3+} ion in the complex . A study based on simultaneous simulation of the X- and Q-band EPR spectra and of the ENDOR Q-band spectrum for ^{55}Mn, replacing Ca^{2+} by Sr^{2+} and DFT modelling, showed that the Mn^{3+} ion is one of the two Mn ions in the Mn_3CaO_4 cubane which do not interact strongly with the fourth Mn [Cox *et al.*, 2011]. The other oxidation states of the active site have also been characterised by EPR. Details of all these studies can be found in [Haddy, 2007; Boussac *et al.*, 2009].

6.5 - Conclusion

The era of the pioneers attempting to determine the structure of the active site and the redox centres of redox enzymes using continuous wave EPR spectra is long past. Spectroscopists now have a range of high-resolution techniques and powerful modelling tools with which they can study the catalytic mechanism of these enzymes in detail. The examples presented in this chapter clearly illustrate the complementary nature of crystallographic and spectroscopic studies:

  ▷ the first supply invaluable information on the architecture of the centres and their organisation in the protein, but they often describe an averaged structure of poorly defined states,

  ▷ the second can characterise well-defined states of the enzyme, and in particular intermediates of the catalytic cycle thanks to time-resolved methods. EPR spectroscopy thus plays an essential role in the characterisation of radicals and metal centres, but it cannot detect non-paramagnetic metal cations which may play a very important role, such as the Fe^{2+} ion $(S = 0)$

in the active site of Ni–Fe hydrogenases and the Ca^{2+} ion in the active site of photosystem II.

We will finish this chapter with a few words about current research into the mechanisms of redox enzymes. Biological evolution selected very efficient enzymes, built from abundant elements found in the Earth's crust. Knowledge of the catalytic mechanism will allow us to better understand how living organisms function, but it is also required if we are to develop a new generation of catalysts. Indeed, some redox enzymes catalyse processes that are very important at industrial scale. For example, in figure 6.15a we illustrated an O_2/H_2 fuel cell in which laccases and hydrogenases are used as catalysts, and in figure 6.15b we showed a system which would be able to produce hydrogen from water and light.

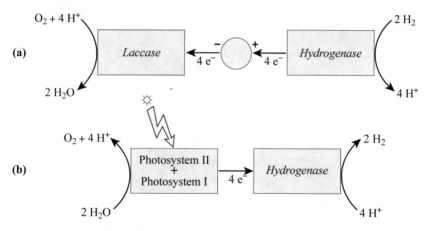

Figure 6.15 - The building blocks for redox enzymes:
(a) O_2/H_2 fuel cell, **(b)** hydrogen production from water and light.

These systems are not pure imagination, since a similar assembly to that shown in figure 6.15a has already been shown to work in laboratory conditions [Vincent *et al.*, 2005]. Hydrogen has also been produced from light energy using photosystem I and a hydrogenase, but using a gold electrode as a source of electrons [Krassen *et al.*, 2009]. These types of assembly are not necessarily transferable at industrial scale, but we can be inspired by the catalytic mechanism of these enzymes to develop very different catalysts from the traditional, generally not very specific and potentially toxic, rare metal-based catalysts, which in addition are very expensive.

Acknowledgements

This chapter benefitted from attentive re-reading and suggestions from Geneviève Blondin (iRTSV/LCBM, CEA, Grenoble), Alain Boussac (CNRS-LBMP, CEA Saclay) and Christophe Léger (BIP CNRS Université Aix-Marseille, Marseille). Simulations of spectra were produced by Emilien Etienne.

Complement 1 - Interpreting the EPR properties of mononuclear copper-containing centres in proteins

The specific spectroscopic properties of the Cu^{2+} centres contained in blue proteins can be attributed based on their structure and data obtained by X-ray absorption spectroscopy. Indeed, these data show that the Cu–$S(Cys)$ bond is highly covalent, in agreement with SCF-Xα-SW-type calculations which indicate that the spin density is equally distributed between the $3d_{x^2-y^2}$ orbital of Cu^{2+} and $3p\pi$ orbital of $S(Cys)$ [Solomon and Lowery, 1993]. This strong covalence explains the high intensity of the absorption at 600 nm, which is due to a charge transfer transition $S(Cys)\pi \to Cu\ 3d_{x^2-y^2}$. It also explains the low values of the hyperfine constants. Indeed, these are generally written:

$$A = A_s + A_{dip}^{spin} + A_{dip}^{orb}$$

The first two terms, due to core polarisation and to spin dipolar interactions, respectively, are negative for component $A_{//}$ and they determine its sign. The third term, due to orbital dipolar interaction via spin-orbit coupling, is positive [Volume 1, appendix 2]. The covalence of the Cu–$S(Cys)$ bond reduces the spin density on the Cu^{2+} ion and consequently the $|A_s|$ and $|A_{dip}^{spin}|$ quantities, which in turn reduces $|A|$ [Solomon and Lowery, 1993]. A more recent DFT study provided a different explanation: the specific structure of the copper-containing centre in blue proteins (figure 6.5a) favours spin-orbit coupling effects, which increases the A_{dip}^{orb} term [Remenyi *et al.*, 2007].

If we now consider the T_1 centre of laccases, the absence of the axial $S(Met)$ ligand (figure 6.5b) slightly increases the covalence of the S–Cys bond, but it also increases the energy of the $d \to d$ transitions, decreasing the effects of spin-orbit coupling. As a result, $g_{//}$, $g_\perp$ and the A_{dip}^{orb} contribution decrease, and the absolute value of the hyperfine constant $|A|$ increases [Palmer *et al.*, 1999].

Complement 2 - Iron-sulfur centres and their EPR properties

Iron-sulfur centres are polynuclear complexes found in numerous redox enzymes and proteins; they were first identified by EPR in 1960 [Beinert and Sands, 1960]. They contain a variable number of iron atoms bridged by sulphide ions, and each iron is generally coordinated to the protein by sulfur atoms from cysteine residues.

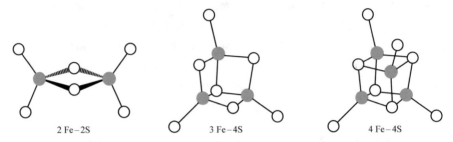

2 Fe–2S   3 Fe–4S   4 Fe–4S

Figure 6.16 - Structure of iron-sulfur centres. Fe atoms are shown in grey, S atoms in white.

These centres have two redox states corresponding to the valence states of the iron ions indicated in Table 6.1.

Table 6.1 - Formal valence of the iron ions and spin S of the ground state of the different redox states of iron-sulfur centres. By convention, we have indicated the centre's charge when considering only the iron ions and the sulphides.

	2Fe–2S	**3Fe–4S**	**4Fe–4S** [a]
Oxidised state	$[2Fe-2S]^{2+}$	$[3Fe-4S]^{1+}$	$[4Fe-4S]^{2+}$
	$2Fe^{3+}$	$3Fe^{3+}$	$2Fe^{3+}, 2Fe^{2+}$
	$S = 0$	$S = \frac{1}{2}$	$S = 0$
Reduced state	$[2Fe-2S]^{1+}$	$[3Fe-4S]^{0}$	$[4Fe-4S]^{1+}$
	Fe^{3+}, Fe^{2+}	$2Fe^{3+}, Fe^{2+}$	$Fe^{3+}, 3Fe^{2+}$
	$S = \frac{1}{2}$	$S = 2$	$S = \frac{1}{2}, \frac{3}{2}$

[a] Some redox proteins from bacterial photosynthetic systems contain $[4Fe-4S]^{3+,2+}$ centres which are paramagnetic in the oxidised state ($3Fe^{3+}, Fe^{2+}$) with $S = \frac{1}{2}$. Their EPR spectrum is characterised by three g values slightly greater than 2.

For each redox state, we have indicated the spin S of the ground state which results from the combined effects of antiferromagnetic exchange coupling and the phenomenon of valence delocalisation [Guigliarelli and Bertrand, 1999].

$[2Fe-2S]^{1+}$ and $[4Fe-4S]^{1+}$ centres of spin ½ have two g values slightly less than 2 and one value slightly greater than 2 [Volume 1, figure 7.11]. Some $[4Fe-4S]^{1+}$ centres have a ground state with $S = \frac{3}{2}$. Their very anisotropic EPR spectrum, characterised by an effective g with a maximal value close to $g \approx 5$ [Volume 1, Figure 6.12], is much more difficult to detect than that of centres of spin ½ [Lanciano *et al.*, 2007]. The $[3Fe-4S]^{1+}$ centres in which the three iron ions are ferric produce a quasi-isotropic spectrum centred near $g \approx 2.01$. In the reduced state $[3Fe-4S]^{0}$ of spin $S = 2$, a broad line is often observed at very low-field. Its detection is facilitated by the use of a dual-mode cavity [Volume 1, section 6.7].

The "cubane" structure of 4Fe–4S centres is particularly stable. In some proteins, these centres even play roles other than that of a redox centre [Guigliarelli and Bertrand, 1999; Beinert, 2000]. Genetic [Eck and Dayhoff, 1966] and biochemical [Martin and Russell, 2007] reasoning suggests that they appeared very early during evolution: nature has been "playing with cubanes" for a long time!

References

ABOU HAMDAM A. *et al.* (2013) *Nature Chemical Biology* **9**: 15-17.

ADMAN E.T. *et al.* (1978) *Journal of Molecular Biology* **123**: 35-47.

ALBRACHT S.P.J. *et al.* (1982) *FEBS Letters,* **140**: 311-313.

ANDROES G.M. & Calvin M. (1962) *Biophysical Journal* **2**: 217-258.

AUGUSTINE A.J. *et al.* (2010) *Journal of the American Chemical Society* **132**: 6057-6067.

BABCOCK G.T. & SAUER K. (1973) *Biochimica et Biophysica Acta* **325**: 483-503.

BABCOCK G.T. & SAUER K. (1975) *Biochimica et Biophysica Acta* **376**: 315-328.

BARRY B.A. & BABCOCK G.T. (1987) *Proceeding of the National Academy of Sciences of the USA* **84**: 7099-7103.

BEINERT H. (2000) *Journal of Biological Inorganic Chemistry* **5**: 2-15.

BEINERT H. & SANDS R.H. (1960) *Biochemical and Biophysical Research Communications* **3**: 41-46.

BERTRAND G. (1895) *Comptes Rendus de l'Académie des Sciences* (Paris) **120**: 266-269.

BERTRAND P. (1991) *Structure and Bonding* **75**: 1-47.

BERTRAND P. (2014) « Molecular modelling of proton transfer kinetics in biological systems » in *Reaction rate constant computations: theories and applications*, K. Han and T. Chu, eds, RSC, Cambridge.

BLANKENSHIP R.E. & SAUER K. (1974) *Biochimica et Biophysica Acta* **357**: 252-266.

BLANKENSHIP R.E. *et al.* (1975) *FEBS Letters* **51**: 287-293.

BLUMBERG W.E. & PEISACH J. (1966) *Biochimica et Biophysica Acta* **126**: 269-273.

BONVOISIN J. *et al.* (1992) *Biophysical Journal* **61**: 1076-1086.

BOUSSAC A. & ETIENNE A-L. (1984) *Biochimica et Biophysica Acta* **766**: 576-581.

BOUSSAC A. *et al.* (1989) *Biochemistry* 28: 8984-8989.

BOUSSAC A. *et al.* (2009) *Journal of the American Chemical Society* **131**: 5050-5051.

BRECHT M. *et al.* (2003) *Journal of the American Chemical Society* **125**: 13075-13083.

BROGIONI B. *et al.* (2008) *Physical Chemistry Chemical Physics* **10**: 7284-7292.

BROMAN L. *et al.* (1963) *Biochimica et Biophysica Acta* **75**: 365-376.

BURLAT B. (2004) Thèse de doctorat, *Etude fonctionnelle des hydrogénases à centres Ni-Fe*, Université de Provence.

CAMMACK R. *et al.* (1987) *Biochimica et Biophysica Acta* **912**: 98-109.

CAREPO M. *et al.* (2002) *Journal of the American Chemical Society* **124**: 281-286.

CARRELL T.G. *et al.* (2002) *J. Biol. Inorg. Chem.* **7**: 2-22.

CHARLOT M-F. *et al.* (2005) *Biochimica et Biophysica Acta* **1708**: 120-132.

COLMAN P.M. *et al.* (1978) *Nature* **272**: 319-324.

COMMONER B. *et al.* (1957) *Science* **126**: 57-63.

Cox N. *et al.* (2011) *Journal of the American Chemical Society* **133**: 3635-3648.

Debus R.J. *et al.* (1988) *Proceeding of the National Academy of Sciences of the USA* **85**: 427-430.

Dementin S. *et al.* (2004) *Journal of Biological Chemistry* **279**: 10508-10513.

Dementin S. *et al.* (2011) *Journal of the American Chemical Society* **133**: 10211-10221.

Dismukes C.C. & Siderer Y. (1981) *Proceeding of the National Academy of Sciences of the USA* **78**: 274-278.

Dole F. *et al.* (1996) *Biochemistry* **35**: 16399-16406.

Dole F. *et al.* (1997) *Biochemistry* **36**: 7847-7854.

Dole F. *et al.* (1997) *Journal of the American Chemical Society* **119**: 11540-11541.

Dorlet P. *et al.* (1999) *Journal of Physical Chemistry* B **103**: 10945-10954.

Dorlet P. *et al.* (2000) *Biochemistry* **39**: 7826-7834.

Ducros V. *et al.* (1998) *Nature Structural Biology* **5**: 310-316.

Eck R.V. & Dayhoff M.O. (1966) *Science* **152**: 363-366.

Fan C. *et al.* (1991) *Journal of the American Chemical Society* **113**: 20-24.

Fee J.A. *et al.* (1969) *Journal of Biological Chemistry* **244**: 4200-4207.

Fernandez V.M. *et al.* (1985) *Biochimica et Biophysica Acta* **832**: 69-79.

Fernandez V.M. *et al.* (1986) *Biochimica et Biophysica Acta* **883**: 145-154.

Ferreira K.N. *et al.* (2004) *Science* **303**: 1831-1838.

Flores M. *et al.* (2008) *Journal of the American Chemical Society* **130**: 2402-2403.

Foerster M. *et al.*(2005) *Journal of Biological Inorganic Chemistry* **10**: 51-62.

Garavaglia S. *et al.* (2004) *Journal of Molecular Biology* **342**: 1519-1531.

Garcin E. *et al.* (1999) *Structure* **7**: 557-566.

Garret T.P.J. *et al.* (1984) *Journal of Biological Chemistry* **259**: 2822-2825.

Guigliarelli B. *et al.* (1995) *Biochemistry* **34**: 4781-4790.

Guigliarelli B. & Bertrand P. (1999) *Advances in Inorganic Chemistry* **47**: 421-497.

Guss J.M. & Freeman M.C. (1983) *Journal of Molecular Biology* **169**: 521-563.

Guss J.M. *et al.* (1986) *Journal of Molecular Biology* **192**: 361-387.

Haddy A. (2007) *Photosynthetic Research* **92**: 357-368.

Hakulinen N. *et al.* (2002) *Nature Structural Biology* **9**: 601-605.

Hasegawa K. *et al.* (1999) *Chemical Physics Letters* **7**: 9-19.

Higuchi Y. *et al.* (1999) *Structure* **7**: 549-556.

Hill R. & Bendall F. (1960) *Nature* **186**: 136-137.

Hofbauer W. *et al.* (2001) *Proceeding of the National Academy of Sciences of the USA* **98**: 6623-6628.

Hoganson C.W. & Babcock G.T. (1992) *Biochemistry* **31**: 11874-11880.

Huyett J.E. *et al.* (1997) *Journal of the American Chemical Society* **119**: 9291-9292.

JOLIOT P. *et al.* (1969) *Photochem. Photobiol.* **10**: 309-329.

KOK B. *et al.* (1970) *Photochem. Photobiol.* **11**: 457-475.

KRASSEN K. *et al.*(2009) *ACS Nano* **3**: 4055-4061.

KUNAMNENI A. *et al.* (2008) *Recent Patents on Biotechnology* **2**: 10-24.

LAMLE S. *et al.* (2004) *Journal of the American Chemical Society* **126**: 14899-14909.

LANCIANO P. *et al.* (2007) *Journal of Physical Chemistry* B **111**: 13632-13637.

LEE S.K. *et al.* (2002) *Journal of the American Chemical Society* **124**: 6180-6193.

LOLL B. *et al.* (2005) *Nature* **438**: 1040-1044.

LUBITZ W. *et al.* (2007) *Chemical Review* **107**: 4331-4365.

MALKIN R. & MALMSTRÖM B.G. (1970) *Advances in Enzymology* **33**: 177-244.

MALMSTÖM B.G. *et al.* (1968) *Biochimica et Biophysica Acta* **156**: 67-76.

MARTIN W. & RUSSELL M.J. (2007) *Philosophical Transactions of the Royal Society B* **362**: 1887-1925.

MATSUOKA H. *et al.* (2006) *Journal of Physical Chemistry* B **110**: 13242-13247.

MOURA J.J.G. *et al.* (1982) *Biochemical and Biophysical Research Communications* **108**: 1388-1393.

PALMER A.E. *et al.* (1997) *Journal of the American Chemical Society* **121**: 7138-7149.

PELOQUIN J.M. *et al.* (2000) *Journal of the American Chemical Society* **122**: 10926-10942.

PELOQUIN J.M. & Britt R.D. (2001) *Biochimica et Biophysica Acta* **1503**: 96-111.

PIONTEK K. *et al.* (2002) *Journal of Biological Chemistry* **277**: 37663-37669.

RAPPAPORT F. & DINER A.D. (2008) *Coordination Chemistry Reviews* **252**: 259-272.

REICHARD P. & EHRENBERG A. (1983) *Science* **221**: 514-519.

REMENYI C. *et al.* (2007) *Journal of Physical Chemistry* B **111**: 8290-8304.

RIGBY S.E.J. *et al.* (1994) *Biochemistry* **33**: 1734-1742.

ROUSSET M. *et al.* (1998) *Proceeding of the National Academy of Sciences of the USA* **95**: 11625- 11630.

RUTHERFORD A.W. (1985) *Biochimica et Biophysica Acta* **807**: 189-201.

SOLOMON E.I *et al.* (2008) *Dalton Transactions* **30**: 3909-4056.

SOLOMON E.I. & LOWERY M.D. (1993) *Science* **259**: 1575-1581.

STADLER C. *et al.* (2002) *Inorganic Chemistry* **41**: 4424-4434.

STEIN M. & LUBITZ W. (2004) *Journal of Inorganic Biochemistry* **98**: 862-877.

SUURNÄKKI A. *et al.* (2010) *Enzyme and Microbial Technology* **46**: 153-158.

TOMMOS C. *et al.* (1995) *Journal of the American Chemical Society* **117**: 10325-10335.

TROFANCHUK O. *et al.* (2000) *Journal of Biological Inorganic Chemistry* **5**: 36-44.

UMENA Y. *et al.* (2011) *Nature* **473**: 55-60.

VAN DER ZWAAN J.W. *et al.* (1990) *Biochimica et Biophysica Acta* **1041**: 101-110.

VAN GASTEL M. *et al.* (2006) *Journal of Biological Inorganic Chemistry* **11**: 41-51.

VERMAAS W.F.J. *et al.* (1988) *Proceeding of the National Academy of Sciences of the USA* **85**: 8477-8481.

VINCENT K.A. *et al.* (2005) *Proceeding of the National Academy of Sciences of the USA* **102**: 16951-16954.

VOLBEDA A. *et al.* (1995) *Nature* **373**: 580-587.

Volume 1: BERTRAND P. (2020) *Electron Paramagnetic Resonance Spectroscopy - Fundamentals*, Springer, Heidelberg.

WARNCKE K. *et al.* (1994) *Journal of the American Chemical Society* **116**: 7332-7340.

WHITEHEAD J.P. *et al.* (1993) *Journal of the American Chemical Society* **115**: 5629-5635.

YOON J. *et al.* (2005) *Journal of the American Chemical Society* **127**: 13680-13693.

ZHENG M. *et al.* (1994) *Inorganic Chemistry* **33**: 382-387.

Chapter 7

Seeking the origins of life: primitive carbonaceous matter

Gourier D. and Binet L.

Chimie-ParisTech Research Institute, CNRS UMR 8247,
Chimie-ParisTech, Paris.

7.1 - Introduction

Where, when and how did life emerge on Earth? Are there any traces of this primitive life in the oldest sedimentary rocks? In what form? All these questions address numerous disciplines from Earth sciences (geology, geochemistry, geophysics, palaeontology, etc.) to physics, chemistry and the life-sciences. *Exobiology* is the interdisciplinary science investigating the emergence of life on Earth 3 to 4 billion years ago [Ga], and the possibility of extinct or current extraterrestrial life on other planets, Mars being the most favourable and accessible. We can very roughly distinguish two distinct and complementary approaches that can be used when attempting to elucidate the origins of life: an experimental approach, which aims to determine the physical, chemical and biological mechanisms responsible for the appearance of the first molecular systems that we can qualify as living; and a field approach, which consists in analyses to identify fossil or physico-chemical (biomarker) traces in the most ancient materials, in particular sedimentary rocks, likely to have preserved them. The second approach is the subject of this chapter, which describes how EPR can be used to detect traces of fossilised *Carbonaceous Matter* (CM) in the most ancient know materials on earth (cherts aged between 2 and 3.5 Ga) and CM present in the most ancient materials from the solar system (carbonaceous

© Springer Nature Switzerland AG 2020
P. Bertrand, *Electron Paramagnetic Resonance Spectroscopy*,
https://doi.org/10.1007/978-3-030-39668-8_7

meteorites) formed 4.5 Ga ago, shortly before the formation of the Earth. In this chapter, we will move back in time to the search for the origins of primitive CM.

7.1.1 - Primitive terrestrial carbonaceous matter

When living beings develop and die in a sea or lake, their remains are progressively buried in sediments. Once the water has been removed through withdrawal, emersion or formation of a mountain range, these sediments progressively become compact sedimentary rocks such as clay, limestone or sandstone, which can retain biological, mineral or organic traces of these organisms. This type of fossil is often found in sedimentary rocks dated at less than 540 million years [Ma] old, the eon from which animals with shells and skeletons became abundant (Phanerozoic, see figure 7.1).

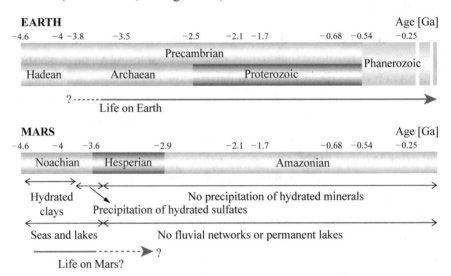

Figure 7.1 - Simplified diagram showing the major Terrestrial and Martian geological phases. Life emerged on Earth in the Archaean eon, between ~ −3.9 Ga and −3.5 Ga (first bacterial fossils). Terrestrial rocks older than 3.5 Ga are highly metamorphosed or have disappeared due to plate tectonics. Conditions on Mars were favourable to the appearance of life during the Noachian eon, from which numerous geological formations remain in the high plateaus of the southern hemisphere.

However, for the oldest eons for which dated rocks still exist (Archaean, figure 7.1), the traces of former life are much more difficult to detect due first to the fact that the most primitive life was solely bacterial, and second to the fact that the oldest sedimentary rocks likely to contain fossil traces, aged between around 3.8 and 3.5 Ga, are rare and often highly modified as a result of pressure and temperature (metamorphism). Traces of life have thus generally disappeared, except in very rare sites in South Africa and north-western Australia where geological formations containing little-metamorphosed siliceous rocks exist. In these formations, carbonaceous microstructures attributed to the fossil remains of primitive bacteria are preserved. These rocks are *cherts*, almost exclusively composed of microcrystalline silica; and therein lies the debate. Are these traces of biological origin or derived from abiotic reactions such as the Fischer-Tropsch reaction or decomposition of carbonates? If they are of biological origin, are they the contemporary biological remains of sedimentation beneath the seas, i.e., true fossils of primitive bacteria, or are they due to later contamination occurring hundreds of millions, or even billions, of years after sedimentation, from fluids enriched in organic matter or by bacteria living in the rocks? This type of *endolithic* bacteria effectively leaves carbonaceous traces similar to those left by primitive bacteria contemporary to the sedimentary deposit [Westall and Folk, 2003; Brocks *et al.*, 2003].

7.1.2 - Carbon before the formation of the Earth

Going further back in time, carbonaceous meteorites are the remains of planetary bodies a few tens of km in diameter which formed at the very start of the solar system's history (– 4.56 Ga), but were then fragmented before evolving like Mercury, Venus, Earth and Mars. Their very primitive nature is reflected in their chemical composition. Indeed, apart from volatile elements like H and He, which escaped, carbonaceous meteorites have the same chemical composition as the Sun [Lodders, 2003]. They contain 3 % carbon, mainly in the form of hydrogenated CM which is chemically quite similar to terrestrial coal, except for its significant deuterium enrichment [Robert and Epstein, 1982]. Although extraterrestrial, this amorphous polymerised CM is interesting from an exobiological point of view as it encases numerous molecules of biological interest including amino acids, carboxylic acids, purines and pyrimidines, carbohydrates [Botta and Bada, 2002; Ehrenfreund *et al.*, 2006]. The Earth was intensely bombarded by comets, dusts and meteorites over a period between – 4.5 and around – 3.9 Ga (figure 7.2). Estimations indicated that the amount of carbonaceous matter and organic molecules which seeded the Earth during this period could be greater than its current biomass [Chyba and Sagan, 1992; Maurette *et al.*, 2006]. There is therefore a strong probability that extraterrestrial organic matter contributed to the emergence of life on Earth [Ehrenfreund *et al.*, 2006; Pizzarello, 2007].

Better knowledge of this CM from the primitive solar system could therefore provide information on the origins of life on Earth. Other planets in the solar system were also subjected to these types of bombardment. Among them, Mars had conditions similar to those on Earth (atmosphere, presence of water, collisions with comets and meteorites, etc.) during the first billion years of its existence. However, as Mars is smaller than the Earth it cooled much faster, and this cooling had several consequences. The loss of its magnetic field due to the solidification of its metallic core led to the loss of its atmosphere under the influence of solar winds, resulting in the disappearance of the greenhouse effect which maintained the conditions appropriate for the existence of liquid water and the development of life. However, Mars probably never had plate tectonics, which has led to the recycling and resorption of the oldest rocks on Earth. The study of very ancient sedimentary rocks (older than 3.5 Ga), which are abundant and well-preserved on Mars, but absent or highly metamorphosed on Earth, could thus provide precious information on our earliest origins.

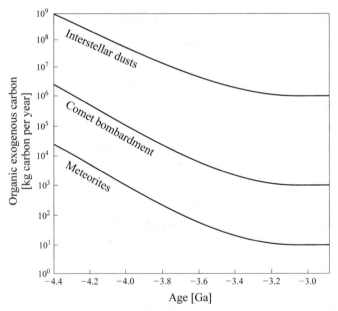

Figure 7.2 - Estimation of the annual carbon influx to Earth due to impact with comets, dusts and meteorites. [Adapted from Chyba and Sagan, 1992]

7.1.3 - Carbon before the solar system

The universe is a maximum of 13.7 Ga years old. All of the chemical elements composing our solar system, born 4.56 Ga ago, were synthesised during this long interval of 9 Ga. In the first minutes following the *Big Bang*, a first nucleosynthesis formed hydrogen (H) and helium (He), as well as traces of other elements such as D, T, Li and Be. Due to rapid cooling and the drop in density, no other element could be synthesised in the rapidly expanding universe. The other elements formed in the core of the stars, which form and die continuously within galaxies. Thus, carbon, the basic element making up organic molecules and living organisms, is produced by the $3 \ ^4He \rightarrow \ ^{12}C$ reaction. These processes occur in the core of stars larger than 0.5 solar masses. The heaviest elements (up to ^{56}Fe) are produced in the cores of more massive stars. At the end of its life, the star explodes producing elements heavier than ^{56}Fe. In total, for an equivalent of 100 hydrogen molecules, the universe contains only 10 He and 0.083 C. Other elements are even less abundant. Deuterium is a special case as it was produced in low amounts during the *Big Bang* and has since been consumed in the stars. All the chemical elements dispersed in interstellar space are once again incorporated into micronic molecules or dusts which go on to form new stars and their planetary systems. This process led to formation of the solar system 4.56 Ga ago. Large interstellar clouds, veritable star nurseries, contain a large variety of carbonaceous molecules and macromolecules such as polyaromatic hydrogenated hydrocarbons (PAH), carbonated chains, diamonds, amorphous hydrogenated (or not) carbon, and complex carbonated systems such as kerogene. PAH are among the most abundant organic molecules in space [Ehrenfreund *et al.*, 2000]. Although the abundance of deuterium in the universe is very low $(D/H \sim 3 \times 10^{-5})$ and is continuously decreasing, ion-molecule interactions at very low temperatures result in an *increase in the D/H ratio in interstellar molecules*. This ratio can thus reach values of up to ~ 0.01, or even more [Millar *et al.*, 2000; Roberts *et al.*, 2003]. We will see that this enrichment in deuterium, veritable marker of low-temperature chemistry in the interstellar medium, is detectable in a complex and partial manner in the carbonaceous matter found in meteorites.

7.1.4 - What can carbon EPR contribute to exobiology?

The structure of primitive CM, whether terrestrial or extraterrestrial, can be studied by analytical chemistry methods. These methods include an acidic demineralisation step, which may be followed by pyrolysis of the extracted CM, separation and analysis of the fragments by gas chromatography combined with mass spectrometry. These degradative techniques can identify all the basic building blocks in carbonaceous matter, but cannot determine its *structure*. *Spectroscopic techniques* are less invasive, they can probe the *local structure* of the CM *in situ* or after demineralisation. In particular, Raman spectroscopy of primitive carbon has been used in numerous studies as it presents a good compromise between sensitivity, spectral resolution, and even spatial resolution in Raman imaging. It could be part of the miniaturised instrument panel implemented on robots during explorations of Mars aiming to discover fossil traces or current bacterial life [Marshall *et al.*, 2010]. However, this spectroscopy cannot clearly discriminate between CM of biological or abiotic origin. NMR is also well adapted to the study of primitive CM, as it can detect the resonance of nuclei such as ^1H and ^{13}C at high resolution and it can characterise various types of chemical bonds involving these elements [Gardinier *et al.*, 2000]. However, the low sensitivity of NMR means that it requires large samples of pure demineralised CM, which can be an issue. In addition, the quality of the NMR signals can be significantly altered by the presence of paramagnetic defects which are frequent in primitive terrestrial CM.

For exobiology applications, EPR is entirely complementary to Raman and NMR spectroscopies due to the good compromise it offers between sensitivity and resolution, and its ease of use (no sample preparation required). This chapter aims to demonstrate the potential of carbon EPR applied to exobiology and cosmochemistry, focusing in particular on continuous wave EPR. Although the spectrum for primitive carbonaceous matter is reduced to a single unstructured line, the detailed study of its shape, the temperature-dependence of its intensity and its relaxation characteristics provides very important information. We will see that additional information can be obtained by exploring the unresolved hyperfine structure using ESEEM and HYSCORE spectroscopies – pulsed EPR techniques – (see appendix 2 to this volume), and by applying EPR imaging. Section 7.2 is devoted to the study of the oldest extraterrestrial carbonaceous matter in the solar system, which is trapped in meteorites known as *carbonaceous chondrites*. Section 7.3 describes characterisation of the oldest terrestrial carbonaceous matter, which is contemporaneous to or slightly older than life on Earth. We conclude in section 7.4 on the possibility of using EPR in the context of the search for traces of extinct life on Mars. As very few teams use the full potential of EPR in this field, significant proportion of the studies described relate to our own work.

7.2 - Primitive carbonaceous matter from the solar system: where and how did it emerge?

7.2.1 - Structure of the carbonaceous matter in meteorites

Numerous studies have been performed the world over since the 1970s to elucidate the structure of CM in carbonaceous meteorites. By combining chemical methods based on chemical or thermal degradation, NMR, FTIR, XANES spectroscopies and electron microscopy [Gardinier *et al.*, 2000; Cody and Alexander, 2005; Ehrenfreund *et al.*, 1992; Derenne *et al.,* 2005, Derenne and Robert, 2010], we have obtained an idea of the average structure of this very primitive CM. This structure, in the case of the Murchison meteorite, is represented in figure 7.3b [Derenne and Robert, 2010]. It can be described as composed of small aromatic entities linked together by short and highly-branched aliphatic chains. This CM is quite hydrogen-rich ($\sim$ 70 %), but also contains heteroelements such as oxygen ($\sim$ 20 %), nitrogen ($\sim$ 3 %) and sulfur ($\sim$ 2 %). It therefore corresponds to a relatively immature CM which has not undergone extensive thermal evolution. It is characterised by a significant deuterium enrichment compared to CM from the emerging solar system, for which the D/H ratio is $\sim 3 \times 10^{-5}$ [Robert and Epstein, 1982].

Very recently, the complexity of the history of this CM was revealed by NanoSIMS (Nano Secondary Ion Mass Spectrometry) experiments, showing the existence of strong concentration heterogeneities in the form of strongly deuterium-enriched micronic *hot spots* [Busemann *et al.*, 2006; Remusat *et al.*, 2009].

For example, in the case of CM from the Orgueil meteorite, which we will discuss further below, the average D/H ratio is around 3×10^{-4}, whereas the *hot spots* are 10 to 20 times more enriched [Remusat *et al.*, 2010]. Nevertheless, this enrichment remains much weaker than that measured for some carbonaceous molecules present in interstellar space (D/H $\sim$ 0.06 to 0.2) [Geiss and Gloecker, 1998; Robert, 2002]. Meteoritic CM therefore retains traces of an interstellar influence, but what influence, and at what stage in its history was this trace acquired?

(a)   (b)

Figure 7.3 - **(a)** Photograph of a fragment of the Orgueil meteorite which originally weighed 14 kg. It contains up to 3 % carbon.**(b)** Average structure of the carbonaceous matter in the Murchison meteorite. [**(b)** Pizzarello S. (2007) *Chemistry & Biodiversity* **4**: 680-693 © 2007 Verlag Helvetica Chimica Acta AG, Zürich]

7.2.2 - EPR of meteoritic carbonaceous matter

We studied carbonaceous matter from three famous carbonaceous meteorites, which fell in Orgueil (Tarn-et-Garonne, France, 14 May 1964) (figure 7.3a), Murchison (Victoria, Australia, 28 September 1969) and Tagish Lake (British Colombia, Canada, 18 January 2000). The CM was extracted from the mineral matrix by HF/HCl treatment [Durand and Nicaise, 1980]. The EPR spectra for the CM from these three meteorites were identical and composed of a narrow *Lorentzian* line (0.4 – 0.5 mT) centred at $g = 2.0031$ (figure 7.4) [Binet *et al.*, 2002, 2004a].

This line is superposed onto a broad ferromagnetic resonance signal (FMR, see chapter 12) due to mineral residues (chromite, magnetite, spinel, etc.) [Binet *et al.*, 2002]. The Lorentzian shape observed at all frequencies (from 4 GHz to 285 GHz) [Binet and Gourier, 2006] might suggest that the line is *homogeneous*, i.e., has no underlying structure [Volume 1, complement 3 to chapter 5]. However, the presence of spin echoes in pulsed EPR (see appendix 2) shows that this line is in fact *inhomogeneous*, and its width, although small, is due to unresolved hyperfine interactions with hydrogen nuclei present in the CM [Gourier *et al.*, 2008].

It might be expected that a single unstructured line at $g \approx 2.00$ provides little information. However, we will see that the detailed study of its characteristics, in particular the temperature-dependence of its intensity and its relaxation properties, associated with analysis of the unresolved hyperfine structure by pulsed EPR, provides a great deal of useful information on the history of the most primitive organic matter in the solar system.

◇ g factor and signal intensity

We can first compare the EPR spectrum for meteoritic CM with that of *terrestrial carbon* with a similar chemical composition. Type III coals derived from the decomposition of plant remains from continental forests has the appropriate composition. Samples of three reference coals aged from 14 Ma (noted A1 and A2) and 320 Ma (noted A3) were studied. These samples are therefore very young compared to CM from meteorites aged 4.5 Ga. Deviation of the g factor for meteoritic CM relative to 2.0023, which is due to the presence of oxygen within the CM [Volume 1, section 4.2.2], was found to be similar to that of terrestrial coals with equivalent O/C ratios. Maturation of terrestrial coals involves both loss of oxygen and loss of hydrogen, and variations in the g factor and the H/C ratio correlate, as shown in figure 7.5 for type I, II and III coals.

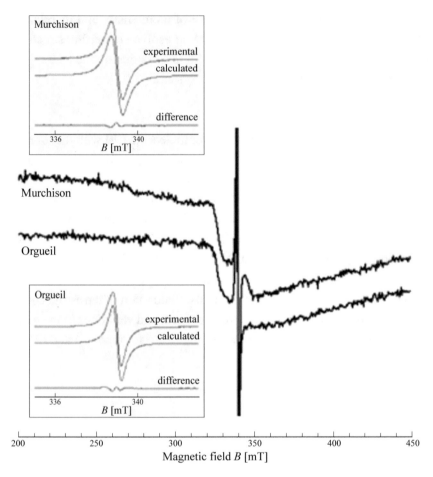

Figure 7.4 - X-band EPR spectra of carbonaceous matter of fragments (~ 4 mg) from the Orgueil and Murchison meteorites, recorded at room temperature. Frequency: 9.514 GHz, power: 30 mW. The main lines recorded at 0.2 mW are well simulated by Lorentzians (inserts). [Adapted from Binet *et al.* (2002) *Geochimica & Cosmochimica Acta* **68**: 4177-4186, *Heterogeneous distribution of paramagnetic radicals in insoluble organic matter from the Orgueil and Murchison meteorites* © 2002, with permission from Elsevier]

Evolution from the least mature coals (the youngest) to the most mature (the oldest) involves loss of oxygen (reduction of *g*) and hydrogen (reduction of H/C). The *g* factor for meteoritic CM has the typical characteristics of type III coal, which is *young and not very mature*, indicating that this CM has not evolved under the influence of temperature and pressure since its formation 4.5 Ga ago.

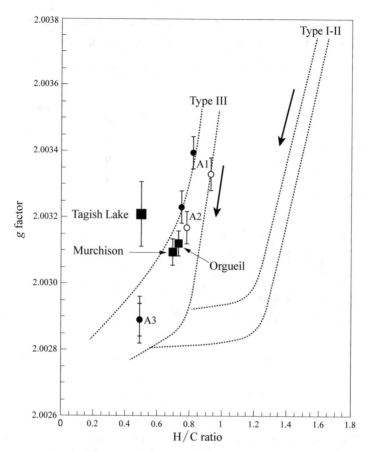

Figure 7.5 - g factor as a function of the H/C ratio for A1, A2 and A3 coal samples and CM samples from the Orgueil, Murchison and Tagish Lake meteorites. The dashed curves show the variations for type I, II and III coals. The arrows correspond to a trip back in time. [Adapted from Binet *et al.* (2002) *Geochimica & Cosmochimica Acta* **68**: 4177-4186, *Heterogeneous distribution of paramagnetic radicals in insoluble organic matter from the Orgueil and Murchison meteorites* © 2002, Elsevier]

A first difference between meteoritic CM and terrestrial coals relates to the concentration in paramagnetic entities, expressed in spins/gram. As coal maturation causes loss of oxygen and hydrogen alongside an increase in the number of paramagnetic centres, the O/C ratio and spin concentration correlate as shown in figure 7.6. In contrast, spin concentrations for meteoritic CM appear very high given its oxygen content. This result indicates that the entities bearing excess oxygen in meteoritic CM are spatially distant (> 10 nm) from spin-bearing radical entities. However, we will see that a more significant difference between primitive CM from the solar system and its terrestrial homologues can be found in the *temperature-dependence* of the EPR spectrum intensity.

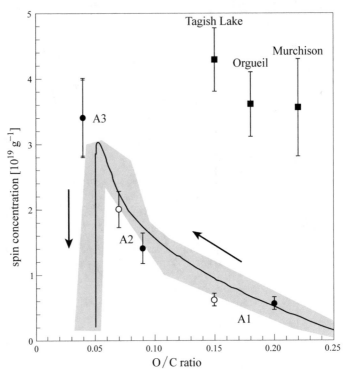

Figure 7.6 - Spin concentration, measured at room temperature, of A1, A2, A3 coals (black and white circles) and CM samples from meteorites (vertical bars) as a function of the oxygen to carbon ratio. The curve and the greyed-out zone represent the values obtained for type III kerogenes from the Douala basin (Cameroun). The arrows indicate the evolution of kerogenes during catagenesis which leads to the production of petrol. Spin concentrations in meteorites are the *local* concentrations assessed as indicated in the legend to figure 7.8. [Adapted from Binet *et al.* (2002) *Geochimica & Cosmochimica Acta* **68**: 4177-4186, *Heterogeneous distribution of paramagnetic radicals in insoluble organic matter from the Orgueil and Murchison meteorites* © 2002, Elsevier]

◇ *Diradical singlets: extraterrestrial markers?*

The intensity, I, of the spectrum of paramagnetic species of spin $S = \frac{1}{2}$ diluted in a diamagnetic matrix is proportional to N/T, where N is the number of centres and T is the temperature, and the product $I(T)T$ is therefore constant [Volume 1, chapter 5]. This is what we expect for CM, whether of terrestrial or extraterrestrial origin, due to its low spin concentration. Figure 7.7 shows the temperature-dependence of the $I(T)T$ product normalised to 1 at $T = 100$ K for four meteorite samples (Orgueil, Murchison, and two samples from Tagish Lake labelled TL1, TL2), coal A3 and two chert samples containing fossil CM of bacterial origin dated at 1.9 Ga (Gunflint) and 3.5 Ga (Warrawoona).

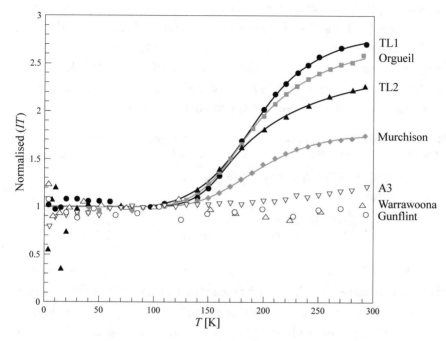

Figure 7.7 - Temperature-dependence of the *IT* product normalised to 1 at *T* = 100 K for four meteorite samples (Orgueil, Murchison, TL1, TL2), coal A3 and two chert samples containing fossil bacteria (Gunflint 1.9 Ga, and Warrawoona 3.5 Ga). The continuous-line curves were calculated by applying equation [7.1] and the parameters from table 7.1.[Adapted from Binet *et al.* (2004) (a) *Meteoritics & Planetary Science* **39**: 1649-1654 © 2004, Meteoritical Society]

The *IT* product can be seen to be constant for terrestrial CM, but it increases above 130 K in the case of meteoritic CM. This phenomenon has been attributed to the presence of two types of paramagnetic entities in meteoritic CM:

▷ centres of spin *S* = ½ (number N_{mono}) similar to those of terrestrial CM, responsible for the constant *IT* product at low temperatures,

▷ *diradical* entities with an *S* = 0 ground state (diradical singlets, number N_{di}), which have an *S* = 1 excited state populated at temperatures above approximately 130 K.

This model leads to the following expression of the *IT* product [Binet *et al.*, 2004b] (figure 7.7):

$$I(T)\,T \propto N_{mono} + \frac{8}{3}N_{di}\frac{1}{1 + \exp(-\Delta\sigma/k_B)\exp(\Delta E/k_B T)} \qquad [7.1]$$

where ΔE is the singlet-triplet gap, $\Delta\sigma$ is the entropy difference between the two states and k_B is Boltzmann's constant. The factor $\frac{8}{3}$ is the ratio of the $S(S+1)$ products for $S = 1$ and $S = \frac{1}{2}$ [Volume 1, section 6.4.2]. The temperature-dependence of the IT product is well reproduced by equation [7.1] for the four samples of meteoritic CM, with the parameters listed in table 7.1. Diradical singlets for the three meteorites have very similar characteristics, only the relative proportions of the radicals and diradicals change.

Table 7.1 - Singlet-triplet split ΔE, difference of $\Delta\sigma$ entropies between the ground singlet and the triplet, and proportion of singlet diradicals in the CM for the three meteorites.

	ΔE [cm^{-1}]	$\Delta\sigma$ [cm^{-1} K^{-1}]	$N_{di}/(N_{mono}+N_{di})$
Orgueil	815 ± 24	4.2 ± 0.1	$40 \pm 1\,\%$
Murchison	838 ± 32	4.3 ± 0.1	$24 \pm 1\,\%$
Tagish Lake	944 ± 24	4.8 ± 0.1	$42 \pm 1\,\%$

According to this model, the CM in meteorites contains between 20 and 40 % of a paramagnetic entity which can be considered an "extraterrestrial signature". The temperature-dependence observed in figure 7.7 can also be reproduced by models differing from that leading to equation [7.1]. Indeed, the existence of a unique narrow, symmetric line is not in favour of an $S = 1$ spin. To demonstrate the relevance of this model, the value of the spin S for the excited state must be *directly determined*. To do so, the *probability of the EPR transitions* is used, which is proportional to $[S(S+1) - M_S(M_S+1)]$ [Volume 1, section 6.4.2]. This quantity is equal to 1 for a $M_S = -\frac{1}{2} \longleftrightarrow M_S = \frac{1}{2}(S = \frac{1}{2})$ transition and to 2 for $M_S = -1 \longleftrightarrow M_S = 0$ and $M_S = 0 \longleftrightarrow M_S = 1(S = 1)$ transitions. The transition probability can be determined by pulsed EPR applying a resonant microwave pulse to cause the system to oscillate between the M_S and $M_S + 1$ spin states. The square of the frequency of these so-called Rabi (or nutation) oscillations is proportional to the transition probability. Experimentally, two frequencies are observed, at a ratio of $\sqrt{2}$, thus definitively demonstrating the presence of *radicals* and *diradicals* in meteoritic CM, whereas only radicals are present in terrestrial CM [Delpoux *et al.*, 2011]. Other pulsed EPR experiments (measurement of the distance between spins by two-quanta excitation) confirmed this interpretation and showed that the two unpaired electrons are separated by a distance of 2 nm in diradicals [Gourier *et al.*, in preparation].

This large distance explains why the zero-field splitting terms due to dipolar magnetic interactions do not alter the spectrum.

Singlet diradicals are therefore confirmed to exist, but what is their structure and why are they absent from (or present only at very low abundance) in terrestrial CM? Recent calculations on the electronic structure of small sheets of large polyaromatic entities (nanographenes) indicated that for sizes less than around 5×5 cycles, polyaromatic entities are diamagnetic [Wang *et al.*, 2010]. At larger sizes, the polyaromatic sheet becomes a singlet diradical ($S = 0$) with a decreasing singlet-triplet gap as the size increases. The two electrons with antiparallel (ground state $S = 0$) or parallel spins (excited state $S = 1$) are localised on the zigzagged edges of the aromatic sheet [Enoki and Takai, 2009]. Thus, the CM in meteorites must contain such well-defined aromatic sheets at sufficiently large sizes to be diradicals, but sufficiently small to not behave like two isolated single radicals. Their magnetic properties therefore become clearly observable as a result of quantum confinement, as these aromatic sheets are embedded in a diamagnetic carbonaceous matrix and are well separated from each other [Gourier *et al.*, in preparation]. In primitive terrestrial CM composed of larger and very disorganised aromatic entities, this effect of singlet diradicals disappears and only isolated $S = \frac{1}{2}$ spin states are observed (see section 7.3).

◇ *Relaxation: evidence of distribution heterogeneities for radical entities*

Are radicals and diradicals distributed homogeneously within the CM or do they display distribution heterogeneities as a result of a complex history? Partial information on this point can be derived from the value of the $T_1 T_2$ product obtained from progressive saturation experiments, where T_1 is the spin-lattice relaxation time and T_2 the spin-spin relaxation time [Volume 1, chapter 5]. As T_2 is determined by the interactions between paramagnetic centres, its value is expected to decrease when the mean concentration increases if the paramagnetic centres are uniformly distributed throughout the sample [Volume 1, complement 3 to chapter 5]. This is effectively what is observed for coals A1, A2, A3 and other coal samples [Thomann *et al.*, 1988] in figure 7.8, where we have represented the value of the $T_1 T_2$ product measured at room temperature as a function of the mean concentration determined by double integration of the EPR line and comparison to a reference sample [Volume 1, complement 4 to chapter 9]. In contrast, the $T_1 T_2$ product appears abnormally small for CM from Orgueil, Murchison and Tagish Lake meteorites given their mean concentrations. If we consider that the chemical composition of meteoritic CM is very similar to that of coal, there is no reason for it to present different relaxation characteristics.

The data from figure 7.8 rather suggest that in CM from meteorites, the *local concentration* of spin is around 4×10^{19} spins g^{-1}, which is higher than the mean concentration which is around 0.2 to 2×10^{19} spins g^{-1}. This interpretation was confirmed by the study of the relative intensities of the proton-ENDOR signal and the EPR signal (ENDOR gain factor) as a function of the mean radical concentration. These two experiments reveal strong heterogeneities in radical and diradical distribution in meteoritic CM [Binet *et al.*, 2004a, 2004b]. As mentioned in section 7.2.1, micronic heterogeneities (*hot-spots*) were recently observed when studying the *deuterium enrichment* of CM from meteorites [Busemann *et al.*, 2006].

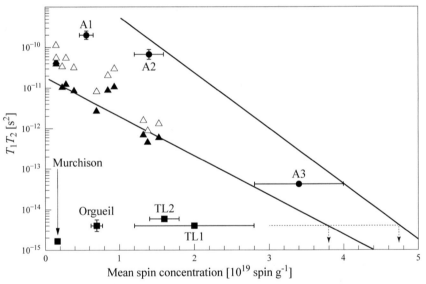

Figure 7.8 - Value of the $T_1 T_2$ product measured at room temperature as a function of the mean spin concentration for coal samples A1, A2, A3 (black circles), other carbons (triangles) [Thomann *et al.*, 1988] and for CM from the Orgueil, Murchison and Tagish Lake meteorites (black squares). The dashed lines indicate how the local concentrations in meteorites can be assessed from the $T_1 T_2$ product. These local concentrations were used in figure 7.6. [Adapted from Binet L. *et al.* (2004) (a) *Meteoritics & Planetary Science* **39**: 1649-1654 © 2004, Meteoritical Society]

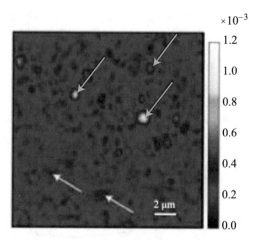

Figure 7.9 - NanoSIMS image of the D/H ratio for carbonaceous matter from the Orgueil meteorite. The green arrows indicate deuterium-rich *hot spots*, the white arrows indicate deuterium-poor regions. [Remusat *et al.* (2009) *The Astrophysical Journal* **698**: 2087-2092 © 2009. The American Astronomical Society]

An example of these deuterium-rich *hot spots* is shown in figure 7.9, which represents the distribution of the D/H ratio determined by NanoSIMS spectroscopy for CM from the Orgueil meteorite [Remusat *et al.*, 2009]. Heterogeneities in the deuterium enrichment are clearly visible. Could there therefore exist a link between these deuterium-rich *hot spots* and the heterogeneities of radical and diradical distribution?

◇ *Where is the deuterium bound?*

The width of the EPR line (0.4 – 0.5 mT) (figure 7.4) is due to unresolved hyperfine interactions with hydrogen nuclei (^{1}H and ^{2}H) and, to a lesser extent, with ^{13}C nuclei. To identify these nuclei and examine their coupling, ENDOR spectroscopy must be used (see appendix 3) along with ESEEM and HYSCORE pulsed EPR experiments (see appendix 2). Figure 7.10 shows the 4-pulse ESEEM spectrum for CM from the Orgueil meteorite.

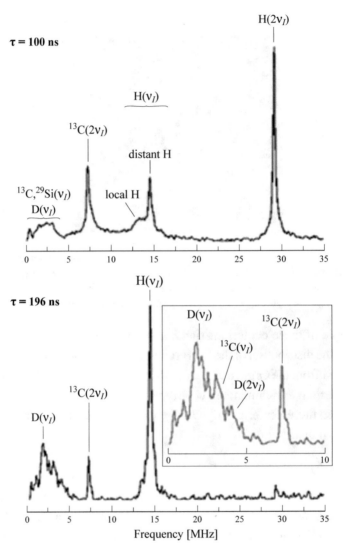

Figure 7.10 - 4-pulse ESEEM spectrum for CM from the Orgueil meteorite recorded at 10 K for two values of the time τ separating the first two microwave pulses.
[© Delpoux, Vezin and Gourier]

As hyperfine interactions are weaker than $h\nu_I$, where ν_I is the Larmor frequency of coupled nuclei, peaks are expected at ν_I and $2\nu_I$ (sections 1.2 and 2 of appendix 2). This is effectively observed for protons ($\nu_I = 14.66$ MHz) in the high-frequency region of the spectrum. The low-frequency region includes an intense peak at a frequency of $\nu_I = 2.25$ MHz for deuterium beside a peak of frequency $v_I = 3.69$ MHz for ^{13}C, which indicates that the radicals are highly enriched in deuterium.

This low-frequency region was studied in detail by HYSCORE spectroscopy (figure 7.11). Two important pieces of information were obtained [Gourier *et al.*, 2008]:

▷ simulation of the deuterium spectrum indicates that the hyperfine interaction is essentially *isotropic*, suggesting that deuterium mainly occupies a benzylic position, i.e., that it is bound to an aliphatic carbon linked to an aromatic carbon,

▷ comparison with a reference spectrum (biphenyl mixed with 1 % deuterated biphenyl) reveals a D/H ratio of 1.5 % for *radicals* from the CM [Gourier *et al.*, 2008].

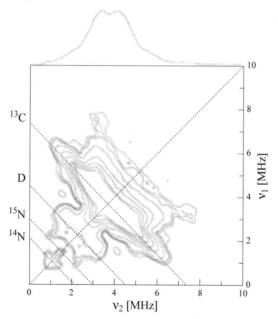

Figure 7.11 - Low-frequency region of the HYSCORE spectrum produced by CM from the Orgueil meteorite recorded at 10 K. [Gourier D. *et al.* (2008) *Geochimica & Cosmochimica Acta* **72**: 1914-1923, *Extreme deuterium enrichment of organic radicals in the Orgueil meteorite: Revisiting the interstellar interpretation?* © 2008, with permission from Elsevier]

Quantitative analysis of the results indicates that the radicals are highly enriched in deuterium, and that their heterogeneous distribution in the CM produces the deuterium-rich *hot-spots* observed by NanoSIMS (figure 7.9) [Remusat *et al.*, 2009].

7.2.3 - Review: the complex history of protosolar organic matter

By combining numerous techniques it was found that meteoritic CM is generally enriched in deuterium (*isotopic mass spectrometry*), that this enrichment is concentrated in micronic *hot-spots* (*NanoSIMS*), which themselves contain radical and diradical aromatic entities that are almost exclusively deuterium carriers (*EPR*), and that deuterium is mainly localised on a single type of C – H bond, which is particularly labile (*HYSCORE*). This multi-scale heterogeneity, from the macroscopic scale to that of the chemical bond, indicates that the CM from meteorites results from the accretion of a *mixture of both deuterium-rich and -poor CM particles*. Figure 7.12 proposes a simplified diagram of the process which may have occurred 4.5 Ga ago in the emerging solar system.

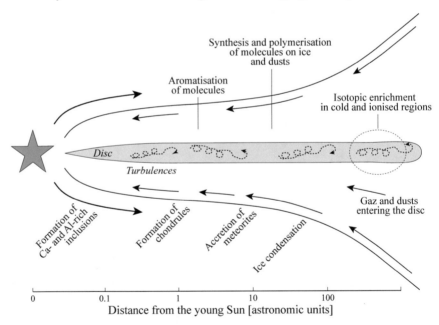

Figure 7.12 - Diagram of the emerging solar system.
In the nebula, the flow of gas and dusts entering under the influence of gravitation is compensated by reverse movement due to turbulence at the centre of the discs where planets will form. The probable localisation of the processes influencing the formation and evolution of the CM in the disc is indicated. The steps leading to formation of meteorites in the disc are also indicated. An astronomic unit corresponds to around 150 million kilometres (distance separating the Earth from the Sun).
[Adapted from Remusat *et al.* (2010) *The Astrophysical Journal* **713**:1048-1058]

This model is based on an influx of gas and dusts attracted towards the young Sun by gravitation, where the temperature increases. This influx is compensated by turbulence which sends particles back towards the periphery of the proto-solar nebula where the temperature is lower. As the D/H ratio increases at low temperature ($D/H \approx 1.5$ % corresponds to $T \approx 35$ K) [Bigeleisen and Mayer, 1947], significant deuterium enrichment could only occur at the periphery of the nebula or in the interstellar medium itself. Thus, the CM in meteorites appears to be derived from the mixture of CM particles synthesised within the proto-solar nebula and dispersed by turbulence. The particles that reach the cold periphery are exposed to irradiation and deuterium-rich ionised gases. Exchange can take place on the *radical aromatic entities* presenting the most *labile* hydrogen atoms (benzylic hydrogens). Finally, mixing of these deuterium-enriched particles with the particles which have remained inside the nebula (thus deuterium-poor) produces CM containing highly deuterium-enriched *hot-spots* [Remusat *et al.*, 2010]. The whole condenses with minerals to form the meteorites' parent bodies, which are then rapidly destroyed by collisions to produce the carbonaceous meteorites which still occasionally fall to Earth.

7.3 - Terrestrial primitive carbonaceous matter: seeking biomarkers of the origins of life

Rocks are complex composite materials which can contain a large variety of paramagnetic species. Even cherts, which are almost exclusively composed of microcrystalline silica, produce EPR signals due to the presence of transition ions such as Mn^{2+}, Fe^{3+}, point defects such as oxygen vacancies (E' centres), aluminium centres (O^- ions linked to Al^{3+} ions in silicium sites, see figure 1.1c in chapter 1), and magnetic inclusions (superparamagnetic or ferromagnetic resonance signals, see chapter 12). In addition, organic radical signals or signals for carbonaceous matter can be detected [Skrzypczak *et al.*, 2008]. Figure 7.13 shows an example of the EPR spectrum produced by a small fragment of cherts aged 3.46 Ga, sampled in the Dresser formation (Warrawoona group, Australia). The cherts in this group contain carbonaceous microstructures which were very early attributed to fossil bacteria [Schopf *et al.*, 2002]. After heated debates on the biological or non-biological origins of these microstructures [Brasier *et al.*, 2002, 2005; Pasteris and Wopenka, 2002; Kazmierczak and Kremer, 2002; Garcia-Ruiz *et al..*, 2003], it has been concluded that this carbonaceous matter is almost certainly of biological origin. Figure 7.13 shows how the different paramagnetic centres producing signals at around $g = 2$ can be identified by successive "zooming", while playing on the temperature, the microwave power and even the in-phase or quadrature phase detection mode (see complement 1 to this chapter). The CM spectrum is readily observed at room temperature, but the $g_{//}$ peak of the signal produced by the E′ centres for silica superposes on it (figure 7.13d). Since these centres have very long T_1 and T_2 relaxation times, their signal can be saturated with a strong microwave power, and it can be selectively recorded thanks to quadrature phase detection (figure 7.13e) [Skrzypczak *et al.*, 2008]. The CM signal is obtained by difference.

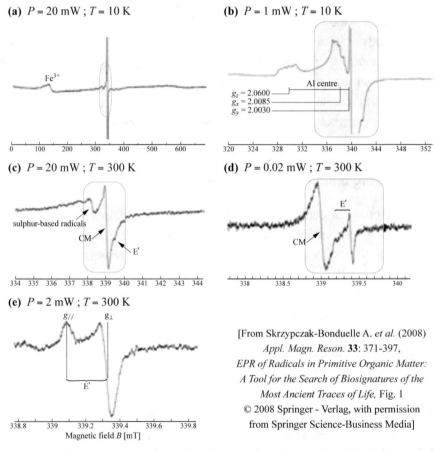

(a) $P = 20 \text{ mW}$; $T = 10 \text{ K}$

(b) $P = 1 \text{ mW}$; $T = 10 \text{ K}$

$g_z = 2.0600$
$g_x = 2.0085$
$g_y = 2.0030$

(c) $P = 20 \text{ mW}$; $T = 300 \text{ K}$

(d) $P = 0.02 \text{ mW}$; $T = 300 \text{ K}$

(e) $P = 2 \text{ mW}$; $T = 300 \text{ K}$

Magnetic field B [mT]

[From Skrzypczak-Bonduelle A. *et al.* (2008)
Appl. Magn. Reson. **33**: 371-397,
EPR of Radicals in Primitive Organic Matter:
A Tool for the Search of Biosignatures of the
Most Ancient Traces of Life, Fig. 1
© 2008 Springer - Verlag, with permission
from Springer Science-Business Media]

Figure 7.13 - EPR spectrum for a chert fragment from Warrawoona (3.5 Ga) recorded at various temperatures and microwave powers, on decreasing field ranges from **(a)** to **(e)**, to cause the CM signal to emerge. Spectra **(a)** to **(d)** are recorded in phase with the modulation. Spectrum **(e)** is recorded in quadrature phase detection.

7.3.1 - Line shape: a tool to date terrestrial carbonaceous matter

The EPR spectrum for the CM encased in a rock can therefore readily be determined at room temperature. Like that of the CM in meteorites (figure 7.4), this spectrum contains only *a single unstructured line*. Figure 7.14 shows spectra produced by coals A1 and A3, defined above, and by cherts aged between 45 Ma and 3.5 Ga. The *g* values and peak-to-peak widths ΔB_{pp} for these signals and those produced by other samples of terrestrial CM are listed in table 7.2.

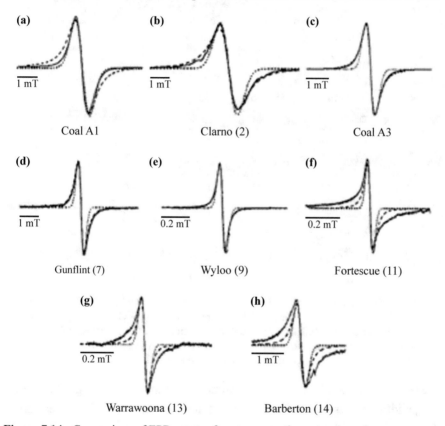

Figure 7.14 - Comparison of EPR spectra for paramagnetic centres in carbonaceous matter from cherts and coal with Gaussian (dotted lines) and Lorentzian (dashed lines) lines. Samples are arranged in increasing order of age and numbered as in table 7.2.

[From Skrzypczak-Bonduelle A. *et al.* (2008) *Appl. Magn. Reson.* **33**: 371-397], *EPR of Radicals in Primitive Organic Matter: A Tool for the Search of Biosignatures of the Most Ancient Traces of Life*, Fig. 3 © 2008 Springer-Verlag, with permission from Springer Science-Business Media]

Table 7.2 - Parameters characterising the EPR line for carbonaceous matter in coals A1, A2, A3 and cherts of various origins. The algebraic shape factor R_{10} is a measure of the difference between the line shape and a Lorentzian (see the text). [From Skrzypczak-Bonduelle A. *et al.* (2008) *Appl. Magn. Reson.* **33**: 371-397]

Sample	Origin	Age [Ma]	g factor	ΔB_{pp} [mT]	R_{10} factor
1	Lost Chicken Creek[a]	2.9	2.002(7)	0.60	nd
A1; A2	Mahakam delta[b]	14	2.003(2)	0.62; 0.67	4.84; 5.06
2	Clarno[c]	45	2.004	0.72	1.74
A3	Solway Basin[d]	320	2.002(9)	0.42	0.63
4	Dengying[e]	655 – 658	2.003(2)	0.40	nd
5	Daongyu[f]	1430 – 1770	2.003(2)	0.22	nd
6; 7	Gunflint[g]	1880	2.003(4)	0.23	0.09; –0.75
8; 9	Wyloo[h]	1850 – 2200	2.00(3)	0.005; 0.034	nd; –0.19
10	Transvaal[i]	2200 – 2340	2.003(3)	0.008	nd
11	Fortescue[j]	2750	2.002(7)	0.04	–2.2
12; 13	Warrawoona[k]	3460	2.003(3)	0.12	–2.43; –1.98
14; 15	Barberton[l]	~ 3400	2.003(2)	0.20; 0.30	nd; –2.16
16	Zwartoppie[m]	3420 – 2450	2.003(1)	0.33	nd

[a] Alaska, USA; [b] Indonesia; [c] John Day Basin, Oregon, USA; [d] United Kingdom; [e-f] China; [g] Schreiber Beach, Ontario, Canada; [h] Duck Creek, Dolomite Wyloo group, Australia; [i] Transvaal supergroup, South Africa; [j] Jeerinah Formation, Australia; [k] Dresser Formation, Warrawoona group, Australia; [l] upper Onverwacht, South Africa; [m] Onverwacht group, South Africa. nd: not determined

The *g* value can be observed to be practically invariant. Although the overall ΔB_{pp} width decreases as the age increases, a net broadening of the spectrum is observed for the oldest CM. *Linewidth* is therefore not a good indicator of age. In contrast, *the line shape* is, as shown in figure 7.14 where the lines are compared to Gaussian and Lorentzian lines. For "young" CM, roughly less than ~ 100 million years old, the shape is between that of a Lorentzian and a Gaussian and tends towards a Lorentzian when the age reaches a billion years. As the age increases up to 3.5 billion years, the line takes the form of a Lorentzian with stretched wings. Such stretched Lorentzians are generally interpreted as the superposition of several Lorentzians with different intensities and widths (proportional to

$1/T_2$). However, the extensive reproducibility of our observations with rocks with different ages and of different origins rather suggests that this stretching is linked to a particular physical phenomenon which occurs as the CM ages at timescales exceeding a billion years. The study of meteoritic CM showed that a narrow line with a Lorentzian shape (true or stretched) is not necessarily the result of exchange narrowing phenomena, as usually considered, since this would produce a *homogeneous line*. The fact that the lines for terrestrial CM produce a spin echo in pulsed EPR (see appendix 2 at the end of this volume) indicates that they are *inhomogeneous*. This inhomogeneity is due to unresolved hyperfine interactions with hydrogen nuclei [Skrzypczak *et al.*, 2008].

◇ *Definition of a line shape parameter*

Due to the continuous variation of the line shape between Gaussian and stretched Lorentzian, a parameter must be defined to reflect the shape of the line without requiring its simulation. It is convenient to use a representation in which a derivative of the *Lorentzian* becomes a *straight line*. The derivative $L(B)$ of a normalised Lorentzian line can be expressed as follows [Volume 1, section 9.4.1]:

$$L(B) = -\frac{16}{3\pi\sqrt{3}\,\Delta B_{pp}^3} \frac{B - B_{res}}{\left[1 + \frac{4}{3}\frac{(B - B_{res})^2}{\Delta B_{pp}^2}\right]^2} \qquad [7.2]$$

where $B_{res} = h\nu/g\beta$ is the magnetic field at the centre of the line and ΔB_{pp} is the peak-to-peak width. If we set

$$x = \left(\frac{b}{\Delta B_{pp}}\right)^2, \quad y = \left(\frac{a_{pp}b}{\Delta B_{pp}a}\right)^{1/2} \qquad [7.3]$$

where $a_{pp} = \dfrac{2\sqrt{3}}{\pi\Delta B_{pp}^2}$ is the peak-to-peak amplitude of $L(B)$ and the new variables a and b are defined on the inset in figure 7.15c, equation [7.2] becomes:

$$y = \sqrt{\frac{9}{8}}\left(1 + \frac{4}{3}x\right)$$

In the transformation defined by equations [7.3], a Lorentzian becomes a straight line, a Gaussian becomes a curve with strong positive concavity, and a stretched Lorentzian becomes a curve with negative concavity. Figure 7.15 shows the transformations obtained by digitising the EPR lines for cherts from Clarno (45 Ma), Gunflint (1.9 Ga) and Warrawoona (3.5 Ga). The shape of the line can be clearly seen to change with age, from a Gaussian-Lorentzian (Clarno) to a quasi-Lorentzian (Gunflint) and then to a stretched Lorentzian (Warrawoona).

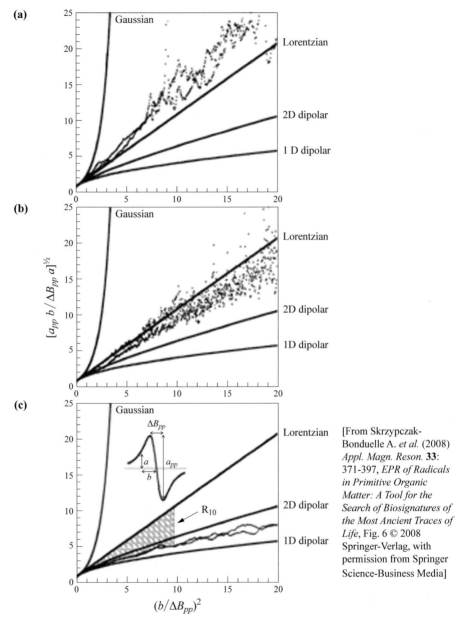

Figure 7.15 - Analysis of the shape of the EPR lines for carbonaceous matter in a representation where the derivative of a Lorentzian line is a straight line. The Gaussian and stretched Lorentzian lines (2D Dipolar and 1D Dipolar) are also represented. The experimental lines correspond to those for cherts from **(a)** Clarno (45 Ma), **(b)** Gunflint (1.9 Ga) and **(c)** Warrawoona (3.5 Ga). The structures apparent in **(a)** are due to the existence of other EPR signals for organic radicals. Figure **(c)** shows the algebraic area used to define the R_{10} factor and the inset defines the a and b variables used in equation [7.3].

◇ *Origin of the stretching of the Lorentzian line*

Magnetic dipolar interactions between fixed paramagnetic centres produce an inhomogeneous *Lorentzian* line when these centres are diluted in a 3-dimensional system. When they are distributed in a system of dimension smaller than 3, the line is no longer Lorentzian. Its shape can be determined by calculating he Fourier transform of the free induction decay signal $S(t)$ which describes the decrease of the transverse magnetisation over time (see appendix 1 at the end of this book), for which the general expression is as follows [Fel'dman and Lacelle, 1996]:

$$S(t) = S(0)\exp[-(t/T_2)^{D/3}]$$ [7.4]

where $D = 1, 2, 3$ is the dimensionality of the spatial distribution of the spins. When the spins are isotropically diluted in space ($D = 3$), the Fourier transform of $S(t)$ is a Lorentzian. When the spins are distributed in a plane ($D = 2$) or along a curve ($D = 1$), the Fourier transform of $S(t)$ has no closed form expression, but it can be numerically calculated and the transformation defined by equations [7.3] is then performed. This produces the curves with negative concavity denoted "2D Dipolar" and"1D Dipolar" in figure 7.15 [Skrypczak *et al.*, 2008]. The line for the oldest CM can be observed to have a shape corresponding to $1 < D < 2$ (figure 7.15c). The variation of the dimensionality of the distribution of the spins can be quantified using the *algebraic area* between the straight line corresponding to a Lorentzian and the experimental curve [Skrypczak *et al.*, 2008]. Figure 7.15c shows an example of measurement limited to the abscissa $(b/\Delta B_{pp})^2 = 10$. The shape factor R_{10} thus defined is equal to 0 for a Lorentzian, it is positive for a Gaussian and negative for a stretched Lorentzian. Its value has been calculated for many samples, as indicated in the last column in table 7.2. Unlike the ΔB_{pp} linewidth, the R_{10} parameter happens to be a good indicator of the age of the carbonaceous matter. Figure 7.16a shows how it evolves as a function of age for carbonaceous cherts and coals. The fact that the Lorentzian becomes stretched (D drops from 3 to 2 then tends towards 1) as the age increases can be interpreted as the result of increasing order within the CM as it ages. When the CM is young (i.e., aged less than ~ 1 Ga!) the carbon is disordered, the spin-bearing aromatic entities are small in size and distributed isotropically within the CM ($D = 3$). As long as the CM contains sufficient hydrogen (age ≤ 500 Ma), the unresolved hyperfine interactions produce an EPR line with a Gaussian-Lorentzian shape (Voigt profile, $R_{10} > 0$).

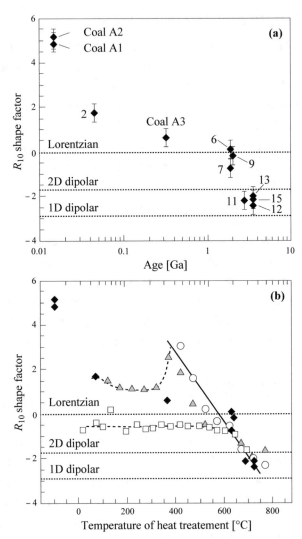

Figure 7.16 - (a) Variations of the shape factor R_{10} as a function of age for a series of chert and coal samples. **(b)** Variation of R_{10} *as a function of temperature* of isochronic heat treatments for cherts from Clarno (grey triangles), Rhynie (white circles) and Gunflint (white squares). The upper logarithmic abscissa, identical to the abscissa in figure **(a)**, and the black symbols show the correlation with the age of ancient cherts. The straight line represents the relation between R_{10} and the age of the carbonaceous matter, equation [7.5]. [From Skrzypczak-Bonduelle A. *et al.* (2008) *Appl. Magn. Reson.* **33**: 371-397], *EPR of Radicals in Primitive Organic Matter: A Tool for the Search of Biosignatures of the Most Ancient Traces of Life*, Fig. 7 © 2008 Springer-Verlag, with permission from Springer Science-Business Media]

When the broadening due to hyperfine interactions drops below the broadening due to dipolar interactions between electron spins due to significant hydrogen release (age $\sim$ 1 to 2 Ga), the line becomes Lorentzian ($D = 3$, $R_{10} = 0$). As the age increases further, the aromatic entities grow and their paramagnetic defects tend to distribute within graphene planes ($D = 2$, $R_{10} = -1.7$) and then to concentrate in their periphery ($D = 1$, $R_{10} = -2.9$) as these graphenes become increasingly ordered.

As the chemical reactions we have just described are thermally activated, the effect of ageing can be simulated by applying isochronic heat treatments (15 minutes per stage) in 50 °C steps, and by recording the EPR spectra at room temperature between each temperature step. The grey and white symbols in figure 7.16b show the results obtained with cherts from Clarno (-45 Ma) and Rhynie (-396 Ma) (the initial EPR spectra for these cherts are shown in figures 7.14b and 7.17a, respectively). Comparison of figures 7.16b and 7.16a indicates that after treatment at 750 °C, the line shape for these two samples is exactly the same as for the CM in samples 13 and 15 in figure 7.16a, which is 3.5 Ga old. It is therefore possible to correlate time (at low temperature) and temperature (at short times) using the R_{10} parameters measured at room temperature and after treatment at 750 °C. This correlation was used in figure 7.16b, where we reported the R_{10} parameters for coals and natural untreated cherts (black symbols) together with those for cherts from Clarno (grey triangles) and Rhynie (white circles) following heat treatment. The relevance of this time-temperature correlation is confirmed by the results of the isochronic heat treatment of cherts from Gunflint aged 1.9 Ga (white squares in figure 7.16b). As long as the temperature for the heat treatment remains below the "equivalent age" of $\sim$2 Ga, the R_{10} factor remains unchanged. As soon as the equivalent age of $\sim$ 2 Ga is reached (around 650 °C), the R_{10} factor decreases and follows the same variation as that of the other cherts. This is thus a *method to date CM* relative to the host rock. In this case, the R_{10} shape factor can be linked to the real age t_{geol} of the CM by the following relation [Skrzypczak et al., 2008]:

$$R_{10} \approx -5,32(\pm 0,36)\log(t_{geol}) + 48,9(\pm 0,9) \qquad [7.5]$$

which is valid between around 600 Ma and 3.5 Ga (thick line in figure 7.16b). This method should make it possible to readily detect whether the CM is from fossilised bacteria in ancient sediments or if it is the result of later contamination by carbonaceous infiltrations or endolithic bacteria. Let us assume, for example, that massive contamination took place around 2 Ga after formation of a 3.5 Ga-old chert. Heat-treatment of these cherts would produce an R_{10} factor that remains constant until an age equivalent to ~ 1.5 Ga (around 600 °C), and would then follow a normal curve after this equivalent age ($T > 600$ °C).

7.3.2 - EPR imaging of carbonaceous matter

Fossil CM containing the centres producing EPR spectra is heterogeneously distributed within the rock. EPR imaging provides access to this distribution. Generally speaking, imaging techniques are very useful when analysing a heterogeneous object. Depending on their resolution and sensitivity, spatial structures that are invisible to the naked eye can be revealed (because the object is opaque or because these structures are very small), and when these techniques are chemically selective, they can also be used to map the spatial distribution of specific species. The images collected are often information-rich and complementary to purely spectroscopic analyses, which only provide access to the physico-chemical characteristics of the species detected. In the case of rocks, optical, Raman or electron microscopy are commonly used to visualise and identify both the different mineral phases and carbonaceous phases. These techniques generally combine good spatial resolution (from the micrometre to the nanometre, depending on the technique), good sensitivity and good chemical selectivity. However, due to the low depth of penetration of optical radiation and the electron beams used, these techniques can only analyse the *superficial layer* of samples. It is therefore necessary to perform numerous thin sections if an in-depth study is to be performed. In addition, when imaging carbonaceous matter, the laser or electron beams deposit enough energy to cause *in situ* chemical and structural modifications of the carbon during analysis. EPR imaging (iEPR), which can be used to view the spatial distribution of electron spins, does not share these disadvantages. Indeed, the microwave radiation used interacts very little with the matter (apart from conducting or highly magnetic materials) and can thus penetrate deep into the samples, over several millimetres or even centimetres. iEPR is therefore compatible with analysis *in the mass* of centimetre-scale objects without requiring specific preparation. Similarly, the low energy deposited when performing EPR leaves the physico-chemical properties of the samples unchanged. However, iEPR has limited resolution (sub millimetre) compared to Raman imaging (a few microns). Its principle is described in complement 2 to this chapter.

◇ *Application to the detection of carbonaceous matter in siliceous rocks*

In rocks, several EPR signals are superposed, which complicates the application of iEPR. Indeed, the principle of iEPR assumes that the sample contains only a single paramagnetic species, making the Fourier transform-based deconvolution method generally used applicable. This method no longer works when the EPR signals for different species overlap. Nevertheless, in a certain number of cases the signal of interest can be isolated by adjusting various instrumental parameters such as the microwave power or the detection phase.

Figure 7.17b shows the EPR image for carbonaceous matter in a fragment of siliceous rock (chert) from the Rhynie formation in Scotland, dated at 396 Ma [Binet *et al.*, 2008]. This rock contains millimetric plant fossils, corresponding to the dark zones in the photos of the faces of the sample. The EPR spectrum for the raw sample presents numerous signals due to molecular radicals and defects in the silica (figure 7.17a, upper spectrum). The signal for the paramagnetic centres in the CM to be mapped by iEPR is not very intense and is masked by the other signals. It is therefore necessary to first heat the sample to 495 °C for 15 minutes to eliminate the signals other than the CM line (figure 7.17a, lower spectrum). The slope of the baseline, which is a broad signal produced by ferromagnetic particles, can be eliminated digitally. The 3D spatial distribution of the paramagnetic centres of the CM obtained by iEPR (figure 7.17b) should reflect the distribution of this matter in the sample. The image shows areas of spin isoquantities. Relatively good correspondence is observed between the shapes obtained by iEPR and those of plant fossils visible in the optical photographs of the faces (dark zones). Thus, the carbonaceous matter is concentrated in the fossils [Binet *et al.*, 2008].

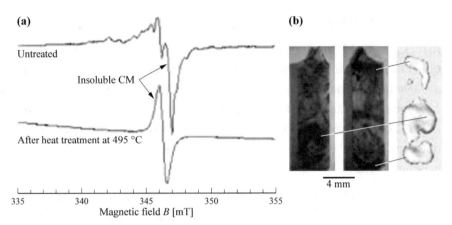

(a)

Untreated

Insoluble CM

After heat treatment at 495 °C

(b)

4 mm

335   340   345   350   355
Magnetic field B [mT]

Figure 7.17 - (a) EPR spectrum for a chert from Rhynie (396 Ma) showing a variety of radicals and paramagnetic centres (upper spectrum). Heat treatment (see text) increases the level of carbonisation and results in a narrow CM signal. **(b)** The photos of the two faces of the sample are compared to the EPR image of the distribution of the CM in the chert. The photos show dark carbon-rich zones representing fossils, separated by lighter zones with a lower carbon content. The optical images only show the features apparent at the surface, whereas the EPR image **(b)** shows the distribution throughout the body of the sample. [From Binet L. *et al.* (2008) *Earth Planet. Sci. Lett.* **273**: 359-366, *Potential of EPR imaging to detect traces of primitive life in sedimentary rocks* © 2008, with permission from Elsevier].

Another example of EPR imaging of siliceous rocks is given in figure 7.18 for a chert from the Gunflint formation in Canada, dated at 1.9 Ga. In this sample, the carbonaceous matter corresponds to a fossilised layer of primitive bacteria. The specificity of this sample is that it is totally opaque and visually homogeneous. The EPR spectrum recorded at low microwave power shows two overlapping signals in the $g = 2$ region (figure 7.18a) produced by the paramagnetic centres in the CM and the E' centres (oxygen vacancies in the silica). As indicated at the start of section 7.3, the signal for the E' centres is saturated at high microwave power, and thus it is possible to isolate the spectrum for the centres of the carbonaceous matter which are much more difficult to saturate (figure 7.18b). The 3D EPR image of the centres in the carbonaceous matter, represented in figure 7.18d, indicates inhomogeneous distribution of this fossil CM within the sample. This distribution would have been much more difficult to observe by other imaging techniques. The EPR spectrum for the E' centres in the silica, detected using quadrature phase detection (complement 1 to this chapter), is represented in figure 7.18c. The corresponding EPR image (figure 7.18e) reproduces the shape of the sample, thus, unlike the CM, the E′ centres are homogeneously distributed throughout the silica [Binet *et al.*, 2008].

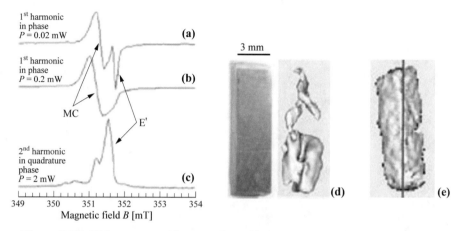

Figure 7.18 - EPR spectra and images of Gunflint cherts (1.9 Ga). **(a)** and **(b)** spectra recorded in normal mode (1st harmonic in phase with the modulation) with a power of **(a)** 0.02 mW, **(b)** 0.2 mW. **(c)** 2nd harmonic detected in quadrature phase at 2 mW. The images of the CM **(d)** and the E′ centres **(e)** correspond to spectra **(b)** and **(c)**. [From Binet L. *et al.* (2008) *Earth Planet. Sci. Lett.* **273**: 359-366, *Potential of EPR imaging to detect traces of primitive life in sedimentary rocks* © 2008, with permission from Elsevier]

7.3.3 - Seeking nuclear biosignatures

Primitive carbonaceous matter, whether of confirmed biological or of abiotic origin (e.g. meteoritic CM), always produces a relatively narrow line. The absence of structure of this line is a handicap to determining its biological or non-biological origin. As the difference between the two categories of CM is revealed by the proportions of C – H bond types (aromatic, benzylic, aliphatic), hyperfine interactions with ^{1}H and ^{13}C nuclei should reflect these differences. As underlined in section 7.2, these unresolved hyperfine interactions can be studied by ENDOR or HYSCORE spectroscopy (see appendices 2 and 3 to this volume). As an example, we present the HYSCORE spectra produced by two cherts. In one, the CM is definitely of biological origin; the other contains CM from a meteorite. Figure 7.19 shows these spectra represented in 3 dimensions and projected onto the frequency plane.

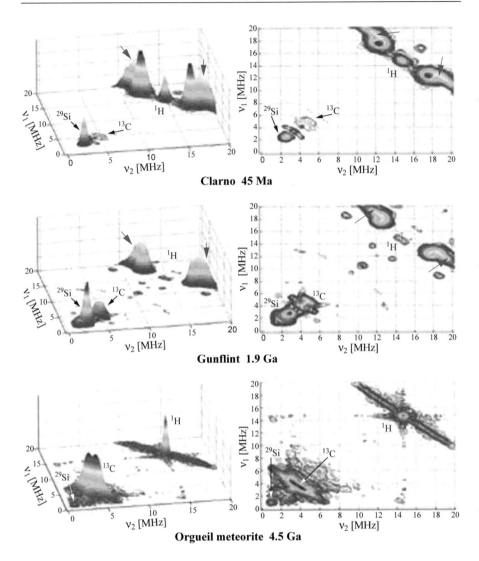

Figure 7.19 - Example of HYSCORE spectrum at 9 K for cherts from Clarno and Gunflint, and CM from the Orgueil meteorite. The spectra on the right are intensity contour maps. [From Gourier D. *et al.* (2013) *Astrobiology* **13**: 932-947 © 2013 Mary Ann Liebert, Inc.]

The spectra for the ^{1}H, ^{13}C and ^{29}Si nuclei for the CM in the cherts are clearly different from those in the meteoritic CM. The spectrum for ^{13}C is intense in the meteoritic CM, and weak in the CM from the cherts. This does not mean that there is less carbon in the CM from the cherts, rather it indicates that *the structure of the aromatic entities* carrying the electron spins is different in the two cases. Similarly, the spectrum for the protons has a linear shape in the meteoritic CM, whereas it takes the form of "commas" in the CM from cherts. This "comma" shape reflects the presence of abundant aromatic hydrogens (presence of a dipolar component in the hyperfine interaction), whereas the linear form reveals the presence of mainly benzylic hydrogens (dominant isotropic hyperfine interaction), and thus highly branched aromatic entities [Gourier *et al.*, 2008]. The hidden hyperfine structure of the EPR line could thus constitute the "fingerprint" of the origin of the CM.

7.4 - Perspectives: destination Mars?

The experiments described in this chapter clearly illustrate the potential of EPR when studying the most primitive carbonaceous matter, and show that this technique is perfectly adapted to the search for the first carbonaceous traces left by emerging life forms. However, the answers we are seeking may not come from Earth. We now know that water flowed abundantly on Mars during the first hundreds of millions of years of its existence. The *Mars Express* European probe effectively discovered abundant clay deposits in the high plains of Mars' southern hemisphere, aged at least 3.7 Ga, i.e., older than the oldest terrestrial cherts containing fossilised bacterial remains [Poulet *et al.*, 2005]. The presence of abundant hydrated minerals in these very ancient rocks, covered by lava flow 3.5 Ga ago, indicates that Mars could have been partially covered by seas before it lost its atmosphere and the planet cooled [Carter *et al.*, 2010]. Conditions compatible with life could therefore have existed on Mars between -4 Ga and -3.7 Ga. The major meteoritic impacts which seeded Earth and Mars with the same carbonaceous and organic components, could have provided the two planets with the same building blocks for the emergence of life [Maurette *et al.*, 2006]. It is therefore in the oldest sedimentary rocks on Mars that we may one day find traces of the most primitive life in the solar system, and be able to determine the conditions and mechanisms which contributed to the emergence of life. The presence of fossil traces on Mars at such an old epoch will help us to better understand the origins of life on Earth, where sedimentary rocks dating from the first 600 million years of its existence are no longer available. If we can show that such traces exist on Mars, or that they are conclusively absent, will obviously alter our understanding of how life emerged. If traces exist, it means that in similar initial conditions, life could have emerged on different planets. If such traces are conclusively absent from Mars, the differences between the two planets will give indications to help identify the specific conditions required for the emergence of life [Hoehler and Westall, 2010]. The search for mineral [Banfield *et al.*, 2001] or organic biomarkers [Marshall *et al.*, 2010] of ancient terrestrial life (around − 3.5 Ga) by various analysis techniques should make it possible to determine the "observables" of even more ancient life forms likely to be discovered on Mars.

Two types of spatial missions are considered for this objective. First, a mission to bring back Martian samples, which presents considerable technical problems and for which a date has not yet been set with any degree of certainty. Alternatively, exploratory missions may be performed by *roving* all-terrain laboratories. With the first approach, very strict quarantine conditions must be respected to avoid cross-contamination (particularly if bacteria or related organisms are still present on Mars), but samples can be studied by a whole range of laboratory techniques. With the second approach, instruments must be miniaturised to an extreme extent and they must be capable of withstanding very harsh conditions, such as very low temperatures and large temperature variations, in addition to constant exposure to ionising radiation. Based on the success of the *Viking* 1976 and *Pathfinder* 1997 missions which validated the landing and roving technology on Mars, the *Mars Exploration Rover* (MER) mission in 2004 landed two rovers (*Spirit and Opportunity*) which discovered geological traces of the presence of water on Mars and evidence of a CO_2-rich atmosphere in the past. The routemap for the MER mission was *Follow the water*, i.e., show that conditions could have been compatible with the emergence of life. That of the *Mars Science Laboratory* (MSL) mission, which landed the *Curiosity* rover on Mars in August 2012, is *Follow the carbon*, i.e., look for carbonaceous molecules which could be linked to life itself. The 70 kg of scientific material in this *rover* include, among other things, a laser spectrometer (ChemCam), an X-ray diffractometer, an X-ray fluorescence spectrometer, a microscope, an X-ray and α particle spectrometer (APXS), a chromatograph, an IR spectrometer, a mass spectrometer, a neutron detector and another particle detector. For the next step, which will heavily depend on NASA and ESA funding, the *Exomars* mission should start in 2019 with a Franco-American *rover*. The aim of this mission is to find traces of life and to prepare the return of samples to Earth.

With its sensitivity and resolution, EPR is a technique which could certainly be applied to the study of Martian samples, whether in the context of a sample-return mission or, after miniaturisation and adaptation, in the context of a mission on Mars. The capacity of long-wavelength radiation (microwaves and radiofrequencies) to penetrate insulating and non-magnetic materials, the potential to analyse carbonaceous matter *in situ* without requiring its extraction, the range of EPR methods that can be used, make these techniques a promising means to search for traces of ancient life-forms on Mars. Its capacity to distinguish meteoritic carbon from carbon of biological origin may be precious since, if very old Martian sedimentary rocks may have preserved carbon structures of biological origin, they may also have preserved extraterrestrial carbon which fell in abundance at the same time.

Acknowledgements

Most of this work benefitted from financial support from the CNES and CNRS (national programme for planetology and Origins of planets and Life programme). Several of the studies described in this chapter were performed in collaboration with Hervé Vezin (Infrared and Raman Spectrochemistry Laboratory, CNRS, Lille 1 University, Villeneuve d'Ascq), Sylvie Derenne (Biogeochemistry and Ecology of Continental Milieux, CNRS, Université Pierre et Marie Curie, Paris) and François Robert (Laboratory of Mineralogy and Cosmochemistry, CNRS, French National Museum of Natural History, Paris).

Complement 1 - Principle of quadrature phase detection

To increase the signal-to-noise ratio, EPR spectra are generally recorded by superposing a modulation $b(t) = \dfrac{B_m}{2}\cos(2\pi v_m t)$ on the static magnetic field B, with a frequency modulation v_m generally equal to 100 kHz. The signal arriving at the detector is then written

$$s(t) = f\left[B(t) + \frac{B_m}{2}\cos(2\pi v_m t)\right]$$

where the function f describes the shape of the absorption spectrum and $B(t)$ represents the slow variation of the "static" field during acquisition of the EPR spectrum [Volume 1, complement 3 to chapter 1]. As the f function is not linear, the $s(t)$ signal contains, along with the $\cos(2\pi v_m t)$ term at the v_m frequency, harmonics in $\cos(2\pi n v_m t)$, where n is a natural integer [Volume 1, section 9.2.2]. Because of relaxation phenomena, the system of electron spins does not instantaneously react to variations in the magnetic field set by the modulation. The $s(t)$ signal thus also contains terms in $\sin(2\pi n v_m t)$ which are in "quadrature phase" relative to the modulation. Thus, the signal arriving at the detector takes the form

$$s(t) = \sum_{n=0}^{\infty} \left[A_n \cos(2\pi n v_n t) + B_n \sin(2\pi n v_n t)\right]$$

For a Lorentzian line, the A_n and B_n amplitudes are given by [Kälin *et al.*, 2003]:

$$A_n = \sum_{k=-\infty}^{+\infty} \frac{T_2\omega_1 J_{-k}(\omega_2/\omega_m)[J_{n-k}(\omega_2/\omega_m) + J_{-n-k}(\omega_2/\omega_m)]}{1 + \omega_1^2 J_k^2(\omega_2/\omega_m) T_1 T_2 + (\Omega - k\omega_m)^2 T_2^2}$$

$$B_n = \sum_{k=-\infty}^{+\infty} \frac{(\Omega - k\omega_m) T_2^2 \omega_1 J_{-k}(\omega_2/\omega_m)[J_{n-k}(\omega_2/\omega_m) - J_{-n-k}(\omega_2/\omega_m)]}{1 + \omega_1^2 J_k^2(\omega_2/\omega_m) T_1 T_2 + (\Omega - k\omega_m)^2 T_2^2}$$

The $J_m(z)$ quantities are the Bessel functions of the first kind. We set $\omega_2 = \gamma B_m/2$, $\omega_m = 2\pi v_m$, $\omega_1 = \gamma B_1/2$ where B_1 is the amplitude of the microwave field, and $\Omega = \gamma B - 2\pi v$, where v is the microwave frequency. When the modulation frequency v_m and the modulation amplitude B_m, are small relative to the linewidth, expressed as a frequency and a magnetic field, respectively, the amplitude A_1 of the in-phase first harmonic is proportional to the first derivative of the absorption spectrum. If the relaxation time T_2 is not too long, the quadrature phase terms are generally negligible, and synchronous detection produces the usual EPR spectrum. However, the B_n/A_n ratio varies as $v_m T_2$ and for paramagnetic species with a sufficiently long relaxation time T_2 (typically around 10 to 100 μs), signals in quadrature phase can be detected. In an EPR spectrometer, the phase of the modulation can be varied and the desired harmonic selected [Volume 1, complement 3 to chapter 1 and section 9.2.2]. We can therefore detect the 1st or 2nd harmonic in quadrature phase, and thus *selectively* detect paramagnetic species with a long relaxation time.

Complement 2 - Principle of EPR imaging

The principle of EPR imaging (iEPR) consists in *encoding the position* of a paramagnetic centre in the value of its *resonance field*. In conventional EPR spectroscopy, a *uniform* magnetic field **B** is applied, and the B_{res} value of the resonance field is independent of the position of the centre in the sample. In iEPR, a static field parallel to its direction is added to **B**. The intensity of this field varies linearly with the position X. The total field is thus written $B(X) = B + GX$, where $G = dB/dX$ is the *field gradient* (figure 7.20). This field gradient is created with specially designed electromagnetic coils which are added to a classical EPR spectrometer. For a paramagnetic centre at position X, the resonance condition is written $h\nu = g\beta B(X)$ and resonance occurs for the value of B given by:

$$B(X) = \frac{h\nu}{g\beta} - GX$$

Thus, when the EPR spectrum for a sample is recorded under a gradient of magnetic field, the signal for a centre appears at a resonance field determined by the position which it occupies in the sample (figure 7.17).

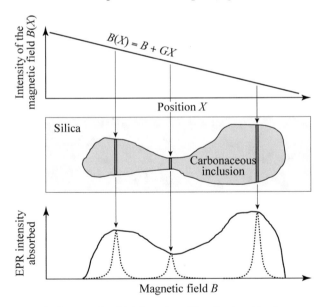

Figure 7.20 - Block diagram for EPR imaging. The abscissa for the upper and central figures is the *X position*. In the lower figure, the dashed lines represent the spectra corresponding to the three sections of the central figure. The broad spectrum represents the superposition of all the lines.

The intensity of the signal in $B(X)$ is proportional to the number of centres having X for abscissa, which are localised in the plane at position X, perpendicular to the direction of the magnetic field gradient. The variation in the EPR intensity of the spectrum under a gradient reflects the variation of the number of paramagnetic centres in this direction and thus constitutes a one-dimensional (1D) image. In practice, the intrinsic shape of the EPR signal for the paramagnetic species must be considered (i.e., that of the signal obtained in the absence of magnetic field gradient), as a result the spectrum under a gradient is in fact the *convolution product* of the spatial distribution of the species and of the intrinsic signal shape. *Deconvolution* must therefore be performed to obtain the 1D image. Two-dimensional (2D) images corresponding to the distribution of the paramagnetic species projected on a given plane, or three-dimensional (3D) images giving a volumic representation of these species can be obtained from a series of 1D images corresponding to different directions of the gradient scanning a plane for a 2D image, or the whole space for a 3D image.

Details on this technique and its applications can be found in [Blank *et al.*, 2003; Subramanian and Krishnan, 2008].

References

BANFIELD J.F. *et al.* (2001) *Astrobiology* **1**: 447-465.

BIGELEISEN J. & MAYER M.G. (1947) *Journal of Chemical Physics* **15**: 261-267.

BINET L. *et al.* (2002) *Geochimica et Cosmochimica Acta* **68**: 4177-4186.

BINET L. *et al.* (2004) (a) *Meteoritics & Planetary Science* **39**: 1649-1654.

BINET L. *et al.* (2004 (b) *Geochimica et Cosmochimica Acta* **68**: 881-891.

BINET L. & GOURIER D. (2006) *Applied Magnetic Resonance* **30**: 207-231.

BINET L. *et al.* (2008) *Earth and Planetary Science Letter* **273**: 359-366.

BLANK A. *et al.* (2003) *Journal of Magnetic Resonance* **165**: 116-127.

BOTTA O. & BADA J. (2002) *Surveys in Geophyics* **23**: 411-467.

BRASIER M.D. *et al.* (2002) *Nature* **416**: 76-81.

BRASIER M.D. *et al.* (2005) *Precambrian Research* **140**: 55-102.

BROCKS J.J. *et al.* (2003) *Geochimica et Cosmochimica Acta* **67**: 4289-4319.

BUSEMANN H. *et al.* (2006) *Science* **312**: 727-730.

CARTER J. *et al.* (2010) *Science* **328**: 1682-1686.

CHYBA C. & SAGAN K. (1992) *Nature* **355**: 125-132.

CODY G.D. & ALEXANDER C.M.O'D (2005) *Geochimica et Cosmochimica Acta* **69**: 1085-1097.

DELPOUX *et al.* (2011) *Geochimica et Cosmochimica Acta* **75**: 326-336.

DERENNE S. *et al.* (2005) *Geochimica et Cosmochimica Acta* **69**: 3911-3918.

DERENNE *et al.* (2008) *Earth and Planetary Science Letter* **272**: 476-480.

DERENNE S. & ROBERT F. (2010) *Meteoritics & Planetary Science* **45**: 1461-1475.

DURAND B. & NICAISE G. (1980) "Procedures for Kerogen isolation" in *Kerogen: Insoluble Organic Matter from Sedimentary Rocks*, Durand B. ed., Technip, Paris.

EHRENFREUND P. *et al.* (1992) *Advances in Space Research* **12**: 53-56.

EHRENFREUND P. *et al.* (2000) *Annual Review of Astronomy and Astrophysics* **38**: 427-483.

EHRENFREUND P. *et al.* (2006) *Astrobiology* **6**: 490-520.

ENOKI T. & TAKAI K. (2009) *Solid State Communications* **149**: 1144-1150.

FEL'DMAN E.B. & LACELLE S. (1996) *Journal of Chemical Physics* **104**: 2000-2009.

GARCIA-RUIZ J.M. *et al.* (2003) *Science* **302**: 1194-1197.

GARDINIER A. *et al.* (2000) *Earth and Planetary Science Letter* **184**: 9-21.

GEISS J. & GLOECKER G. (1998) *Space Science Review* **84**: 239-250.

GOURIER D. *et al.* (2008) *Geochimica et Cosmochimica Acta* **72**: 1914-1923.

GOURIER D. *et al.* (2013) *Astrobiology* **13**: 932-947.

HOEHLER T.M. & WESTALL F. (2010) *Astrobiology* **10**: 859-867.

KÄLIN M. *et al.* (2003) *Journal of Magnetic Resonance* **160**: 166-182.

KAZMIERCZAK J. & KREMER B. (2002) *Nature* **420**: 477-478.

LODDERS K. (2003) *The Astrophysical Journal* **591**: 1220-1247.

MARSHALL P. *et al.* (2010) *Astrobiology* **10**: 229-243.

MAURETTE L. *et al.* (2006) *Advances in Space Research* **38**: 701-708.

MILLAR T.J. *et al.* (2000) *Philosophical Transactions of the Royal Society of London* **A358**: 2535-2547.

PASTERIS J.D. & WOPENKA B. (2002) *Nature* **420**: 476-477.

PIZZARELLO S. (2007) *Chemistry & Biodiversity* **4**: 680-693.

POULET F. *et al.* (2005) *Nature* **438**: 623-627.

REMUSAT L. *et al.* (2009) *The Astrophysical Journal* **698**: 2087-2092.

REMUSAT L. *et al.* (2010) *The Astrophysical Journal* **713**: 1048-1058.

ROBERT F. & EPSTEIN S. (1982) *Geochimica et Cosmochimica Acta* **16**: 81-95.

ROBERT F. (2002) *Planetary and Space Science* **50**: 1227-1234.

ROBERTS H. *et al.* (2003) *The Astrophysical Journal* **591**: L41-44.

SCHOPF J.W. *et al.* (2002) *Nature* **416**: 73-76.

SKRZYPCZAK-BONDUELLE A. *et al.* (2008) *Applied Magnetic Resonance* **33**: 371-397.

SUBRAMANIAN S. & KRISHNA M.C. (2008) *Magnetic Resonance Insight* **2**: 43-74.

THOMANN H. *et al.* (1988) *Energy & Fuels* **2**: 333-339.

Volume 1: BERTRAND P. (2020) *Electron Paramagnetic Resonance Spectroscopy - Fundamentals*, Springer, Heidelberg.

WANG J. *et al.* (2010) *Physical Chemistry Chemical Physics* **12**: 9839-9844.

WESTALL F. & FOLK R.L. (2003) *Precambrian Research* **126**: 313-330.

Chapter 8

Using paramagnetic probes to study structural transitions in proteins

Belle V. and Fournel A.

Laboratory of Bioenergetics and Protein Engineering, UMR 7281,
Institute of Microbiology of the Mediterranean,
CNRS & Aix-Marseille University, Marseille.

8.1 - Introduction

Thanks to technical progress in structural biology, a large number of three-dimensional structures for proteins with a wide variety of functions have been determined over the last two decades. These structures provide very important information, helping us to understand how these complex macromolecules work. However, as their dynamic properties often play an essential role, this information can be insufficient. Indeed, structural flexibility is central to numerous biological processes such as enzymatic catalysis, cellular signalling and molecular recognition, as well as the assembly of complex structures. Among the techniques that have been developed to obtain information in this field, "spin labelling" combined with EPR spectroscopy, a technique known as SDSL-EPR (for *Site-Directed Spin Labelling*) has emerged as a powerful tool for the study of structural changes occurring within proteins (see appendix 4 to this volume for a description of protein structure). SDSL-EPR consists in "attaching" one or several paramagnetic probes to well-defined sites in a protein and analysing their EPR spectrum to glean information on their environment. These probes are generally stable nitroxide-type radicals with a functional group allowing its binding to a *cysteine* residue in a protein. Although the first applications of spin labelling to proteins were performed in the 1960s in pioneering work by

H. McConnell at Stanford University [Stone *et al.*, 1965; Griffith and McConnell, 1966], the technique has recently been the focus of a spectacular renewal of interest thanks to progress made in molecular biology tools, allowing any amino acid in a protein to be replaced by a cysteine, and to the development of pulsed EPR techniques. Given the local nature of the information provided by paramagnetic probes, the size of the biological system studied is not a limitation, and SDSL-EPR can be applied across a vast range of systems. For example, this technique has been used to obtain structural information on large membrane-bound protein complexes which cannot be studied by conventional techniques, such as X-ray diffraction crystallography or high-resolution NMR. It is also ideally suited to the study of structural changes occurring within a protein, whether induced by an external physical process, such as illumination, or by its interaction with a physiological partner (protein, ligand, substrate). This type of application, in particular as developed by W.L. Hubbell and his group at the University of California and Los Angeles (UCLA) [Altenbach *et al.*, 1989; Altenbach *et al.*, 1990], will be dealt with in this chapter.

After a review of the EPR properties of nitroxide radicals in section 8.2, we will present the different methods used to study conformational changes by SDSL in section 8.3. In sections 8.4 and 8.5, we describe the application of these methods to two very different proteins, an enzyme: human pancreatic lipase; and a protein with an intrinsically disordered domain: the measles virus nucleoprotein. Recent developments related to other applications of SDSL can be found in review articles [Fanucci and Cafiso, 2006; Klug and Feix, 2008; Klare and Steinhoff, 2009].

8.2 - EPR spectrum and mobility of nitroxide radicals

The stability, ease of synthesis, chemical properties and EPR characteristics of nitroxide radicals explain why they are used in so many applications in EPR. Thus, they appear in this volume in chapter 3 as adducts in radical traps, and in chapter 9 as "building blocks" for the development of new magnetic materials. We will briefly review the properties that make them suitable for use as *probes* in biological macromolecules. Nitroxide radicals are π radicals in which the unpaired electron is approximately equally delocalised over the nitrogen and oxygen atoms of the N–O bond. The principal values of the $\tilde{g}$ matrix are very close to $g_e = 2.0023$ and they are generally ranked in the order $g_x > g_y > g_z$ where $\{x, y, z\}$ are the principle axes represented in the inset in figure 8.1. The interaction between the single electron and the ^{14}N nucleus ($I = 1$) produces a very anisotropic hyperfine $\tilde{\mathbf{A}}$ matrix with $A_x \approx A_y \ll A_z$ [Volume 1, exercise 4.6 and appendix 3].

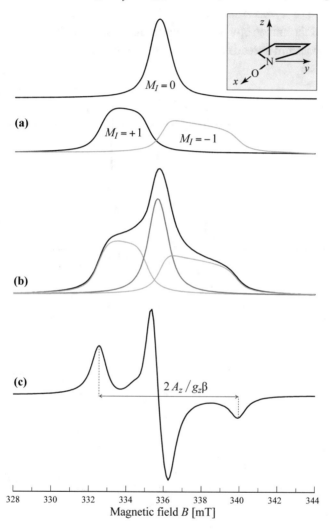

Figure 8.1 - X-band spectrum for a frozen solution of nitroxide radicals.
(a) The components corresponding to the 3 values of M_I are calculated for
$\nu = 9.428$ GHz, $(g_x, g_y, g_z) = (2.0089, 2.0064, 2.0027)$ and
$(A_x/g_x\beta, A_y/g_y\beta, A_z/g_z\beta) = (0.49, 0.49, 3.51)$ mT. **(b)** Sum of the 3 components.
(c) Derivative of the absorption signal. The (x, y, z) directions are defined in the inset.

This interaction can be considered a perturbation relative to the Zeeman term and the spectrum is the result of superposition of the three components corresponding to $M_I = -1, 0, 1$ [Volume 1, section 4.4.2]. Its shape is highly dependent on the mobility of the molecules in the sample. The following two limit situations can be defined:

▷ In the case of a *frozen solution*, the three components are clearly separated because of the anisotropy of the hyperfine constants [Volume 1, section 4.4.3] and the spectrum spreads over a field range of around $2A_z/g_z\beta$ (figure 8.1).

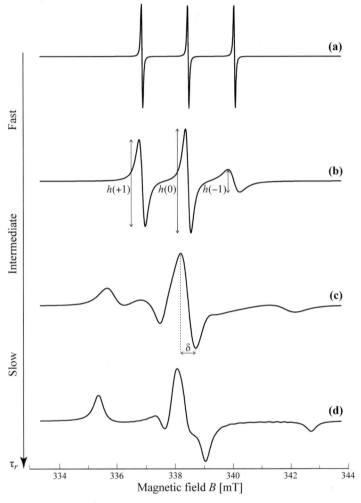

Figure 8.2 - X-band spectrum produced by a solution of nitroxide radicals undergoing isotropic movement. The spectra illustrating the different mobility regimes were calculated using *EasySpin* software [Stoll and Schweiger, 2006] for $\nu = 9.501$ GHz, $(g_x, g_y, g_z) = (2.0089, 2.0064, 2.0027)$; $(A_x/g_x\beta, A_y/g_y\beta, A_z/g_z\beta) = (0.49, 0.49, 3.51)$ mT and the following τ_r correlation times **(a)** 10^{-12} s, **(b)** 10^{-9} s, **(c)** 10^{-8} s, **(d)** 10^{-5} s. The shape of the spectrum shown in **(d)** differs from that in figure 8.1c because of its narrower linewidth. The semi-quantitative indicators used to characterise the shape of the spectrum are indicated (see section 8.3.1).

▷ In a *liquid solution*, the molecules rapidly explore all the orientations relative to the applied field with an equal probability (isotropic regime), the anisotropic effects disappear and each component reduces to a single line. The spectrum is thus a simple pattern of 3 equidistant narrow lines separated by $A_{iso}/g_{iso}\beta$, centred at $B = h\nu/g_{iso}\beta$ [Volume 1, section 2.3.3], where:

$$g_{iso} = \frac{g_x + g_y + g_z}{3} \;;\; A_{iso} = \frac{A_x + A_y + A_z}{3}$$

Between these two limit situations, the calculation is more complicated, but as long as the motion of the radicals remains *isotropic*, the shape of the spectrum depends on only one parameter: the *rotational correlation time* τ_r. Some free software can calculate the EPR spectrum in this situation.

Figure 8.2 illustrates four spectral shapes calculated for four different values of τ_r:

▷ $\tau_r < 10^{-11}$ s: isotropic regime (figure 8.2a).

▷ 10^{-11} s $< \tau_r < 10^{-9}$ s: the three lines broaden without changing position (fast motion regime, figure 8.2b).

▷ 10^{-9} s $< \tau_r < 10^{-6}$ s: the spectrum broadens and spreads (intermediate regime, figure 8.2c).

▷ $\tau_r > 10^{-6}$ s: all the effects of anisotropy are observed on the spectrum (slow motion regime, figure 8.2d).

When the movement of the nitroxide radicals is not isotropic, several parameters must be used to characterise it and the spectrum is more difficult to calculate (see section 8.3.1).

8.3 - Studying structural transitions by spin labelling

Before describing the various methods used to obtain information on conformational changes occurring in proteins by SDSL, we will indicate how the labelling operation itself is performed. The nitroxide radical most frequently used to label proteins is the commercially-available compound MTSL (1-oxyl-2,2,5,5-tetramethyl-δ^3-pyrroline-3-methyl-methanethiosulfonate) which binds covalently to *cysteine* amino acid residues, forming a disulfide bridge (figure 8.3).

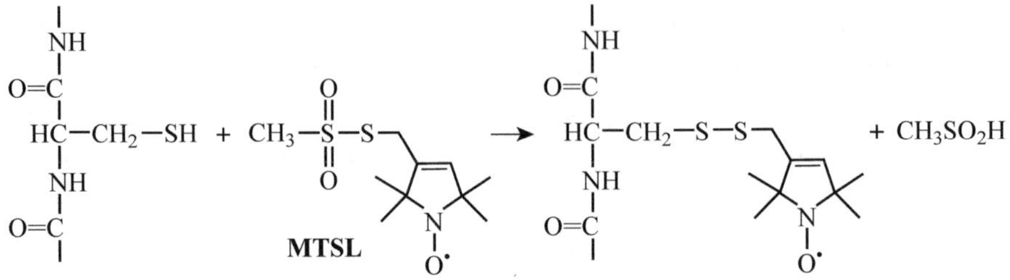

Figure 8.3 - Labelling of a cysteine with the MTSL radical.

Generally, the labelling reaction is performed by allowing the cysteine-bearing protein to react for a few hours with the label, which is present in excess. The excess unreacted label is then eliminated by gel filtration. Labelling yields obtained with this method are generally around 80 %. To label a well-defined position in the polypeptide chain, it is not sufficient to replace the amino acid located at this position by a cysteine; in addition, the cysteine residues naturally present and likely to be accessible to MTSL must be replaced by other residues. These operations are performed by "site-directed mutagenesis", i.e., by introducing appropriate modifications to the gene sequence coding for the protein (see appendix 4 to this volume). These modifications and the binding of the probe could be thought to perturb the structural changes under investigation. However, the presence of the probe generally has no, or very little, effect on the structure or function of the protein studied. This is due to the flexibility of the side chain carrying the label and the relatively low steric hindrance of the pyrrolidinyl cycle. It is nevertheless very important to verify whether the modifications affect the mutated protein, both before and after labelling (see sections 8.4 and 8.5).

In figure 8.4 we have summarised the various strategies used to obtain information on conformational changes by SDSL:

  ▷ The first involves analysis of *changes in mobility* for a probe bound to a region undergoing a structural transition [Steinhoff *et al.*, 1994; Thorgeirsson *et al.*, 1997; Klug *et al.*, 1998; Perozo *et al.*, 1999; Kim *et al.*, 2007; Belle *et al.*, 2008].

  ▷ Another, complementary, approach consists in revealing *variations in the accessibility* of the probe through the addition of relaxing agents. This technique is of particular interest for the study of membrane protein topology

[Hubbell *et al.*, 1998; Fanucci *et al.*, 2003; Altenbach *et al.*, 2005; Pyka *et al.*, 2005].

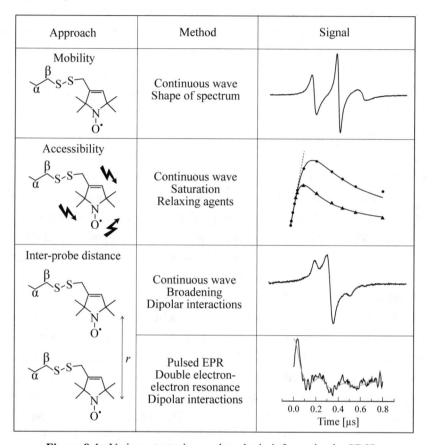

Approach	Method	Signal
Mobility	Continuous wave Shape of spectrum	
Accessibility	Continuous wave Saturation Relaxing agents	
Inter-probe distance	Continuous wave Broadening Dipolar interactions	
	Pulsed EPR Double electron-electron resonance Dipolar interactions	

Figure 8.4 - Various strategies used to obtain information by SDSL.

▷ SDSL can also be used to measure *variations in the distance* between two sites in a protein. For these measurements, a label is bound to each site and the distance between the labels is determined by analysing their dipole-dipole interaction [Volume 1, section 7.2.4]. The development of pulsed EPR techniques considerably broadened the field of application of this method [Jeschke *et al.*, 2005; Altenbach *et al.*, 2008; Drescher *et al.*, 2008; Sugata *et al.*, 2009; Ranaldi *et al.*, 2010].

8.3.1 - Analysis of probe mobility

The EPR spectrum recorded at room temperature for a solution of labelled proteins is determined by the motion of the radical probes relative to the applied magnetic field. This motion is the result of the composition of the "local" motion of the radical relative to the protein, and the rotational Brownian motion of the protein molecules in solution. When the molar mass of the proteins exceeds around 50 kDa, the protein moves much more slowly than the radical, and the proteins can be considered to be immobile and randomly oriented relative to the magnetic field. In this case, the shape of the EPR spectrum is only determined by the steric restrictions affecting the radical bound to the protein. When these restrictions are stringent, the spectrum is close to that of a frozen solution of radicals (figure 8.2d).

General rules for the qualitative interpretation of the EPR spectrum of labelled proteins were established in 1996 by Hubbell and collaborators. These authors bound a probe to various sites on T4 lysozyme, the structure of which was known, and assessed its mobility using "shape indicators" measured directly on the spectrum: width δ of the central line (figure 8.2c) and second moment of the spectrum [Hubbell *et al.*, 1996; Mchaourab *et al.*, 1996]. The results indicated that the mobility of the radical increases in the following order: buried sites, sites involving interactions with neighbouring structural elements, sites accessible to the solvent at the surface of α helices, sites located on loops at the protein's surface (see appendix 4). When probes are bound to the surface of an α helix, examination of the EPR spectra suggests that the probe's mobility is practically independent of the nature of the amino acids neighbouring the binding site, and that it is mainly determined by interaction between the S–S bond and the polypeptide chain (figure 8.3). This interpretation was confirmed by analysis of the crystal structures of the various labelled forms of T4 lysozyme [Langen *et al.*, 2000]. This study also revealed that the side chain bearing the probe can take several orientations relative to the peptide chain, with each position producing a spectral component corresponding to a structural sub-group known as a *rotamer*.

Since these first studies, the influence of fluctuations of the polypeptide chain on the EPR spectra for bound nitroxide radicals has been studied in detail [Columbus *et al.*, 2001; Columbus and Hubbell, 2002]. The consequences of the existence of several rotamers of the side chain on the shape of the spectrum have also been analysed [Guo *et al.*, 2007; Guo *et al.*, 2008; Fleissner *et al.*, 2009]. These consequences are essential to consider when using the SDSL technique to reveal conformational changes in proteins. Indeed, the spectrum produced by a single protein conformation in which the probe can explore two different environments is identical to that which would be produced by two different conformations in which the probe explores the same environment. To distinguish between these two situations, various methods have been proposed, based on molecular dynamic calculations [Pistolesi *et al.*, 2006; Ranaldi *et al.*, 2010] or measurement of the spin-lattice relaxation time T_1 [Bridges *et al.*, 2010].

The studies mentioned above show that using indicators of the shape of the spectrum gives ready access to important information. In the *fast regime*, the best parameters to describe mobility of the radical are the ratios between the peak-to-peak amplitudes for the lateral lines labelled $h(M_I = \pm 1)$ and that of the central line $h(M_I = 0)$, which actually reflect the differences between the *widths* of the lines (figure 8.2).

▷ When the radical performs *isotropic* movement, the linewidth takes the form $\Delta B = W + a + bM_I + cM_I^2$, where W is the residual width and a, b, c are coefficients proportional to the rotational correlation time, which depends on the principal values of the $\tilde{g}$ and $\tilde{A}$ matrices [Goldman *et al.*, 1972] (also see Volume 1, section 5.4.1). In this case, the best indicator of the mobility of the label is the $h(-1)/h(0)$ ratio [Qu *et al.*, 1997].

▷ *Anisotropic* radical movement can be considered to result from rapid rotational movement around the y axis (inset in figure 8.1) characterised by a rotational correlation time $\tau_{//}$, and a slower rotational motion around an axis perpendicular to y, characterised by $\tau_{\perp}$ [Goldman *et al.*, 1972; Marsh *et al.*, 2002]. In this case, the $h(+1)/h(0)$ ratio is a better indicator of radical mobility than the $h(-1)/h(0)$ ratio [Morin *et al.*, 2006; Belle *et al.*, 2009]. We will describe an example of the use of these semi-quantitative parameters in section 8.5.1.

To interpret the EPR spectrum in a more quantitative manner, it must be simulated from an appropriate model. Several models have been proposed and we will briefly describe two. In the MOMD (*Microscopic Order Macroscopic Disorder*) model developed by J. Freed and collaborators, motion of the side chain to which the radical is bound is characterised by two parameters: an order parameter linked to *the amplitude of the motion* and an effective correlation time associated with the *rotation rate* [Freed, 1976; Budil *et al.*, 1996; Barnes *et al.*, 1999]. In another model, the rotational correlation time is considered to always be very short and the variations of the spectrum for the radical mainly reflect those of the *steric hindrance* of the movement [Stopar *et al.*, 2005;

Strancar *et al.*, 2005]. This model describes the *conformational space* explored by the probe as an asymmetric cone defined by two angles corresponding to the amplitude and the anisotropy of the movement. An example of application of this model will be given in section 8.5.2.

8.3.2 - Determining probe accessibility

The addition of a "relaxing agent" (metal complexes or O_2 molecule) to a solution of labelled proteins results in accelerated relaxation of the radicals. The acceleration is increased when the labelled site is accessible to the relaxing agent. We will see that quantitative study of this effect, based on saturation of the spectrum for the radical, can be used to assess its accessibility.

Collisions between molecules of relaxing agent and nitroxide radicals bound to the proteins create transient complexes. In these complexes, strong exchange coupling between the two paramagnetic entities produces a paramagnetic ground state, but its EPR spectrum is not observable at room temperature. When exchange between the complexed and free forms of the nitroxide is slow, the positions of the resonance lines for the radical remain unchanged, but the *lifetime* of its spin states is shortened. If k_{ex} [$M^{-1} s^{-1}$] is the rate constant characterising formation of these complexes, the relaxation rates for the nitroxide radical are given by the following equations [Altenbach *et al.*, 2005]:

$$1/T_1^R = 1/T_1^0 + k_{ex} C_R$$
$$1/T_2^R = 1/T_2^0 + k_{ex} C_R \qquad [8.1]$$

where T_1^0 and T_2^0 are the relaxation times in the absence of relaxing agent, and C_R is the molar concentration of the agent. These expressions have the same form as equation [11.2] in chapter 11, which describes acceleration of the relaxation of protons from water by Gd^{3+} complexes. They indicate that the rate constant k_{ex}, which depends on the *accessibility* of the radical to the relaxing agent, can be deduced from the variation of the relaxation rates. To measure this variation, experimental saturation curves for the EPR signal of the radical are constructed, and they are simulated using an appropriate model [Volume 1, sections 5.3.2 and 9.4]. If the concentration C_R is not excessive, the variation of T_2 is negligible and the increase in power at half-saturation $P_{1/2}$ due to the relaxing agent can be written:

$$\Delta P_{1/2} = P_{1/2}^R - P_{1/2}^0 = \frac{2^{1/\varepsilon} - 1}{\frac{\gamma^2}{4} \lambda^2 T_2} \left[\frac{1}{T_1^R} - \frac{1}{T_1^0} \right] \qquad [8.2]$$

where $P_{1/2}^0$ and $P_{1/2}^R$ are the half-saturation powers in the absence and presence of the relaxing agent, $\gamma = g\beta/\hbar$ is the magnetogyric ratio, and ε is the inhomogeneity factor for the line. ε varies from ¾ for a completely homogeneous line to ½ for a completely inhomogeneous line [Haas *et al.*, 1993]. λ is the "conversion factor" for the cavity defined by $B_1 = \lambda\sqrt{P}$. Its value is provided by the manufacturer or can be measured. Analysis of the linewidth in the spectrum gives T_2, and the $P_{1/2}^R$, $P_{1/2}^0$ and ε parameters can be deduced from simulations of the saturation curves in the absence and presence of a relaxing agent. Equation [8.2] can then be used to calculate the $\dfrac{1}{T_1^R} - \dfrac{1}{T_1^0}$ difference and thus deduce the rate constant k_{ex} for equation [8.1]. Sometimes a dimensionless relative accessibility Π is also used, it is defined as follows:

$$\Pi = \frac{\dfrac{\Delta P_{1/2}}{\delta}}{\left(\dfrac{\Delta P_{1/2}}{\delta}\right)_{ref}} \qquad [8.3]$$

Division by the width δ of the central line normalises the spectrum relative to T_2 (equation [8.2]), and division by the quantity $\left(\dfrac{\Delta P_{1/2}}{\delta}\right)_{ref}$, determined by studying the effect of the relaxing agent on saturation of a reference molecule such as DPPH, eliminates the λ parameter which can be difficult to determine.

To construct the saturation curves, the EPR spectrum for the nitroxide radicals must be sufficiently saturable, even in the presence of agents accelerating relaxation. This is generally impossible at room temperature when using standard cavities, and samples must therefore be analysed in specific "high conversion factor" cavities. Different relaxing agents are used depending on whether the protein studied is membrane-bound or in solution:

▷ in the case of membrane proteins, molecular O_2 is used as it tends to concentrate in membranes due to its apolar nature [Merianos *et al.*, 2000; Marsh *et al.*, 2006],

▷ in aqueous medium, Cr(III) oxalate (abbreviated as Crox) or Ni(II) ethylenediaminediacetate (abbreviated as NiEDDA) are often used [Altenbach *et al.*, 1994; Lin *et al.*, 1998; Doebber *et al.*, 2008].

When studying membrane proteins, two relaxing agents with differing polarities can be used (e.g. O_2 and NiEDDA) to produce opposing concentration gradients between the membrane phase and the aqueous phase. These gradients can be used to calculate a Φ parameter which reflects the *depth* of the paramagnetic probe within the membrane:

$$\Phi = \ln\left[\frac{\Pi(O_2)}{\Pi(NiEDDA)}\right]$$

The relative accessibilities Π are defined by equation [8.3]. The Φ parameter is *positive* when the probes are deeply embedded in the membrane where collisions with O_2 dominate, or *negative* when the probes are located close to the aqueous phase where collisions with the Ni(II) complexes dominate [Altenbach *et al.*, 1994; Kaplan *et al.*, 2000].

8.3.3 - Measuring inter-probe distances

If two radicals are bound to the same protein, the interaction between the two magnetic dipoles causes splitting of the EPR lines [Volume 1, section 7.3.1]. Quantitative study of this effect, which varies with distance according to $1/r^3$, can be used to determine their distance. In practice, the technique used must be adapted depending on the value of r.

▷ When the distance is small enough for the splitting to result in visible broadening of the spectrum ($r \lesssim 20$ Å), the EPR spectrum can be simulated using an appropriate model [Rabenstein and Sgin 1995; Steinhoff *et al.*, 1997]. This method cannot be used over very short distances ($r \lesssim 8$Å) as in these conditions exchange interactions also contribute to splitting of the lines.

▷ When the distance is too great for the effects of the dipolar interaction to be visible on the spectrum, pulsed EPR techniques can be used to extend the range of measurement to around 80 Å. Specific pulse sequences can distinguish between the effects of dipolar interactions and other types of magnetic interactions. Currently, the most frequently used sequence is the 4-pulse DEER (*Double Electron Electron Resonance*) sequence [Pannier *et al.*, 2000; Jeschke, 2002] described in section 3 of appendix 2 to this volume. Analysis of the results is complicated by the fact that inter-centre distances are generally *distributed* in proteins, but several procedures can be used to take this phenomenon into consideration [Jeschke *et al.*, 2006].

The methods described in this section are illustrated hereafter by two studies in which a protein undergoes a conformational change as a result of interaction with its physiological partner(s).

8.4 - Activation of human pancreatic lipase

Lipases are enzymes catalysing lipid hydrolysis. In humans, Human Pancreatic Lipase (HPL) is the key enzyme in fat digestion. HPL alone is soluble in water and cannot interact with its insoluble lipid substrate: the enzyme is inactive. In the presence of a small protein, colipase, and bile salts – amphiphilic molecules which ensure the formation of lipid droplets – the enzyme can adsorb at the water-lipid interface and catalysis takes place: the enzyme is said to be *activated* [Bezzine *et al.*, 1999]. HPL activity can be determined by measuring the quantity of fatty acids released during hydrolysis of a tributyrine emulsion [Thirstrup *et al.*, 1993]. The maximum specific activity of HPL is around 12,500 U/mg (1 U produces 1 µmol of fatty acid per minute).

Several groups have sought to determine HPL's mechanism of action. Significant progress was made when the crystal structures of HPL and the HPL-colipase complex were determined. In the structure of the HPL-colipase complex, the region of the protein interacting with the lipid substrate – the *active site* – is covered by a polypeptide loop known as the *lid*. This lid is open in the structure of a HPL-colipase complex crystallised in the presence of a competitive inhibitor and amphiphilic molecules [van Tilbeurgh *et al.*, 1992; van Tilbeurgh *et al.*, 1993]. These observations led to the proposal of the following mechanism:

▷ when HPL is in solution, alone or in the presence of colipase, the substrate cannot access the active site, which is covered by the lid (HPL is said to be in the *closed* conformation),

▷ in the presence of bile salts, the presence of colipase induces a *conformational change* in the enzyme causing the lid to open (HPL is said to be in the *open* conformation).

Comparison of the two crystal structures suggested this mechanism, but no evidence has yet been obtained proving that it takes place in solution. In addition, the respective roles of colipase and bile salts in the activation process remain to be defined. We therefore undertook the study of this process *in solution* using the SDSL methods described in section 8.3.

8.4.1 - Attributing spectral features to HPL conformations

To demonstrate the conformational change of the lid in solution, a MTSL probe was bound to it, and its mobility analysed. The amino acid corresponding to aspartic acid (code D, see table 1 in appendix 4) in position 249 of the lid was selected as the labelling site, based on information from the crystal structures indicating that this site is accessible to solvent and that it does not interact with the enzyme's core in either the open or closed conformations. As HPL has a naturally accessible cysteine (code C) in position 181, labelling at position 249 requires the construction of a *double mutant* in which D249 is replaced by a cysteine, and C181 is replaced by a tyrosine. The specific activities of the mutated HPL and the labelled mutated HPL were similar to those for native HPL, indicating that the enzyme's catalytic mechanism is affected neither by the mutations nor by the presence of the probe. To define the spectral shapes which characterise the two enzymatic conformations, we examined the effect of colipase and a bile salt (sodium taurodeoxycholate) on the shape of the EPR spectrum for the MTSL radical recorded at room temperature (figure 8.5).

When alone in solution, labelled HPL produces a "narrow" spectrum corresponding to the *fast motion* regime for the probe (see figure 8.2b), with a separation between lateral lines of 3.22 ± 0.02 mT (figure 8.5a). This spectrum is not affected by the presence of colipase (figure 8.5b). In contrast, a broad component appears when the lipase is exposed to bile salts at 4 mM – a concentration which is greater than the critical micelle concentration (CMC) (figure 8.5c). The CMC, from which bile salt molecules assemble to form micelles of sodium taurodeoxycholate is around 1 mM. By varying bile salt concentrations, a sudden increase in the proportion of the broad component was observed above the CMC [Belle *et al.*, 2007]. Upon addition of colipase, this proportion increases further, reaching 75 % (figure 8.5d). The lateral lines of the broad component determined by subtraction are separated by 6.65 ± 0.05 mT (figure 8.5e), which indicates that the probe is in the *intermediate mobility* regime (see figure 8.2c).

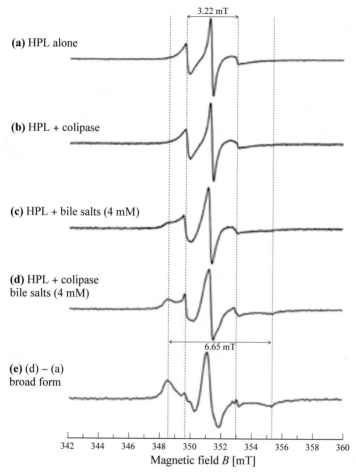

Figure 8.5 - Effect of physiological partners on the X-band spectrum for HPL labelled at position 249. **(a)** Labelled HPL in solution, **(b)** HPL in the presence of colipase (in 2-fold molar excess), **(c)** HPL in the presence of bile salts at supramicellar concentration (4 mM), **(d)** HPL in the presence of both partners, **(e)** spectrum obtained by subtraction of spectrum **(a)** from spectrum **(d)**. Microwave frequency: 9.869 GHz, power 10 mW. Modulation amplitude 0.1 mT.

Complementary experiments showed that the narrow component of the spectrum is produced by the closed conformation, and that the broad component is produced by the open conformation [Belle *et al.*, 2007] (figure 8.6). Only these two conformations were identified by SDSL. All these experiments show that colipase alone cannot induce conformational change in HPL, and that *in solution* an equilibrium between the closed and open conformations is observed in the presence of bile salts at concentrations exceeding the CMC; the presence of colipase shifts this equilibrium to the open conformation.

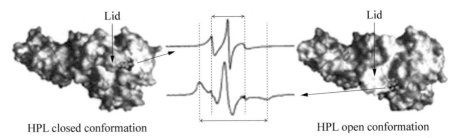

Lid   Lid

HPL closed conformation   HPL open conformation

Figure 8.6 - Attributing spectral features to the two enzymatic conformations. The nitroxide radical is represented.

8.4.2 - Studying probe accessibility

The probe is less mobile when the enzyme is in the open conformation than in the closed conformation, but the reason for this difference is not obvious. To obtain some answers, we assessed *the probe's accessibility to the solvent* in these two conformations using the method described in section 8.3.2, with Cr(III) oxalate as relaxing agent. The samples studied were the following: HPL alone in solution (closed conformation), HPL in the presence of bile salts and colipase (open conformation). Saturation curves for the central line were constructed for a range of concentrations of relaxing agent, using a cavity with a high conversion factor, $\lambda = 2.0 \text{ G W}^{-\frac{1}{2}}$ (figure 8.7). They are well reproduced by the relation

$$a_{pp} \propto \frac{\sqrt{P}}{\left[1 + (2^{1/\varepsilon} - 1)\dfrac{P}{P_{1/2}}\right]^{\varepsilon}} \qquad [8.4]$$

which describes the variation of the peak-to-peak amplitude of the derivative of an inhomogeneous line as a function of the incident microwave power, P [Haas *et al.*, 1993] (also see Volume 1, section 9.4). $P_{1/2}$ is the microwave power at half-saturation, and ε is a number describing the inhomogeneity of the line (section 8.3.2).

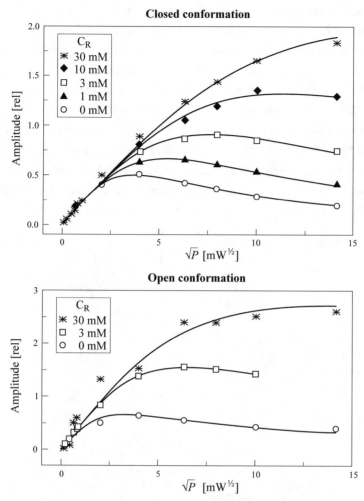

Figure 8.7 - Saturation curves for the MTSL radical bound to position 249 of HPL, in the closed and open conformations, for different concentrations of relaxing agent (C_R). The continuous line corresponds to calculations performed using equation [8.4].

Adjustment of the calculated curves to the experimental points returned $\varepsilon = 1.2$ for the closed conformation and $\varepsilon = 0.7$ for the open conformation. Variation of $P_{1/2}$ as a function of C_R can be used to determine the *exchange rate constant* k_{ex} which characterises the formation of radical-relaxing agent complexes (equations [8.1] and [8.2]). We therefore obtain:

$$k_{ex}^{closed} = 6.6 \times 10^8 \text{ M}^{-1} \text{ s}^{-1}; \ k_{ex}^{open} = 2.1 \times 10^8 \text{ M}^{-1} \text{ s}^{-1}$$

When the enzyme switches from the closed conformation to the open conformation, the exchange rate constant drops around 3-fold. The *reduced mobility* of the probe therefore correlates with a *reduction in its accessibility* to the solvent. Detailed analysis of the crystal structures suggests that aspartic acid D249 is in a less constrained environment in the closed conformation than in the open conformation (figure 8.8). It can therefore be assumed that after mutation to cysteine and binding of the probe, the latter is in a relatively free environment (probe pointing towards the solvent) in the closed conformation and a more restricted environment in the open conformation (probe trapped between two small helices). Analysis of the crystal structures could therefore help to understand the origin of the two spectral shapes observed [Ranaldi *et al.*, 2009].

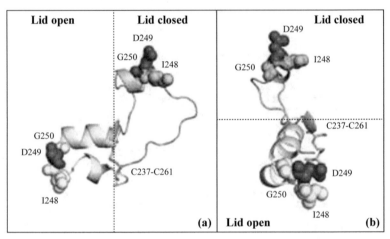

Figure 8.8 - Structure of the lid showing the environment of the aspartic acid D249. Two different views **(a)** and **(b)** are shown. The C237-C261 disulfide bridge serves as a hinge for the conformational change. In the closed conformation, D249 is on a helix exposed to the solvent, whereas in the open conformation it is on a small loop trapped between two helices.

8.4.3 - Assessing the amplitude of the conformational change

To determine whether the conformational change in solution effectively corresponds to that observed on the crystal structures, the variation of the distance between two sites on the protein was estimated using the double-labelling method described in section 8.3.3. The first site, as previously, is position 249 on the lid. The second is cysteine C181, already mentioned in section 8.4.1, which is close to the active site. According to the crystal structures, the environment for this amino acid is unaltered by the conformational change (figure 8.9a). In addition, the specific activity of the doubly-labelled HPL was similar to that of native HPL.

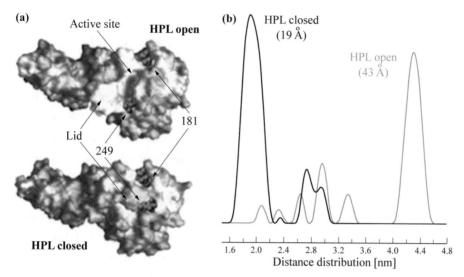

Figure 8.9 - **(a)** Modelling the positions of the two paramagnetic probes bound to positions 249 and 181 on the crystal structures of HPL in the open and closed conformations, **(b)** distribution of the inter-probe distances determined by DEER experiments performed on the closed (black line) and open (grey line) forms of the enzyme. $T = 70$ K, $v_{pump} - v_{obs}$ = 72.25 MHz, pulse duration 12 ns for $\pi/2$ and 24 ns for π (see section 3 in appendix 2).

DEER experiments were performed at 70 K on the HPL samples described previously, with and without the partners (colipase and bile salts at 4 mM) required to induce conformational changes. The results indicate an inter-probe distance of 19 ± 2 Å in the closed conformation and of 43 ± 2 Å in the open conformation (figure 8.9b) [Ranaldi *et al.*, 2010]. These distances are perfectly coherent with those deduced from the crystal structures, demonstrating that these structures effectively reflect the enzyme's behaviour in solution. X-band EPR spectra for the same samples were recorded at room temperature and at 100 K. Analysis of these spectra shows that for some molecules in the closed conformation, the interaction between the two radicals is visible on the EPR spectrum and it has a much larger effect than expected for a distance of 19 Å. This apparent contradiction between the results obtained by pulsed EPR and by continuous wave EPR can be explained if we remember that pulsed EPR only measures distances greater than 15 Å and if we assume that two possible orientations exist for one of the probes (two rotamers, see section 8.3.1). This hypothesis was validated by simulation of the EPR spectra and by molecular dynamic calculations: when the probe is bound to position 249, it can take two orientations in the enzyme's closed conformation [Ranaldi *et al.*, 2010].

8.5 - Folding induced by the measles virus nucleoprotein

"Intrinsically disordered" proteins do not have a stable three-dimensional structure in physiological conditions (see appendix 4). A large number of totally or partially disordered proteins are present in living organisms, where they play essential roles in a variety of processes such as cellular regulation and cell division, molecular recognition or even processes involved in neurodegenerative diseases [Tompa, 2002; Dyson and Wright, 2005; Fink, 2005] (also see chapter 4 in this volume). The total or partial disorder of these proteins makes them highly flexible, and thus capable of interacting with multiple partners. Intrinsically disordered proteins often take a well-defined conformation when they interact with their physiological partners [Dyson and Wright, 2002; Dunker *et al.*, 2005; Uversky *et al.*, 2005]. A type of disorder-order transition, known as *induced folding*, will be dealt with in this section.

Measles remains one of the main causes of infant mortality in developing countries. According to WHO, it was responsible for 158,000 deaths worldwide in 2013. There is currently no treatment for measles, but the protein complex which allows *viral replication* is considered a prime target for the development of antiviral agents.

To better understand how this complex functions, we investigated *nucleoprotein N* – a large number of copies of which encapsidate the viral genome – and its interaction with *phosphoprotein P* – which plays an essential role in viral transcription and replication (figure 8.10). Interaction between the nucleoprotein and the phosphoprotein involves two domains: the C-terminal domain (see appendix 4) of the N protein (amino acids 401 to 525), known as N_{TAIL}, which is intrinsically disordered; and the XD domain of the phosphoprotein [Longhi *et al.*, 2003]. Study of these domains indicated that their interaction results in a part of N_{TAIL} folding into an α helix [Johansson *et al.*, 2003]. NMR, circular dichroism and small-angle X-ray diffusion experiments performed to determine the structure of the (N_{TAIL}-XD) complex only provided very partial information [Bourhis *et al.*, 2005, 2006]. We therefore undertook the study of the N_{TAIL}-XD interaction by the SDSL technique.

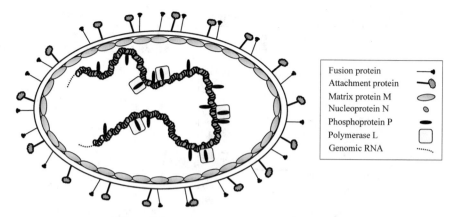

Figure 8.10 - Schematic structure of the measles virus.
Transcription and replication of the virus are ensured by the genomic RNA
and the nucleoprotein N/phosphoprotein P complexes.

8.5.1 - Mapping interaction sites for the (N_{TAIL}-XD) complex

To identify the sites on N_{TAIL} interacting with XD, we developed a series of 14 cysteine mutants of N_{TAIL} to successively probe 14 sites using MTSL. Of these 14 sites, 12 are in the 488–525 region which is known to be involved in interaction with XD, and two (positions 407 and 460) are located outside this region (top of figure 8.11). Circular dichroism experiments were used to verify that the cysteine mutations and binding of the label in the series studied affected neither the overall structure of the protein nor its folding [Belle *et al.*, 2008]. The EPR spectra for the labelled N_{TAIL} domains were recorded at room temperature, alone and in the presence of XD, and the $h(+1)/h(0)$ ratio was used as an indicator of mobility of the radical (see section 8.3.1). The results are presented in figure 8.11b.

▷ As expected, the mobility of the radicals bound to positions 407 and 460 is not affected by the presence of XD.

▷ For the labels bound in the 488–502 region, the mobility indicator decreases from 0.85 to around 0.45 in the presence of XD. As an example, we have represented the spectra produced by MTSL bound to position 496 in figure 8.11a.

▷ For labels bound in the 505–522 region, the mobility indicator shifts from 0.90 to around 0.80.

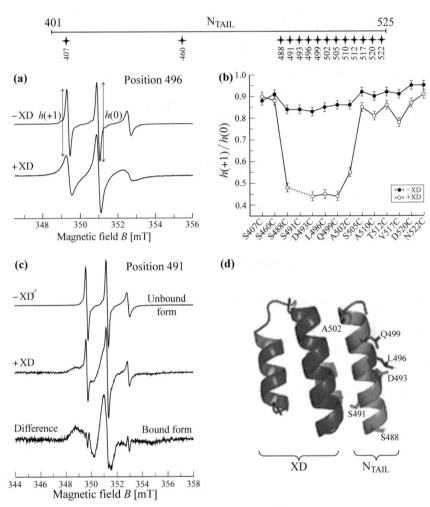

Figure 8.11 - Top: Positions of the 14 labelled sites in the 401–525 amino acid sequence of N_{TAIL}. **(a)** Spectrum for N_{TAIL} labelled at position 496, with and without XD, **(b)** variation of the $h(+1)/h(0)$ ratio as a function of the label's position, **(c)** spectrum for N_{TAIL} labelled at position 491, with and without XD. Spectra were recorded in the following conditions: microwave frequency: 9.855 GHz, power 10 mW. Modulation amplitude: 0.1 mT, **(d)** crystal structure of a chimera of XD and an N_{TAIL} region limited to amino acids 486 to 504, showing the positions of the labelled sites.

Based on these results, the N_{TAIL} region which folds as an α helix is contained between residues 488 and 502. We note that the mobility of the labels is slightly reduced in this region *before the interaction with XD*, suggesting that some *pre-structuration* exists within N_{TAIL}. This observation is compatible with a so-called conformational selection mechanism, whereby the partner of a protein selects that with the best conformation for the interaction [Tsai *et al.*, 2001].

Labelling at position 491 produces unusual results. Indeed, the addition of XD leads to a very significant modification of the spectrum for 75 % of the proteins present, and no modification for the remaining 25 %, which therefore do not interact with XD (figure 8.11c). When the spectrum for the bound form is obtained by calculating the difference, the distance between the lateral lines is 5.95 ± 0.05 mT (figure 8.11c). This value corresponds to the intermediate mobility regime (figure 8.2c), typical of a probe located in a *buried site* [Mchaourab *et al.*, 1996]. In this complex, the probe is therefore effectively located in the region of interaction, but it is directed *towards the XD partner*.

A crystal structure for a chimeric construction resulting from the association of XD with the (486–504) region in N_{TAIL} has been obtained [Kingston *et al.*, 2004]. In this structure, the side chains of all the residues selected as binding sites are directed towards the outside of the complex, except for residue 491 which points towards XD (figure 8.11d). The results obtained by SDSL in the N_{TAIL}-XD complex therefore indicate that this structural arrangement is conserved in solution [Morin *et al.*, 2006; Belle *et al.*, 2008].

8.5.2 - Conformational analysis of the (N_{TAIL}-XD) complex

As indicated in section 8.3.1, *simulation of the EPR spectra* can provide more detailed information than that given by the mobility indicators. The model developed by Strancar and collaborators [Strancar *et al.*, 2005; Stopar *et al.*, 2006], in which rapid rotation of the probe is confined to a space restricted to a cone can be used to determine *the conformational space* of the nitroxide radical at the level of its binding site. A novel approach was developed to study the N_{TAIL}-XD complex. This approach consists in adjusting the conformational space for the probe obtained by molecular modelling, including the local conformation of the polypeptide chain described by the Ramachandran angles and the steric interactions with neighbouring residues, to reproduce the conformational space deduced from simulations of the EPR spectra. This approach confirmed the pre-structuring of an α helix in the 490–500 region in the free form of N_{TAIL}. In the presence of the XD partner, structuring as an α helix extends to the 485–506 region. Modelling also confirmed that the 507–525 region of N_{TAIL} retains a considerable degree of conformational freedom, even when bound to XD [Kavalenka *et al.*, 2010] (figure 8.12).

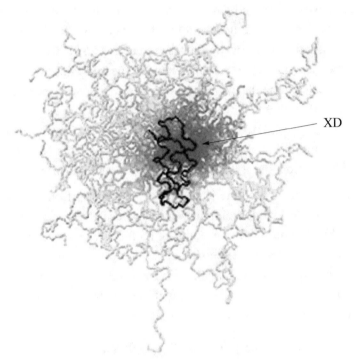

XD

Figure 8.12 - Model of the (488–525) region of N_{TAIL} in complex with XD. The figure represents the fifty best N_{TAIL} structures obtained by adjusting the conformational space calculated by molecular modelling to that which results from simulations of the EPR spectra. The N_{TAIL} models are represented in color describing the concentrations of structures (red = high concentration, yellow = low concentration). XD is represented in black. A significant structural concentration is observed at the level of the induced helix (485–506 region) and a strong structural dispersion is observed for the upstream region (507–525 region) of the induced helix. [From Kavalenka *et al.*, 2010]

8.6 - Conclusion

In this chapter, we presented paramagnetic probe-based methods which can be used to study conformational changes occurring in proteins. These methods can be used to implement a wide range of strategies, as illustrated by their applications to two very different systems:

▷ Activation of human pancreatic lipase requires transition from the closed conformation to the open conformation, a complex phenomenon involving several partners. Study of a probe bound *to a carefully selected site* revealed the conformational change occurring in the enzyme in solution and was used to determine the factors controlling it.

▷ We also investigated the disordered region of a viral nucleoprotein, and in particular its folding following interaction with another protein. In this case, successive labelling of a large number of sites was used to *precisely map* the region undergoing folding into an α helix. The SDSL technique had already been used to study denaturation/renaturation processes in proteins [Kreimer *et al.*, 1994; Qu *et al.*, 1997], but this is the first time that it was applied to study folding induced by another protein.

Acknowledgements

The authors would like to thank all those who contributed to the work presented in this chapter. We are grateful to Sonia Longhi and her team at the "Architecture and Function of Biological Macromolecules" laboratory in Marseille for the study of the measles virus nucleoprotein, and to Frédéric Carrière and Robert Verger from the "Interfacial Enzymology and Physiology of Lipolysis" laboratory, Marseille for the study of human pancreatic lipase. We thank Janez Strancar from the Josef Stefan Institute, Ljubljana for the conformational analysis of the N_{TAIL}-XD complex, and Hervé Vezin from the Infrared and Raman spectroscopy laboratory in Lille for the DEER experiments on human pancreatic lipase. Bruno Guigliarelli, head of our team, Sébastien Ranaldi, David Köpfer and Mireille Woustra are also acknowledged for their contribution to the study of human pancreatic lipase.

References

ALOULOU A. *et al.* (2006) *Biochimica et Biophysica Acta* **1761**: 995-1013.

ALTENBACH C. *et al.* (1989) *Biochemistry* **28**: 7806-7812.

ALTENBACH C. *et al.* (2005) *Biophysical Journal* **89**: 2103-2112.

ALTENBACH C. *et al.* (1994) *Proceeding of the National Academy of Sciences of the USA* **91**: 1667-1671.

ALTENBACH C. *et al.*(2008) *Proceeding of the National Academy of Sciences of the USA* **105**: 7439-7444.

ALTENBACH C. *et al.* (1990) *Science* **248**: 1088-1092.

BARNES J.P. *et al.* (1999) *Biophysical Journal* **76**: 3298-3306.

BELLE V. *et al.* (2007) *Biochemistry* **46**: 2205-2214.

BELLE V. *et al.* (2008) *Proteins: Structure, Function and Bioinformatics* **73**: 973-988.

BELLE V. *et al.* (2009) « Assessing Structures and Conformations of Intrinsically Disordered Proteins » in *Site-directed spin labeling EPR spectroscopy,* Uversky V.N. ed., John Wiley and Sons, New Jersey.

BEZZINE S. *et al.* (1999) *Biochemistry* **38**: 5499-5510.

BOURHIS J.M. *et al.* (2006) *Virology* **344**: 94-110.

BOURHIS J.M. *et al.* (2005) *Protein Science* **14**: 1975-1992.

BRIDGES M.D. *et al.* (2010) *Applied Magnetic Resonance* **37**: 363-390.

BUDIL D.E. *et al.* (1996) *Journal of Magnetic Resonance Series A* **120**: 155-189.

COLUMBUS L. & HUBBELL W.L. (2002) *Trends in Biochemical Sciences* **27**: 288-295.

COLUMBUS L. *et al.* (2001) *Biochemistry* **40**: 3828-3846.

DOEBBER M. *et al.* (2008) *Journal of Biological Chemistry* **283**: 28691-28701.

DRESCHER M. *et al.* (2008) *Journal of the American Chemical Society* **130**: 7796-7797.

DUNKER A.K. *et al.* (2005) *FEBS Journal* **272**: 5129-5148.

DYSON H.J. & WRIGHT P.E. (2002) *Current Opinion in Structural Biology* **12**: 54-60.

DYSON H.J. & WRIGHT P.E. (2005) *Nature Reviews Molecular Cell Biology* **6**: 197-208.

FANUCCI G.E. & CAFISO D.S. (2006) *Current Opinion in Structural Biology* **16**: 644-653.

FANUCCI G.E. *et al.* (2003) *Biochemistry* **42**: 1391-1400.

FINK A.L. (2005) *Current Opinion in Structural Biology* **15**: 35-41.

FLEISSNER M.R. *et al.* (2009) *Protein Science* **18**: 893-908.

FREED J.H. (1976) « Theory of slow tumbling ESR spectra for nitroxides » in *Spin labeling: theory and applications,* Berliner L. J., ed., Academic Press, New York.

GOLDMAN A. *et al.* (1972) *Journal of Chemical Physics* **56**: 716-735.

GRIFFITH O.H. & MCCONNELL H.M. (1966) *Proceeding of the National Academy of Sciences of the USA* **55**: 8-11.

GUO Z.F. *et al.* (2008) *Protein Science* **17**: 228-239.

Guo Z.F. *et al.* (2007) *Protein Science* **16**: 1069-1086.

Haas D.A. *et al.* (1993) *Biophysical Journal* **64**: 594-604.

Hubbell W.L. *et al.* (1998) *Current Opinion in Structural Biology* **8**: 649-656.

Hubbell W.L. *et al.* (1996) *Structure* **4**: 779-783.

Jeschke G. (2002) *ChemPhysChem* **3**: 927-932.

Jeschke G. *et al.* (2005) *Journal of Biological Chemistry* **280**: 18623-18630.

Jeschke G. *et al.* (2006) *Applied Magnetic Resonance* **30**: 473-498.

Johansson K. *et al.* (2003) *Journal of Biological Chemistry* **278**: 44567-44573.

Kaplan R.S. *et al.* (2000) *Biochemistry* **39**: 9157-9163.

Karlin D. *et al.* (2002) *Virology* **302**: 420-432.

Kavalenka A. *et al.* (2010) *Biophysical Journal* **98**: 1055-1064.

Kim M. *et al.* (2007) *Proceeding of the National Academy of Sciences of the USA* **104**: 11975-11980.

Kingston R.L. *et al.* (2004) *Proceeding of the National Academy of Sciences of the USA* **101**: 8301-8306.

Klare J.P. & Steinhoff H.J. (2009) *Photosynthesis Research* **102**: 377-390.

Klug C.S. *et al.* (1998) *Biochemistry* **37**: 9016-9023.

Klug C.S. & Feix J.B. (2008) *Biophysical Tools for Biologists Vol 1 in Vitro Techniques* **84**: 617-658.

Kreimer D.I. *et al.* (1994) *Proceeding of the National Academy of Sciences of the USA* **91**: 12145-12149.

Langen R. *et al.* (2000) *Biochemistry* **39**: 8396-8405.

Lin Y. *et al.* (1998) *Science* **279**: 1925-1929.

Longhi S. *et al.* (2003) *Journal of Biological Chemistry* **278**: 18638-18648.

Marsh D. *et al.* (2006) *Biophysical Journal* **90**: 49-51.

Marsh D. *et al.* (2002) *Chemistry and Physics of Lipids* **116**: 93-114.

Mchaourab H.S. *et al.* (1996) *Biochemistry* **35**: 7692-7704.

Merianos H.J. *et al.* (2000) *Nature Structural Biology* **7**: 205-209.

Morin B. *et al.* (2006) *Journal of Physical Chemistry B* **110**: 20596-20608.

Pannier M. *et al.* (2000) *Journal of Magnetic Resonance* **142**: 331-340.

Perozo E. *et al.* (1999) *Science* **285**: 73-78.

Pistolesi S. *et al.* (2006) *Biophysical Chemistry* **123**: 49-57.

Pyka J. *et al.* (2005) *Biophysical Journal* **89**: 2059-2068.

Qu K. *et al.* (1997) *Biochemistry* **36**: 2884-2897.

Rabenstein M.D. & Sgin Y.K. (1995) *Proceeding of the National Academy of Sciences of the USA* **92**: 8239-8243.

Ranaldi S. *et al.* (2010) *Biochemistry* **49**: 2140-2149.

RANALDI S. *et al.* (2009) *Biochemistry* **48**: 630-638.

STEINHOFF H.J. *et al.* (1994) *Science* **266**: 105-107.

STEINHOFF H.J. *et al.* (1997) *Biophysical Journal* **73**: 3287-3298.

STOLL S. & SCHWEIGER A. (2006) *Journal of Magnetic Resonance* **178**: 42-55.

JONES T.J. *et al.* (1965) *Proceeding of the National Academy of Sciences of the USA* **54**: 1010-1017.

STOPAR D. *et al.* (2005) *Journal of Chemical Information and Modeling* **45**: 1621-1627.

STOPAR D. *et al.* (2006) *Biophysical Journal* **91**: 3341-3348.

STRANCAR J. *et al.* (2005) *Journal of Chemical Information and Modeling* **45**: 394-406.

SUGATA K. *et al.* (2009) *Journal of Molecular Biology* **386**: 626-636.

THIRSTRUP K. *et al.* (1993) *FEBS Letters* **327**: 79-84.

THORGEIRSSON T.E. *et al.* (1997) *Journal of Molecular Biology* **273**: 951-957.

TIMOFEEV V.P. & TSETLIN V.I. (1983) *Biophysics of Structure and Mechanisms* **10**: 93-108.

TOMPA P. (2002) *Trends in Biochemical Sciences* **27**: 527-533.

TSAI C.D. *et al.* (2001) *Critical Reviews in Biochemistry and Molecular Biology* **36**: 399-433.

UVERSKY V.N. *et al.* (2005) *Journal of Molecular Recognition* **18**: 343-384.

VAN TILBEURGH H. *et al.* (1992) *Nature* **359**: 159-162.

VAN TILBEURGH H. *et al.* (1993) *Journal of Molecular Biology* **229**: 552-554.

Volume 1: BERTRAND P. (2020) *Electron Paramagnetic Resonance Spectroscopy - Fundamentals*, Springer, Heidelberg.

Chapter 9

Organic radicals and molecular magnetism

Turek P.

University of Strasbourg, Chemistry Institute
(UMR 7177, UDS-CNRS), Strasbourg

9.1 - Introduction

9.1.1 - Molecular materials

Molecular materials are composed of organic and / or inorganic molecules, the "bricks and mortar" architecture of which is composed of organic components. The basic building blocks are molecules with specific properties (polarity, response to optical excitation, electron delocalisation, etc.). Their organisation within a condensed phase depends on their geometry and their intermolecular interactions (van der Waals interactions, hydrogen bond, π-molecular stacking, etc.) which are mainly determined by the charge distribution specific to the molecules (π-type delocalised molecular orbitals and hybridation). For this reason, molecular materials have often been used as models to study the physical properties of low-dimensional systems. Molecular materials can be distinguished from their inorganic homologues, generally produced as by-products of metallurgy, by the following characteristics: low density, soluble, optically transparent, modulable properties (engineered chemical synthesis), flexible, multifunctional, partially biocompatible, low energy required for production, production at low temperatures, independent of mining resources.

The search for new materials is motivated by economic and strategic reasons, or more simply our thirst for knowledge. Initially, attempts were made to reproduce the optical, electronic, magnetic, or mechanical properties of traditional materials, metals or alloys, with molecular materials. This quest resulted in new fields of study, from molecular conductors and semiconductors in the 1960s, to graphene and conducting polymers (Heeger, McDiarmid and Shirakawa, Nobel Prize for Chemistry 2000) and the unexpected properties of several organised carbon states (Kroto, Curl and Smalley, Nobel Prize for Chemistry 1996; Geim and Novoselov, Nobel Prize for Physics 2010). In parallel, nanosciences and nanotechnologies have taken off, motivated by the emergence of scientific and technological hurdles in fields where miniaturisation is a major objective (e.g. microelectronics and biotechnologies). This has led to the launch of research into molecular materials by the *bottom-up* strategy used in nanosciences, i.e., from the molecule to the material.

9.1.2 - Molecular magnetism

Research into molecular magnetism aims to develop and characterise molecules and molecular materials with specific magnetic properties. Interest in this field has progressed significantly since the 1990s [Kahn, 1993; Miller and Epstein, 1994; Lahti, 1999; Itoh and Kinoshita, 2000; Miller and Drillon, 2001–2005] thanks to the discovery of the first molecular compounds displaying permanent magnet behaviour (ferromagnetic materials) [Chittipedi *et al.*, 1987; Kinoshita *et al.*, 1991; Takahashi *et al.*, 1991; Tamura *et al.*, 1991]. It was also discovered that molecular complexes, polynuclear clusters and nanoparticles of transition metals [Sessoli *et al.*, 1993a, b; Troiani *et al.*, 2010], and more recently rare earths [Bogani and Wernsdorfer, 2008; Kyatskaya *et al.*, 2009; Sorace *et al.*, 2011], display properties of single-molecule magnets (i.e., behave like magnets even though they are molecular objects in an isolated state) with specific properties such as quantum tunnelling of magnetisation [Thomas *et al.*, 1996]. The advantages of multifrequency/high field EPR spectroscopies when studying complexes of transition metals presenting single-molecule magnet behaviour was recently highlighted [Barra, 2005; Fittipaldi *et al.*, 2009]. In addition, multifunctional materials with magnetic properties that can, for example, be modulated by light (photomagnetism), are actively being sought. In summary, interest in this field of study is mainly driven, as in many scientific domains, by the impetus of nanosciences.

◇ *Intramolecular exchange coupling*

Exchange coupling between magnetic sites involves general magnetic mechanisms [Volume 1, section 7.2]: orbital overlap and geometry of the contact orbitals, concepts which have been revised for systems with molecular orbitals [Kahn and Briat, 1976; Kahn, 1993; Onofrio and Mouesca, 2010]. *Intermolecular* exchange propagates by overlap through space, whereas *intramolecular* exchange propagates through chemical bonds. Intramolecular exchange is specific to organic molecular compounds, which are the subjects of this chapter. The objective is to obtain the ground state with the highest spin multiplicity for the system under consideration in order to obtain a ferromagnetic or superparamagnetic material. The general problem addressed can be schematically represented as follows (figure 9.1) in the case of two sites bearing spins S_A and S_B.

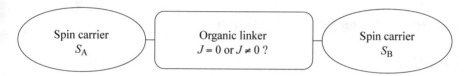

Figure 9.1 - Intramolecular exchange coupling.

Exchange coupling between the two sites is described by a Heisenberg Hamiltonian [Volume 1, section 7.2.2]:

$$\hat{H} = -J\, \mathbf{S}_A \cdot \mathbf{S}_B \qquad [9.1]$$

In this convention, $J > 0$ corresponds to *ferromagnetic* coupling which stabilises the ground state with maximum spin. In the case of organic radicals, the effects of spin-orbit coupling are negligible and the anisotropic components of the exchange interaction can be ignored [Volume 1, section 7.2.3].

It is essential to understand the intramolecular coupling mechanisms if we are to develop molecular components carrying a large number of magnetic sites in mutual interaction, such as linear or branched magnetic polymers (dendrimers). In the field of molecular electronics and nanosciences, another interest lies in the comparison between the propagation of *exchange interactions* [Higashiguchi *et al.*, 2010] and propagation of *electron delocalisation*, which can be studied by measuring conductance at the molecular scale [Choi *et al.*, 2008; Liu *et al.*, 2008]. Several mechanisms have been proposed to describe the propagation of exchange interactions through chemical bonds in organic compounds, and more specifically in conjugated systems [Crayston *et al.*, 2000; Rajca, 1994; Miller and Drillon, 2001–2005]. These mechanisms take into acount the nature of the radical, and the nature and length of the linker, as well as its geometry and connectivity, and its molecular conformation. In summary, the models are based on the topology of the molecular linkers and the nature of the molecular orbitals. These *topological* rules predict *that the sign for the spin density will alternate* [Volume 1, appendix 5] along the conjugated skeleton. This concept is simple to apply, although it should not be generalised, in particular when heteroatoms are present in the conjugated coupler. Spin polarisation [Volume 1, section 2.2.2] assisted by electron delocalisation as a result of conjugation can thus promote propagation of exchange between two distant spin carriers. The sign of the *J* parameter is then determined by parallel (ferromagnetic) or antiparallel (antiferromagnetic) alignment of the spin polarisation on distant radical sites.

9.2 - Molecules and methods of study

The studies presented in this chapter focus on diradicals and triradicals derived from nitronyl-nitroxides and imino-nitroxides. We will briefly present these radical *termini* and then describe how EPR can be used to characterise them.

9.2.1 - Nitronyl-nitroxide and imino-nitroxide radicals

Here, we consider purely organic molecular compounds in which the spin carriers are nitroxide-derived stable radicals (NO•): *nitronyl-nitroxide* (**NN**, figure 9.2a) and *imino-nitroxide* (**IN**, figure 9.2b) [Ullman *et al.*, 1972]. Synthesis of these radicals has been extensively documented because of their use in *spin trapping* (see chapter 3) and *spin labelling* (see chapter 8) methods.

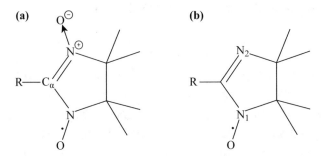

Figure 9.2 - Radicals **(a)** nitronyl-nitroxide (**NN**) and **(b)** imino-nitroxide (**IN**). R is a substituent which can take a variety of forms.

The major advantage of **NN** and **IN** radicals lies in their *great stability*, due in particular to delocalisation of the unpaired electron over the whole ONCNO fragment for **NN** and ONCN fragment for **IN**. Their X-band EPR spectra in the isotropic regime are presented below.

◇ *NN radical (figure 9.3)*

The spectrum for the NN radical is composed of five hyperfine lines centred around $g_{iso} \approx 2.007$, with relative intensities 1:2:3:2:1 reflecting interactions with the two equivalent ^{14}N nuclei ($I = 1$) [Turek *et al.*, 1991]. The hyperfine constant is equal to 21.1 MHz, i.e., half the value of 42 MHz observed for an R–NO• nitroxide radical [Volume 1, section 2.4.1]. The spin polarisation phenomenon results in a negative spin density on the central C_α carbon in the imidazole cycle, since the symmetry of the molecule requires this site to be a node of the wave function for the π-molecular orbital occupied by the unpaired electron [Zheludev *et al.*, 1994].

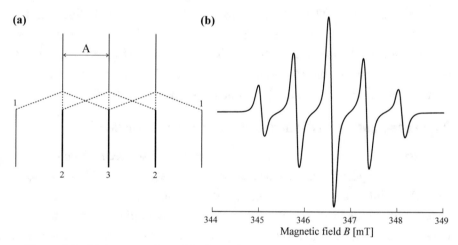

Figure 9.3 - (a) Interpretation of the hyperfine structure of the EPR spectrum for an NN radical, **(b)** Spectrum for an **NN** radical in solution (8×10^{-5} M in CH_2Cl_2) at room temperature. Microwave frequency: 9.734 GHz, power 1.5 mW. Modulation: frequency 100 kHz, peak-to-peak amplitude 0.07 mT.

◇ *IN radical (figure 9.4)*

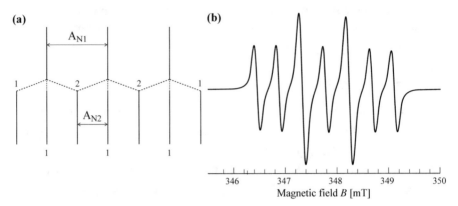

Figure 9.4 - **(a)** Interpretation of the hyperfine structure of the EPR spectrum for an **IN** radical, **(b)** Spectrum for an **IN** radical in solution (4.3×10^{-5} M in a 1:1 $CH_2Cl_2/$ xylene mixture) at room temperature. Microwave frequency: 9.775 GHz, power 1 mW. Modulation: frequency 100 kHz, peak-to-peak amplitude 0.06 mT.

The spectrum for an **IN** radical is composed of seven hyperfine lines centred around $g_{iso} \approx 2.006$ with relative intensities 1:1:2:1:2:1:1, corresponding to two non-equivalent ^{14}N nuclei, with $A_{N1} \approx 2A_{N2} \approx 25.3$ MHz (figure 9.4a). Because of the asymmetry of the **IN** molecule, the spin polarisation phenomenon is weaker than observed with the **NN** radical [Bonnet *et al.*, 1995].

9.2.2 - Contribution of EPR: characterisation of isolated molecules

This chapter describes how the magnetic properties of isolated molecules can be characterised by EPR spectroscopy, which requires the use of *dilute* solutions. The UV-visible absorption spectrum for the solution can be used to verify the absence of aggregation, which frequently occurs with planar aromatic molecules in solution. For the information provided by the hyperfine structure of the EPR spectrum to be exploitable, the width of the lines must be as narrow as possible. Solutions are therefore *diluted* to limit broadening as a result of intermolecular dipolar interactions, and they are *degassed* to limit interactions with the molecular oxygen triplet. Concentrations between 10^{-3} and 10^{-4} M are usually sufficient, and degassing is generally performed by bubbling argon through the solution or performing freeze-pump-thaw cycles under vacuum. Solutions prepared in this way can be studied by EPR in both the liquid and frozen state to glean complementary information:

▷ In *liquid solution*, all the effects of anisotropy disappear from the spectrum in the isotropic regime [Volume 1, section 2.2.3]. In particular, this is the case for effects due to dipolar interactions [Volume 1, section 6.2.1]. The shape of the spectrum is determined by the hyperfine structure produced by the *isotropic components* of the hyperfine interactions. Within the limit of *strong exchange coupling* [Volume 1, section 7.4], this structure reflects the distribution of the spin density over the molecule and the value of the total spin *S* at room temperature. The spectrum can be used, first, to assist chemical synthesis by identifying the molecule during its preparation. Determining the concentration of paramagnetic species (integrated intensity measurement, see Volume 1, chapter 9) can also help, in that it can be used to determine the quantitative yield when preparing a radical. For molecules that carry several radicals, the spectrum simply and rapidly reveals the existence of intramolecular exchange coupling (section 9.3). When weak hyperfine coupling is observed (with a limit of around 0.3 MHz), analysis of the EPR spectrum can produce a detailed map of the distribution of the spin density over the molecule by applying the McConnell relation [Cirujeda *et al.*, 1995]. These studies can be efficiently completed by electron-nucleus double resonance experiments (ENDOR, see appendix 3 at the end of this volume) to measure very weak coupling constants [Takui *et al.*, 1995; Hirel *et al.*, 2005].

▷ The shape of the spectrum for a *frozen solution* of molecules with several radicals interacting through exchange interactions is mainly determined by the value of the total spin *S*, and by the zero-field splitting terms due to interactions between the electrons magnetic dipoles. Indeed, given the large distances between radical sites in the compounds studied, the "high-field" limit is always reached at X-band [Volume 1, section 6.5]. It should be remembered that line splitting due to dipolar interaction between two radicals is equal to around $(\mu_0/4\pi)g\beta/r^3 \approx 2000$ mT/$\mathring{A}^3$, i.e., 2 mT at a distance of $r = 10$ $\mathring{A}$ [Volume 1, section 7.2.4]. The hyperfine structure is unresolved on spectra for frozen solutions because of extensive broadening of the lines due to distribution of the zero-field splitting parameters. It is interesting to seek for the half-field resonance [Volume 1, section 6.5.3], as it can directly indicate the existence of a spin multiplicity larger than or equal to three ($S \geq 1$) due to exchange coupling between at least two sites. Finally, a major aspect of the EPR signal is its *integrated intensity* which is proportional to the magnetisation produced by diradicals and triradicals in the sample [Volume 1, section 5.5.1]. The main advantage is to be able to measure the temperature-dependence of the susceptibility. Experimental results can be compared to predictions from exchange Hamiltonian models, which can be used to determine the spin of the ground state and the value(s) of exchange parameter(s). The sensitivity of EPR can thus be exploited to study isolated molecules and very small systems, such as thin layers and single crystals [Brinkmann *et al.*, 1997, 2004; Gallani *et al.*, 2001].

In the following sections we are interested in oligoradical systems with two or three **NN**- and/or **IN**-type radicals connected by a conjugated organic coupler. The coupling unit is phenylene ethynylene, and its *length* can be modulated in oligomeric units (**OPE**) (figure 9.5). The coupling *topology* can be modified by substituting radicals *meta* or *para* on the linker. The **OPE** moiety was selected for its rigidity and the cylindrical symmetry provided by the ethylene bond, allowing the conjugation to remain unbroken over the whole length of the chain. The same arguments are presented in the context of molecular electronics [Yoon *et al.*, 2010; Lu *et al.*, 2008, 2009].

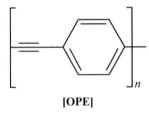

[OPE]

Figure 9.5 - The radical coupling unit: phenylene ethynylene.

Figure 9.6 summarises how EPR will be used in this chapter.

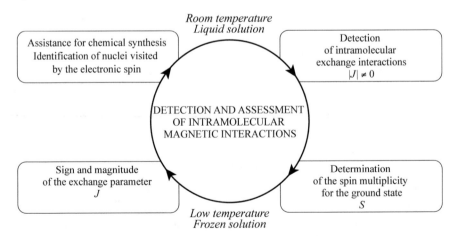

Figure 9.6 - Using EPR: from assistance for synthesis to determining the magnitude and sign of the exchange parameter.

9.3 - Study of liquid solutions of diradicals and triradicals: revealing intramolecular exchange

9.3.1 - Diradicals

Several series of compounds can be used to study different aspects of the propagation of the exchange interaction in diradicals.

◇ *Effect of the length of the coupler on exchange interactions*

Up to what distance can exchange interaction between two nitroxide radicals be observed through the phenyl ethynyl moiety (figure 9.5)? To answer this question, the series of *linear diradicals* **np-IN** (n = 2, 3, 5) represented on the left in figure 9.7 was synthesised. Only the structure of the **3p-IN** compound could be determined from X-ray diffraction experiments on a single crystal. The distances between radicals considered to be point radicals, as determined by molecular modelling, were as follows: 1.5 nm for **2p-IN**, 2.1 nm for **3p-IN** and 3.6 nm for **5p-IN**. These linear molecules with numerous aromatic nuclei tend to form aggregates [Breitenkamp and Tew, 2004; Chu and Pang, 2003]. It was thus necessary to use low concentrations, and a xylene/CH_2Cl_2 mixture as solvent. The appearance of narrow hyperfine lines and the UV-visible spectra produced confirm the absence of aggregates in these conditions.

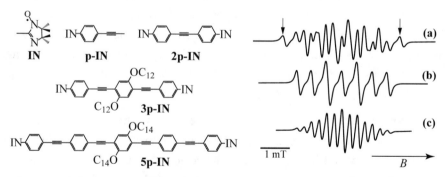

Figure 9.7 - X-band EPR spectra of linear diradicals in a 1:1 xylene/CH_2Cl_2 mixture. **(a)** 5p-IN (6.3×10^{-5} M), **(b)** p-IN (4.3×10^{-5} M), **(c)** 3p-IN (1.1×10^{-5} M). The molecular structures are represented on the left ($OC_{12} = OC_{12}H_{25}$, $OC_{14} = OC_{14}H_{29}$). Spectra **(a)** and **(c)** are centred at the central field of **(b)**, and the slight differences in g_{iso} are taken into account. The spectrum for **2p-IN** is similar to that for **3p-IN (c)**. The arrows indicate lines which are discussed in the text. Microwave **(b)**: frequency 9.775 GHz, power 1 mW. Modulation: frequency 100 kHz, peak-to-peak amplitude 0.06 mT. Same power and modulation for **(a)** and **(c)**. [From Wautelet P. *et al.* (2003) *Journal of Organic Chemistry* **68**: 8025–8036 © 2003 American Chemical Society]

The EPR spectrum directly provides information on the magnitude of the exchange coupling between radical entities. Indeed, in the absence of interaction, the spectrum for the **IN** monoradical should be observed (figure 9.4b) which is recalled in figure 9.7b (above) for the **p-IN** monoradical, radical synthon for the **2p-IN**, **3p-IN** and **5p-IN** diradicals. If the radicals interact in the limit of strong coupling, the spectrum should be that of a bis-imino diradical with spin $S = 1$ and isotropic hyperfine coupling A_{di} two-fold smaller than that observed for the A_{mono} monoradical. Indeed, the ^{14}N nuclei located on either side of the linker in this case are equivalent, and the spin density is shared to an equivalent extent between the nuclei of both radicals [Volume 1, section 7.4.3]. It should be noted that the limit of strong coupling is reached as soon as $|J| \gg |A|$, where A is around 10 MHz. This limit is readily reached when effective spin conjugation and polarisation mechanisms come into play. This is effectively observed with **2p-IN** and **3p-IN** compounds for which the spectrum (figure 9.7c) shows a hyperfine structure with 13 lines corresponding to two groups of two equivalent ^{14}N nuclei, with hyperfine constants $A_{N1}/2$ and $A_{N2}/2$ (see figure 9.4).

The spectrum for the **5p-IN** compound is more complex (figure 9.7a). Its hyperfine structure is neither that of a monoradical nor that of a strongly coupled diradical. Additional lines are observed outside the usual range for these compounds in the isotropic regime. Indeed, $|J|$ and $|A|$ are of the same order of magnitude here, and the number and position of the resonance lines can only be determined by full calcuation [Brière *et al.*, 1965; Glarum and March, 1967; Metzner *et al.*, 1977; Hernandez-Gasiò *et al.*, 1994]. This calculation indicates that $|J|/|A| \sim 3.6$, and thus $|J|/|k_B| \sim 2$ mK [Wautelet *et al.*, 2003]. The *sign* of J will be specified in section 9.4.2.

It was impossible to synthesise the same series with the **NN** radical, but an example shows that exchange coupling can also propagate over long distances with these radicals. Figure 9.8 shows the structure of a molecule where two nitronyl-nitroxide radicals are separated by an **OPE**-type linker centred on a Pt atom, and the EPR spectrum given by a dilute solution [Stroh *et al.*, 2004]. Although the **NN** radicals are spaced more than 2 nm apart and are separated by a diamagnetic Pt atom, the EPR spectrum is that of a diradical with strong coupling (interpretation of this 9-line spectrum is detailed in section 7.4.3 of Volume 1). The spin density is therefore transferred to the Pt, probably via the spin polarisation mechanism characteristic of **NN** radicals, assisted by the conjugation inherent to the **PE** linker. This mechanism can be considered analogous to the superexchange mechanism observed in metal oxides [Anderson, 1963].

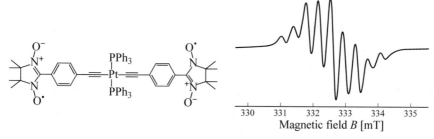

Figure 9.8 - Structure and EPR spectrum for a diradical derived from nitronyl-nitroxide bridged by a Pt atom. $\sim 10^{-4}$ M solution in CH_2Cl_2 at room temperature. Microwave frequency: 9.756 GHz, power 1 mW. Modulation: frequency 100 kHz, peak-to-peak amplitude 0.1 mT.

◇ *Conformational effects*

Figure 9.9 shows three series of molecules where the **IN** and **NN** radicals are bound to phenylene-ethynylene linkers with different geometries. According to the topological rules summarised in section 9.1.2, the sign for *J* should change depending on the position of the substitution (*ortho, meta, para*) of the radical on the linker. Here, we describe the EPR spectra for liquid solutions of the **om-NN** compound, compounds in the **pmp-IN$_R$** series and "9c series" compounds represented in figure 9.9c.

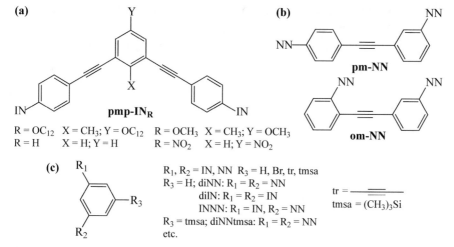

Figure 9.9 - Series of diradical compounds used to study topological effects. **(a)** IN derivatives with *para-meta-para* (**pmp**) topology, **(b)** NN derivatives with *para-meta* (**pm-NN**) and *ortho-meta* (**om-NN**) topology, **(c)** mixed **IN/NN** derivatives (diradicals and triradicals).

◇ *om-NN compound and pmp-INR series*

The **om-NN** compound produces a 9-line spectrum with relative intensities that differ from those on the spectrum in figure 9.8, which is typical of strong coupling. When the temperature is lowered, the slowing of molecular motion results in broadening of the high-field lines which creates asymmetry in the spectrum [Luckhurst and Pedulli, 1971; Alster and Silver, 1986].

A detailed study of the shape of the spectrum as a function of temperature shows that in this compound molecular motion allows interactions to propagate through bonds *and* through space. Derivatives of the **pmp-IN$_R$** series produce the 13-line spectrum of a strongly coupled diradical, identical to that of the linear series (figure 9.7c). This spectrum is not temperature-dependent, which indicates that, thanks to the rigidity of the **pmp** moiety, exchange coupling only propagates through bonds in these compounds [Wautelet *et al.*, 2003].

◇ *Series 9c*

EPR played an important role in assisting the synthesis of *mixed compounds* from the series represented in figure 9.9c, as analysis of the hyperfine interactions with the ^{14}N nuclei confirmed the radical formula. In these compounds, the radical groups are close enough on the *meta*-phenylene connector for the dipolar interactions to cause broadening of the lines in some solvents. Acetone has been observed to produce the best-resolved hyperfine structure. The **diNN-R$_3$** derivatives produce a diradical spectrum in the strong coupling limit with nine lines similar to that shown in figure 9.8, with $A_{di} = A_{mono}/2 = 10.4$ MHz and $g_{iso} = 2.0067$ (figure 9.10a). Similarly, the mixed **INNN-R$_3$** derivatives produce an 8-line spectrum with hyperfine constants twofold smaller than those of monoradicals, with $A_{N1} = 10.4$ MHz, $A_{N2} = 12.65$ MHz, $A_{N3} = 6.18$ MHz, $g_{iso} = 2.0061$ (figure 9.10b).

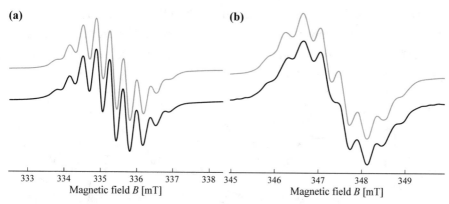

Figure 9.10 - Spectra for **diNN** and **INNN** diradicals in the 9c series solubilised in acetone at room temperature. **(a)** diNN, 10^{-4} M, frequency 9.42 GHz, **(b)** INNN, 10^{-4} M,
frequency 9.754 GHz. Simulations[1] in grey were obtained with the parameters indicated
in the text. For the two spectra: power 1 mW. Modulation: frequency 100 kHz, peak-
to-peak amplitude 0.1 mT. [**(b)** From Catala L. *et al.* (2001) *Chemistry - A European Journal* 7:
2466–2480 © 2001 Wiley-VCH Verlag GmbH, Weinheim, Fed. Rep. of Germany]

The spectra for **diIN-R₃**-type derivatives are more complex. This complexity
is due to the fact that the distribution of the spin density on the **IN** radical is
asymmetric, with approximately ⅔ on N_1 and ⅓ on N_2 (figures 9.4). When a
diradical is constructed on an *m*-phenyl cycle, the NO groups can either be
external or internal (figure 9.11a). The **IN** or **NN** cycles can turn about the
bond with the *m*-phenyl cycle, which can disrupt the intramolecular exchange
through deconjugation. **diIN** compounds are more sensitive to this phenomenon. Indeed, the barrier for rotation around the C_α(radical)-*m*-phenylene bond
is weaker for **IN** than for **NN** and we will see that the exchange coupling is
weaker for **diIN** derivatives than for the other derivatives. The spectrum in
figure 9.11b was simulated by adding:

▷ the spectrum for a **diIN** *diradical* in the strong coupling limit (see figure 9.7c) with A_{N1} = 12.65 MHz, A_{N2} = 6.27 MHz, g_{iso} = 2.0061 (80 %),

▷ the spectrum for a **IN** *monoradical* (figure 9.4) with A_{N1} = 25.28 MHz,
   A_{N2} = 12.65 MHz, g_{iso} = 2.0061 (10 %),

▷ a "dipolar component" due to dipolar interactions between the two radicals
   (Gaussian lines with width 1.2 mT) (10 %).

---

[1]   WINEPR SimFonia software, Brüker Analytische Messtechnik GmbH©, 1995–1996,
      Version 1.25.

The specific conformational effects due to the **IN** radical entity in series 9c, resulting from interactions with its environment, are well illustrated by the case of the **diINtmsa** derivative in tetrahydrofuran (THF) with **tmsa** = trimethylsilylacetylene (see figure 9.9c).

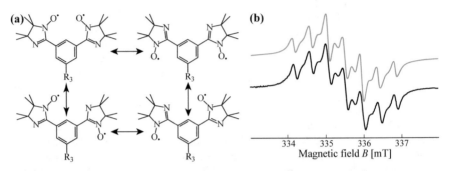

Figure 9.11 - (a) The four possible internal/external conformations for **diIN-R₃** derivatives, **(b)** spectrum for **diIN**, 10^{-4} M in acetone at 295 K. The simulation[1] in grey was obtained by adding the 3 components described in the text. Microwave frequency: 9.4207 GHz,power 1 mW. Modulation: frequency 100 kHz, peak-to-peak amplitude 0.1 mT. [**(b)** From Catala L. *et al.* (2001) *Chemistry - A European Journal* **7**: 2466–2480 © 2001 Wiley-VCH Verlag GmbH, Weinheim, Fed. Rep. of Germany]

Whereas the spectrum for **diNNtmsa** in solution in THF is that of a bis(nitronyl-nitroxide) *diradical* with strong coupling, that for **diINtmsa** looks like a *monoradical* **IN** spectrum (figure 9.4b) on which a broad dipolar component is superposed (figure 9.12a). In addition, the spectrum for the mixed **INNNtmsa** diradical (figure 9.12b) can be simulated by superposing:

▷ the spectrum for an **IN** *monoradical* with $A_{N1} = 25.29\,\text{MHz}$, $A_{N2} = 12.65\,\text{MHz}$, $g_{iso} = 2.0061$.

▷ the spectrum for an **NN** *monoradical*, with $A_N = 20.51\,\text{MHz}$, $g_{iso} = 2.0067$.

▷ a Gaussian dipolar component with $\Delta B_{pp} = 1.2\,\text{mT}$.

It is interesting to note that these effects are not observed in 2-methyl-THF, which is a slightly more bulky solvent with similar characteristics (polarity, chemical nature) [Ulmann *et al* ., 1972]. The influence of *steric factors* on the diradicals' conformation, and subsequently on the intramolecular exchange coupling, are clearly visible from the EPR spectra obtained for liquid solutions.

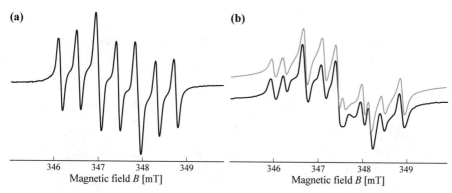

Figure 9.12 - Spectra for the **diINtmsa** and **INNNtmsa** compounds in THF at room temperature. The simulation[1] in grey was obtained as described in the text.
(a) diINtmsa, 10^{-4} M. Microwave: frequency 9.756 GHz, **(b) INNNtmsa**, 10^{-4} M. Microwave: frequency 9.762 GHz. In both cases: power 1 mW. Modulation: frequency 100 kHz, peak-to-peak amplitude 0.1 mT. [**(b)** From Catala L. *et al.* (2001) *Chemistry - A European Journal* 7: 2466–2480 © 2001 Wiley-VCH Verlag GmbH, Weinheim, Fed. Rep. of Germany]

9.3.2 - Triradicals

Triradicals were produced from series 9c (figure 9.13a) and examples of spectra for dilute liquid solutions are shown in figures 9.13b, c, d. Their study reveals one or more intramolecular exchange interactions based on the same principles as those discussed for diradicals. The strong coupling limit involves distribution of the spin density over the three radical fragments.

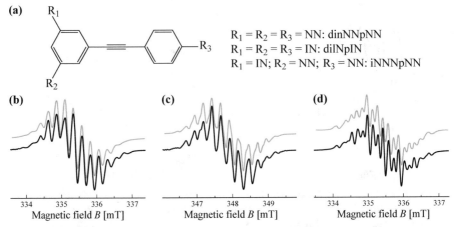

Figure 9.13 - Structure and EPR spectrum for a series of nitroxide-derived triradicals. The simulations [1] in grey were obtained as described in the text. **(a)** Overall structure. The other figures show the spectra for **(b)** diNNpNN, **(c)** diINpIN, **(d)** INNNpNN, at 10^{-4} M in solution in a 1:1 mixture of dichloromethane/xylene at room temperature. Microwave: frequency 9.4213 GHz for **(b)** and **(d)** and 9.767 GHz for **(c)**, power 1 mW. Modulation: frequency 100 kHz, peak-to-peak amplitude 0.1 mT.

Thus, for example, in an **NN** triradical with strong coupling, six equivalent ^{14}N nuclei are present and the hyperfine structure is composed of 13 lines ($2nI + 1$ where $I = 1$ and $n = 6$) separated by $A_{mono}/3 \sim 7$ MHz (figure 9.13b). Similarly, in a triradical composed of two **NN** radicals and one **IN** radical with strong coupling (figure 9.13d), four equivalent ^{14}N nuclei ($A_{mono}/3 \sim 7.03$ MHz) and two non-equivalent ^{14}N nuclei ($A_{N1}/3 \sim 8.43$ MHz; $A_{N2}/3 \sim 4.07$ MHz) are present, and so on for other combinations. Satisfactory simulations of the EPR spectra for the three triradicals were obtained in the strong coupling limit. Nevertheless, to simulate the spectrum for the triradical composed only of **IN** radicals (figure 9.13c), a mixture of molecular conformations involving complete or partial uncoupling of the radicals (mono-, di- and tri-radical) must be considered, as for the **diIN** compounds described above. These observations suggest that the barrier to rotation between the cycles bearing the radicals and the phenyl is sufficiently low for the molecular conformation to be sensitive to the environment. In addition, exchange coupling is itself sensitive to the molecular conformation. We therefore expect to observe dynamic effects in solution, similar to those observed for diradical and triradical derivatives bearing **IN** fragments.

To conclude this section, it should be remembere that the existence of, even very weak ($|J| \geq |A|$), exchange coupling between two organic radicals is readily detectable on the EPR spectrum for a liquid solution, with a hyperfine structure reflecting the distribution of the spin density over the molecule. This can be the result of an exchange mechanism through space (intermolecular) or through bond (intramolecular). One of the major advantages of EPR is that the process studied occurs at the molecular scale, and that it can be detected thanks to its very high sensitivity.

9.4 - Study of frozen solutions

9.4.1 - Information contained in the spectrum

As mentioned in the introduction, the shape of the spectrum for a frozen solution of molecules bearing several radicals coupled by exchange interactions is mainly determined by the value of the total spin S and by the zero-field splitting terms resulting from the electron-electron magnetic dipolar interactions [Volume 1, section 6.5]. The values of $|D|$ and $|E|$ can be deduced from the spectral features, as indicated in chapter 6 in [Volume 1, figures 6.5 and 6.9 for $S = 1$, figure 6.6 for $S = \frac{3}{2}$]. When a triradical is composed of a *diradical fragment* which is relatively well separated from a *monoradical fragment* (see figure 9.16b), the dipolar interactions between the two fragments are significantly weaker than those existing in the diradical. In these conditions, the zero-field splitting parameters for the D_T diradical (T for triplet state) and the D_Q triradical (Q for quartet state) are linked by [Bencini and Gatteschi, 1990]:

$$D_T = 3D_Q \qquad [9.2]$$

In the section above, we showed that analysis of liquid solutions by EPR spectroscopy can readily reveal the *existence* of intramolecular exchange coupling between radicals. However, neither the *nature* nor the *magnitude* of the exchange coupling could be determined. They are generally elucidated by studying the temperature-dependence of the magnetisation with a SQUID magnetometer. In complement 1 we explain why this type of experiment is not appropriate for the study of dilute solutions and we show how the temperature-dependence of the integrated intensity of the EPR spectrum can be used to obtain the desired information.

9.4.2 - Study of diradicals: effect of length and topology of the linker on the J parameter

This study was performed on diradical derivatives of imino-nitroxide from the **np-IN** *linear series* (N =2, 3, 5) (figure 9.7) and the **pmp-IN$_R$** series (R = H, OCH$_3$, NO$_2$) (figure 9.9a); study of these series at room temperature was described in section 9.3.1. In these diradicals, the sites bearing spins are separated by large distances, and magnetic dipolar interactions are therefore very weak. As a result, the features due to the zero-field splitting terms are unresolved on spectra recorded at low temperature. Similarly, the half-field line, the intensity of which varies as the $(D/h\nu)^2$ ratio, cannot be observed [Weissman and Kothe, 1975].

◇ *Method used*

The temperature-dependence of the intensity of the EPR spectrum was monitored between 4 K and 100 K. The singlet-triplet gap J was determined by modelling this dependence using the following relation [Bleaney and Bowers, 1952]:

$$\chi_{di}(T) = \frac{C}{T}\left[\frac{3}{3 + \exp\left(-J/k_B T\right)}\right] \qquad [9.3]$$

In this expression, C is a constant which depends on the units, and k_B is Boltzmann's constant. The ground state is $S = 0$ for $J < 0$ and $S = 1$ for $J > 0$.

We mentioned above (section 9.3.1) that at low temperature the EPR signal for polyradical species resulting from the $\Delta M_S = \pm 1$ transitions is easily saturated at relatively low power. This phenomenon is illustrated in figure 9.14a through the example of the **2p-IN compound.** The saturation curve at 4 K is typical of a homogeneous line [Volume 1, sections 5.3.2 and 9.4.1] with a marked maximum and saturation at relatively low power. Because of saturation, the intensity is underestimated at low temperature, and as a result the $|J|$ parameter is underestimated in the case of a ground triplet, or overestimated in the case of a ground singlet. The effects of saturation on the temperature-dependence of the intensity and its impact on the estimation of the singlet-triplet gap are illustrated in figure 9.14b by the case of **pmp-IN$_H$**. The reduction in the signal due to saturation is observed not only to result in overestimation of the singlet-triplet gap, but also to cause a temperature-dependence that does not fit equation [9.3].

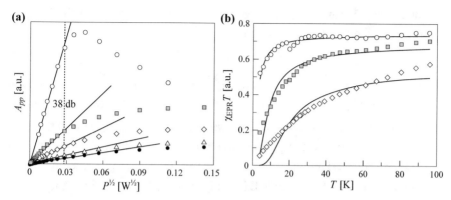

Figure 9.14 - (a) Saturation of the EPR signal for the **2p-IN** compound ($\sim 10^{-4}$ M in 1:1 dichloromethane/xylene). The variation in the peak-to-peak amplitude as a function of the square root of the power is represented for different temperatures: $T = 4$ K (white circles), $T = 14$ K (white diamonds), $T = 40$ K (grey squares), $T = 80$ K (white triangles), $T = 120$ K (black dots). **(b)** Temperature-dependence of the intensity of the EPR signal ($\Delta M_S = \pm 1$ transitions) for the **pmp-IN$_H$** molecule ($\sim 10^{-4}$ M in 1:1 dichloromethane/xylene) for different attenuations of the microwave power: 50 dB (white circles), 40 dB (grey square), 30 dB (white diamonds). The curves with continuous lines were calculated by applying equation [9.2] for a ground singlet with $J/k_B = -4.5 \pm 0.2$ K for 50 dB, $J/k_B = -15.5 \pm 0.9$ K for 40 dB, $J/k_B = -37.0 \pm 2.4$ K for 30 dB.

To conclude this section, it should be noted that due to the easy conformational changes which occur in liquid solution (sections 9.3.1 and 9.3.2), it is likely that several conformations are trapped in frozen solution. As a result, the values of J measured must be considered *mean values*.

◇ *Analysis of results*

Two essential results were obtained with this method [Wautelet, 1996; Wautelet *et al.*, 2001, 2003]:

▷ We saw that at room temperature the compounds in the linear **np-IN** series produce a triplet state spectrum corresponding to the strong coupling limit for $n = 2$ and $n = 3$ (figure 9.7c) and that $|J| \approx |A|$ for $n = 5$ (figure 9.7a). Study of the spectrum at low temperature reveals a ground state with $S = 0$, and exchange coupling is observed up to a distance of around 4 nm, with $J/k_B = -9.7$ K for **2p-IN**, -3.2 K for **3p-IN**, -0.002 K for **5p-IN**. This is a remarkable result given the distance covered by the electronic processes (conjugation and spin polarisation) involved in the intramolecular exchange mechanism. However, if one considers the quasi-planar geometry of the radical substituents relative to the conjugated chain, the coupling appears relatively weak when compared to the values measured with other linkers (phenylene-vinylidene, phenylene). Variation of the J parameter with the length L of the linker is well described by the following relation:

$$J = J_0 \exp(-\beta_{exch}L) \qquad [9.4]$$

where $J_0/k_B = -173 \pm 22$ K, and $\beta_{exch}^{OPE} = 0.19 \pm 0.01$ Å^{-1}. The attenuation factor β_{exch}^{OPE} should be compared to $\beta_{cond}^{OPE} = 0.21 \pm 0.01$ Å^{-1}, which is obtained for the decrease of the *conductance* of similar **OPE** derivatives [Liu *et al.*, 2008], and to $\beta_{cond}^{OPP} = 0.51 \pm 0.01$ Å^{-1}, deduced from a similar study [2] performed on oligo-*para*-phenylene (**OPP**) derivatives with nitronyl-nitroxide radical terminals [Higashiguchi *et al.*, 2010]. As in the case of **OPP** derivatives, the reduction in the *exchange coupling* is observed to be similar to that of the *conductance*, indicating that the processes involved in spin delocalisation on the one hand, charge delocalisation and charge transfer on the other, are the result of related, or even identical, mechanisms. This is an important result for molecular spin electronics as it implies that spin and charge are closely linked at the molecular scale.

▷ **pmp-IN$_\mathbf{R}$**-type derivatives, which produce triplet state spectra at room temperature (section 9.3.1), all have an $S = 0$ ground state (figure 9.15). This result contradicts predictions based on the topological rules set out for *meta*-substituted phenyl diradicals (section 9.1.2).

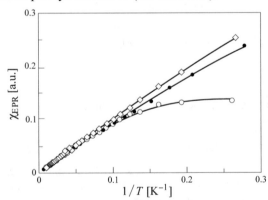

Figure 9.15 - Variation of the EPR signal intensity ($\Delta M_S = \pm 1$ transition) as a function of $1/T$ for a series of compounds substituted in meta to the phenyl: **pmp-IN$_\mathbf{H}$** (white circles), **pmp-IN$_{\mathbf{OC}_{12}}$** (black circles), **pmp-IN$_{\mathbf{NO}_2}$** (white diamonds) (see figure 9.9a for structures). Continuous lines indicate adjustments to equation [9.3] using the J/k_B values indicated in the text. [From Wautelet P. *et al.* (2003) *Journal of Organic Chemistry* **68**: 8025–8036 © 2003 American Chemical Society]

---

2   This study [Higashiguchi *et al.*, 2010] claims to be the first of its type, but we performed and reported similar studies in the previous decade [Wautelet, 1996; Wautelet *et al.*, 2001, 2003].

The J values measured are as follows: $J/k_B = -4.5 \pm 0.2$ K (**pmp-IN$_H$**), -2.6 ± 0.1 K (**pmp-IN$_{OC_{12}}$**), -1.3 ± 0.1 K (**pmp-IN$_{NO_2}$**). The acceptor or donor nature of the substituent on the central phenyl group therefore appears to affect the singlet-triplet gap. However, the weakness of the coupling means we must take care when quantitatively analysing these results. The undeniable result is the "violation" of the topological rules which predict a triplet ground state for a *meta*-phenylene-type conformation. However, other diradicals constructed on a *meta*-phenylene basis also have a *singlet* ground state [Dvolaitzki *et al.*, 1992; Kanno *et al.*, 1993; Silverman and Dougherty, 1993; Yoshioka *et al.*, 1994; Fang *et al.*, 1995; Fujita *et al.*, 1996; Schultz *et al.*, 1997; Dei *et al.*, 2004]. In parallel to these experimental studies, the capacity of the *meta*-phenylene fragment to couple paramagnetic entities and the nature of the resulting ground state have also been the subject of theoretical works [Barone *et al.*, 2011]. Calculations performed on compounds bearing **NN** radicals rather than **IN** radicals and built on a central pyridine nucleus rather than a phenyl, concluded on the existence of an $S = 1$ ground state [Rajadurai *et al.*, 2003]. In addition to the effects of the molecular conformation, in particular of the dihedral angle between the *meta*-substituted radicals and the plane of the central phenyl, our results raise the question of the role played by the nature of the radical substituent, and in particular that of the spin polarisation between an **IN** radical and an **NN** radical, as well as that played by the heteroatom on the central group, which can be compared to that of the donor or acceptor substituents on the central phenyl.

In conclusion, it should be emphasised that EPR can be used to determine the *sign* and the *magnitude* of exchange coupling occurring within diradicals, and at the scale of isolated molecules. A study of the magnetic properties of isolated molecules, with similar objectives and exploiting a range of experimental techniques, including EPR, was performed on nitronyl-nitroxide-derived diradicals bound through a *carotenoid*-type linker [Stroh, 2002; Ziessel *et al.*, 2004; Stroh *et al.*, 2005]. Very strong antiferromagnetic couplings, greater than $k_B T$ at room temperature, were measured in these systems.

9.4.3 - Influence of the nature of the radical substituents on the J parameter in diradicals and triradicals

In this section we will examine how the nature of the radical groups influences the J parameter in diradicals (figure 9.9c) and triradicals (figure 9.13a) from series 9c. Additional details on these studies can be found in [Catala, 1999;

Catala *et al.*, 1999, 2001, 2005]. As diradical entities evidently constitute the building blocks for triradicals, the aim is to determine the J_1 parameter within the diradical fragment and the J_2 parameter between the diradical and the terminal radical in the triradicals (figure 9.16). The studies in dilute liquid solution described in sections 9.3.1 and 9.3.2 indicated that the radicals interact in the strong exchange limit at room temperature.

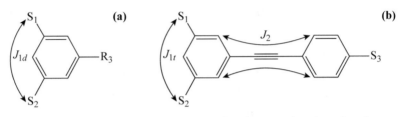

$$\hat{H}_{exch} = -J_{1d}\hat{\mathbf{S}}_1 \cdot \hat{\mathbf{S}}_2 \qquad \hat{H}_{exch} = -J_{1t}\hat{\mathbf{S}}_1 \cdot \hat{\mathbf{S}}_2 - J_2(\hat{\mathbf{S}}_1 \cdot \hat{\mathbf{S}}_3 + \hat{\mathbf{S}}_2 \cdot \hat{\mathbf{S}}_3)$$

Figure 9.16 - Definition of the exchange parameters.
(a) diradical: J_{1d}, **(b)** triradical: J_{1t} and J_2. The Hamiltonians used to calculate the energy levels and the magnetic susceptibility are indicated.

Low-temperature spectra for the **INNN** diradical and the **INNNpNN** triradical are shown in figure 9.17 (the spectrum for a **diNN** diradical is given in figure 6.8 in Volume 1). A half-field line is observed for all the oligoradicals studied. No line at one-third-field ($\Delta M_S = \pm 3$) was observed for triradicals, which is not surprising as the ratio of its intensity to that of the $\Delta M_S = \pm 1$ transitions is around $(D/h\nu)^4$.

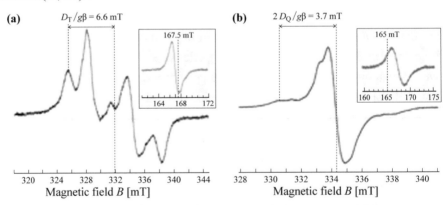

Figure 9.17 - (a) Spectrum produced by the **INNN** diradical (2.5×10^{-4} M in a xylene / CH_2Cl_2 mixture) at $T = 10$ K, **(b)** spectrum for the **INNNpNN** triradical (5×10^{-4} M in an acetone/CH_2Cl_2 mixture) at $T = 12$ K. The insets show the half-field lines. Microwave frequency: 9.39 GHz, power 0.1 mW for $\Delta M_S = \pm 1$ transitions, 20 mW for $\Delta M_S = \pm 2$ transitions. Modulation: frequency 100 kHz, peak-to-peak amplitude 0.3 mT.

The $D_T/g\beta$ and $3D_Q/g\beta$ values measured on the spectra are indicated in table 9.1. Relation [9.2] holds for the compound bearing only **NN** fragments, but not quite as well for compounds bearing one or more **IN** fragments. For these compounds, the geometry of the diradical fragment of the triradical compound could differ from that of the pure diradical. This observation may be linked to the effects of "disorder" induced by the different possible positions of the NO group in an imino-nitroxide in liquid solution (figure 9.11a).

Table 9.1 - Zero-field splitting parameters determined for the diradical fragment of triradicals ($3D_Q$) and the corresponding diradical (D_T).

Diradical	$D_T/g\beta$ [mT]	Triradical	$3D_Q/g\beta$ [mT]
diNN	10 ± 1	diNNpNN	9.6 ± 1.2
INNN	6.6 ± 0.4	INNNpNN	5.6 ± 0.3
diIN	9.0 ± 0.5	diINpNN	7.4 ± 0.4
		diINpIN	7.4 ± 0.4

The temperature-dependence of the susceptibility was monitored using different quantities measured on the EPR spectrum (integrated intensity of the $\Delta M_S = \pm 1$ signal, integrated intensity of the $\Delta M_S = \pm 2$ signal, peak-to-peak amplitude multiplied by the square of the linewidth) to determine the confidence intervals for the results obtained when the theoretical curves are fitted to the data (figure 9.18).

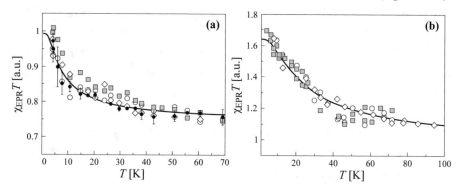

Figure 9.18 - Temperature-dependence of the product of EPR susceptibility and temperature for **(a)** the **diIN** diradical, **(b)** the **diNNpNN** triradical. Susceptibility was monitored by measuring the integrated intensity of the $\Delta M_S = \pm 1$ signal (white diamonds), the integrated intensity of the $\Delta M_S = \pm 2$ signal (grey squares), the peak-to-peak amplitude of the $\Delta M_S = \pm 2$ signal (white circles), the mean value of all experimental data (black circles). The continuous line curves were calculated by applying equation [9.3] for **(a)**, equation [9.5] for **(b)**, with the parameters listed in table 9.2.

[Adapted from Catala L. *et al.* (2001) *Chemistry - A European Journal* 7: 2466–2480 © 2001 Wiley-VCH Verlag GmbH, Weinheim, Fed. Rep. of Germany]

To describe the temperature-dependence of triradical susceptibility, we used the "isosceles triangle"-type Heisenberg Hamiltonian from figure 9.16. This produces two doublets ($S = \frac{1}{2}$) and a quartet ($S = \frac{3}{2}$) in an order depending on the sign of J_{1t}. The temperature-dependence of the magnetic susceptibility is described by the following relation [Belorizki, 1993; Belorizki and Fries, 1993; Fujita *et al.*, 1996]:

$$\chi(T) = \frac{C}{T}\left[\frac{1 + \exp(-\Delta_1/k_B T) + 10\exp(-\Delta_2/k_B T)}{1 + \exp(-\Delta_1/k_B T) + 2\exp(-\Delta_2/k_B T)}\right] \qquad [9.5]$$

In this expression, we set $\Delta_1 = 3J_2/2$ and $\Delta_2 = (2J_{1t} + J_2)/2$ (figure 9.16). The corresponding J values are indicated in table 9.2.

Table 9.2 - Exchange parameters (J_{1t}, J_2) for triradicals and (J_{1d}) for the corresponding diradical. Solution ($10^{-3} \sim 10^{-4}$ M) in a 1:1 mixture of (xylene/CH_2Cl_2) for diradicals and (acetone/CH_2Cl_2) for triradicals.

Exchange parameter	diNNpNN	lNNpNN	dilNpNN	dilNpIN
J_2/k_B [K]	$+7 \pm 2$	$+7 \pm 2$	$+1.0 \pm 0.5$	$+0.7 \pm 0.1$
J_{1t}/k_B [K]	$+23 \pm 4$	$+15 \pm 1$	$+6.5 \pm 4.0$	$+6.5 \pm 1.0$
	diNN	lNN	dilN	
J_{1d}/k_B [K]	$+23 \pm 5*$	$+19 \pm 6$	$+7.6 \pm 0.2$	

* $J_{1d}/k_B = +36 \pm 10$ K in a polystyrene matrix.

All the diradicals are observed to have an $S = 1$ ground state, and all the triradicals have an $S = \frac{3}{2}$ ground state, stabilised by *ferromagnetic* exchange interactions. This result correlates with the prediction of the alternation of spin density rule for the "*meta*" topology adopted. Within the limits of experimental uncertainty, the values of J_{1t} for the diradical fragments are equal to the values of J_{1d} estimated for isolated diradicals. The differences in conformation suggested by the data presented in table 9.1 for derivatives with an **IN** fragment are therefore undetectable. The nature of the substituted radical has a significant effect on the intensity of the exchange coupling:

▷ within a diradical series, thus for a *meta*-phenylene, a net decrease in J_{1d} is observed when **NN** is replaced by **IN**, in the order $J(\mathbf{diNN}) > J(\mathbf{INNN}) > J(\mathbf{diIN})$,

▷ within a triradical series, the exchange coupling (J_2) through the phenylene ethynylene linker (figure 9.16) is greater when it involves **NN** rather than **IN** radicals.

The decrease in the interactions when shifting from **NN** to **IN** can be attributed to spin polarisation effects which contribute to spin delocalisation on the molecule. These effects are known to be more important for **NN** than for **IN**. The factors determining the nature of the ground state and the origin of the coupling have been studied by theoretical calculations performed on several series of compounds; some of the results are presented in complement 2.

9.5 - Conclusion

Naturally, EPR spectroscopy cannot be presented as the universal tool to answer all the questions raised in the field of molecular magnetism. For example, magnetisation should always be measured by SQUID magnetometry when possible. Before performing these measurements, of course, it is necessary to determine the physical and chemical nature of the objects studied. This characterisation requires a large number of tools, even in its simplest form (chemical microanalysis, optical UV-visible and infrared spectroscopy, NMR, mass spectrometry, etc.). The structure or organisation of these objects if it exists, can be determined by X-ray diffraction, electron microscopy, scanning probe microscopy (AFM, STM). EPR should only be used after all these characterisation steps, otherwise the signals observed will be uninterpretable.

The relevance and power of EPR are due to it compatibility with a wide variety of samples (liquid or frozen solutions, powders, crystals) and its high sensitivity when examining the behaviour of isolated molecules through the study of dilute solutions. Among the topics not addressed here, we can mention the use of different frequencies of study which allow different aspects of a paramagnetic centre's local environment to be examined [Barra *et al.* 2005; Gatteschi *et al.*, 2006]. Finally, access to pulsed EPR and ENDOR spectroscopy further extends the orders of magnitude of the magnetic quantities accessible with this powerful technique (see appendix 2 and 3 at the end of this volume).

Acknowledgements

The results presented in this chapter are the fruit of PhD studies performed under the shared direction of chemists and physicists. I honour my former "PhD" companions, now colleagues [Wautelet, 1996; Catala, 1999; Stroh, 2002], who gave so much of their time to deal intelligently with the basic questions that are quite simple to formulate, as you have seen, using EPR spectroscopy as the main tool to probe their molecular constructs.

I also thank my collaborators at the Institut de Chimie, Strasbourg (POMAM laboratory, UMR 7177 University of Strasbourg-CNRS), for their critical reading of this manuscript: Sylvie Choua, Nathalie Parizel, Jérôme Tribollet and Bertrand Vileno. I particularly acknowledge Maxime Bernard, CNRS research engineer, all-weather, all-terrain EPR handler, experienced emergency interventionist, without whom much less would have been possible.

Complement 1 - Temperature-dependence of the susceptibility of dilute solutions monitored by EPR

Magnetisation is generally measured as a function of temperature (typically between 2 K and 300 K), and sometimes as a function of the magnetic field, using a SQUID magnetometer. However, for technical and sensitivity reasons, this instrument is not easily adapted for measurement of the magnetisation of dilute solutions of organic radicals. Technically, there is a high risk of breaking the sample-holder in the cryostat as a result of fusion/solidification of the solvent, which would require the measurement chamber to be dismantled for cleaning, resulting in immobilisation of the instrument. In addition, the magnetisation measured includes a paramagnetic component and a diamagnetic component. For organic compounds, these contributions are of the same order of magnitude. But because of the solvent, in dilute solutions the diamagnetic component is much greater than the paramagnetic contribution of the molecules. To avoid these pitfalls, the magnetic properties are often determined with polycrystalline powders. In polycrystals, molecular stacking often creates intermolecular interactions. These interactions, even when weak, make it impossible to determine the magnetic state of isolated molecules. Indeed, their effects appear at low temperature where the ground state of molecular systems, in which intramolecular magnetic interactions are generally weak, reveals itself.

EPR can be used to study isolated molecules in sufficiently dilute solutions to allow intermolecular interactions to be neglected, and the integrated intensity of the spectrum can be used to monitor the temperature-dependence of the magnetisation specifically produced by the paramagnetic entities. A number of experimental problems can be encountered with this type of experiment. For example:

▷ precise temperature measurement,

▷ problems posed by saturation of the EPR signal at low temperature,

▷ errors linked to the two successive integrations of the signal,

▷ limitation of the accessible temperature range due to the point of fusion for the solvent.

With regard to the accessible temperature range, if necessary the solution can be diluted in a polymer matrix and the mixture freeze-dried. As a result, the temperature range can be extended beyond the point of fusion for the solvent.

Determining the temperature of the sample is a real problem when working with a helium-flux cryostat (e.g. *Oxford Instruments'* ESR900/910) in which temperature gradients can become significant when the temperature increases due to a concomitant decrease in helium flux. The temperature probe (thermocouple, thermistance, etc.) in the regulation system is located relatively far from the sample support, and differences of several degrees are frequent beyond 30 K. In the laboratory, we have opted for a separate measurement system with, if possible, a probe immersed in the measurement solution. The problem arises less, or not at all, with cryostats in which the whole resonator is immersed in the helium bath. Ceramic resistors (e.g. Cernox™ from *Lake Shore*) installed in new generations of cryostats are also easier to use than systems with differential thermocouples. Study of the susceptibility deduced from the EPR signal requires a few precautions to ensure reliable estimation of the exchange coupling [Rajca, 1994; Berson, 1988; Borden, 1994]. The signal saturation problem [Volume 1, section 5.3], results from the fact that the signal produced by polyradical samples involving spins greater than ½ are often easily saturated at low temperature, typically at less than 20 K. As a result, the intensity decreases and sometimes the signal becomes distorted which can seriously hamper measurement of its integrated intensity. Measurements must therefore be performed at very low microwave power, which reduces the signal/noise ratio. This problem can be partially overcome by using the half-field signal for "forbidden" transitions ($\Delta M_S = \pm 2$) [Volume 1, section 6.5.3]. Indeed, this signal is generally saturated at a higher power than the "normal" signal due to the $\Delta M_S = \pm 1$ transition. However, it is much weaker, and must generally be overmodulated while performing accumulations [Weissman and Kothe, 1975]. The difficulties presented by the numerical integration of potentially weak signals, which rarely have linear baselines (empty resonator) devoid of parasitic signals can be resolved by determining the peak-to-peak amplitude for the signals observed, after having ensured that their shape and width do not depend on temperature. Indeed, the intensity obtained by two successive integrations of an EPR line is proportional to the product of the peak-to-peak amplitude A_{pp} multiplied by the square of the width ΔB_{pp} [Volume 1, section 5.2.1]:

$$\chi_{RPE} \propto \int_{B_{min}}^{B_{max}} \left[\int_{B_{min}}^{B} \frac{dY}{db}(b - B_0)\,db \right] dB \propto A_{pp}\Delta B_{pp}^2$$

If the shape and linewidth are constant, which is generally the case for isolated molecules in frozen solution, A_{pp} is proportional to the integrated intensity.

To finish, we would like to highlight that several modes of representation are possible to reveal the deviation of the susceptibility with respect to Curie's law (χ proportional to $1/T$), and consequently the existence of exchange coupling. The most frequently used are the following:

▷ the inverse of the susceptibility as a function of temperature ("physicists'" representation),

▷ the product of temperature and susceptibility as a function of temperature ("chemists'" representation), which reveals ferromagnetic or antiferromagnetic coupling depending on whether the curve is concave or convex with respect to the constant defined by Curie's law,

▷ the susceptibility as a function of the inverse of temperature. In this representation, a lack of deviation from Curie's law indicates that the ground state is a triplet with a significant triplet-singlet gap, or that singlet and triplet states are quasi degenerate (figure 9.15) [Rajca, 1994; Borden *et al.*, 1994; Berson, 1998; Matsuda and Iwamura, 1998].

Complement 2 - What molecular orbital calculations provide

The description of the intramolecular exchange interaction mechanism generally involves electron spin delocalisation on the molecule under investigation. The underlying concepts are linked to spin polarisation and conjugation, mechanisms which are sensitive to the molecular conformation (e.g. dihedral angles, ortho/meta/para substitution topology). Molecular orbital calculations are thus often useful, sometimes essential, when estimating the distribution of the spin density and exchange interaction, along with the effects of the molecular geometry. These calculations have the advantage of being capable of simulating situations which would be difficult to study experimentally and/or implement in terms of chemical synthesis. This approach is illustrated here through examples of calculations performed on the series of oligoradicals studied in this chapter. The results are presented without mention of the approximations used or the tests performed to validate these calculations. Further details can be found in [Catala, 1999; Wautelet *et al.*, 2001; Catala *et al.*, 2001, 2005].

Methods

The molecular geometry, distribution of the spin density, nature of the ground state and the barriers to rotation are strongly dependent on the dihedral angle between the phenyl group and the groups bearing radical sites. The nature of the ground state and the intensity of the exchange coupling were thus determined for all the oligoradicals studied. For the **pmp-IN$_R$** and ***n*p-IN** series, semi-empirical methods (MOPAC software [Stewart, 1990, 1993]) were used which allow, for example, the length of the linker to be extended to $n = 10$ for ***n*p-IN** [Wautelet *et al.*, 2001]. In this case, the energy levels were calculated with configuration interaction (CI), i.e., taking stabilisation of the ground state by the excited states into account. For the series of mixed diradicals and triradicals, calculations were performed by density functional theory (DFT) using GAUSSIAN 98 software [Frisch *et al.*, 1998] and exchange coupling was estimated using the "broken symmetry" method [Noodleman, 1981; Noodleman and Davidson, 1986; Noodleman and Case, 1992].

Results

Figure 9.19a shows the energy of the rotation barrier involving the dihedral angle between the **IN** radical fragment and the phenyl cycle for **pmp-IN$_{OCH_3}$**

and **np-IN** derivatives. Calculations indicate that a broad range of molecular geometries are accessible at room temperature ($k_B T \approx 0.58$ kcal mol^{-1} for $T = 300$ K) in a range covering rotation over more than 100 degrees. This result confirms that the exchange parameters deduced from the intensity measurements in frozen solution must be considered *average values* (methods subsection in section 9.4.2). The effect of excited states on stabilisation of the ground state is very important as the spin of the ground state shifts from $S = 1$ to $S = 0$ when the dimension of the space used to calculate the configuration interaction is increased (figure 9.19b) [Bieber *et al.*, 2003]. This dimension is represented here by the number CI = n for a space of configurations $\{n, n\}$ including the two singly-occupied orbitals in the radicals (SOMO) as well as the $(n-2)/2$ doubly-occupied molecular orbitals, and the $(n-2)/2$ unoccupied molecular orbitals. It is interesting to note that these calculations also account for the notion of superexchange over a long distance with an exponential decrease in the value of J for **np-IN** where n varies from 1 to 10 (equation [9.4]), as well as the influence of the nature of the radical (**IN** or **NN**) on the value of J [Bieber *et al.*, 2003].

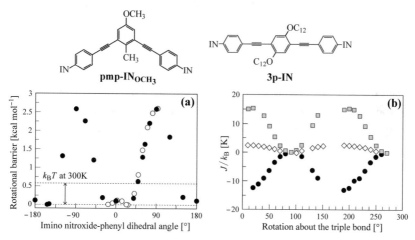

Figure 9.19 - Results of semi-empirical calculations.
(a) Energy of the rotation barrier as a function of the dihedral angle between the **IN** cycle and the phenyl cycle for **pmp-IN$_{OCH_3}$** (black circles) and **3p-IN** (white circles) compounds. The value of $k_B T$ at 300 K is indicated.
(b) Calculation of the exchange parameter J/k_B as a function of the angle of rotation around the triple bond for the **pmp-IN$_{OCH_3}$** compound for various levels of configuration interactions identified by their CI number. The triplet state is stabilised for CI = 4 (grey squares) and CI = 6 (white diamonds), whereas the singlet state is stabilised for CI = 8 (black circles). [From Bieber *et al.*, 2003]

For the series of mixed diradicals (figure 9.9c) and triradicals based on **IN** and **NN** (figure 9.13a), exchange coupling was calculated by applying the DFT (B3LYP) method (table 9.3). Qualitatively, the results agree with the experimental results presented in section 9.4.3 and predict, in particular, a ground state with $S = 1$ for all the diradicals studied. Like in the case of semi-empirical calculations (figure 9.19b), the calculated values are greater than the experimental values (table 9.3). Nevertheless, these calculations correctly reproduce the hierarchy of the interactions: replacement of an **NN** radical by an **IN** radical within a diradical or a triradical systematically leads to a reduction in the interactions (table 9.3).

Table 9.3 - Exchange coupling calculated using the GAUSSIAN code for mixed **IN/NN** diradical and triradical compounds. ΔE_{D-Q} is the difference between the $S = \frac{1}{2}$ and $S = \frac{3}{2}$ levels in triradicals, ΔE_{S-T} is the difference between the $S = 0$ and $S = 1$ levels in diradicals corresponding to the diradical fragment of the triradical. Corresponding experimental values (table 9.2) are indicated in brackets.

	diNNpNN	INNNpNN	diINpNN	diINpIN
ΔE_{D-Q} [K] *	67 (27)	40 (19)	8 (7)	16 (7)
	diNN	INNN	diIN	
ΔE_{S-T} [K] *	94 (23)	38 (19)	11(8)	

* The ground state is $S = \frac{3}{2}$ for triradicals, $S = 1$ for diradicals.

The role of spin polarisation in coupling is observed through the correlation between the value of J and the spin density calculated for the carbon atom noted C2 in the inset to figure 9.20a. The alternation of the sign for the spin density along the conjugated linker is also revealed (figure 9.20b).

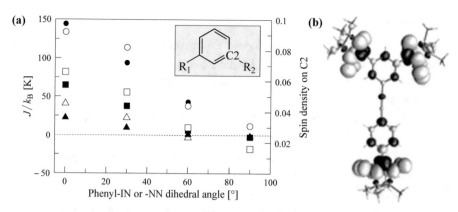

Figure 9.20 - (a) Dependence of the value of J (filled symbols, left-hand scale) and the spin density on carbon C2 in the m-phenylene (open symbols, right-hand scale) on the dihedral angle between the phenyl and the **IN** or **NN** radical, for **diNN** (circles), **INNN** (squares) and **diIN** (triangles) diradicals.

(b) Distribution of the spin density on the **diINpNN** triradical represented by the surfaces with a density of 0.002 atomic units (1 atomic unit of electron density is equivalent to around 6.7 electrons $\overset{\circ}{A}^{-3}$). Positive values are indicated in white, negative values in black. DFT calculations were performed using GAUSSIAN 98. [From Catala L. *et al.* (2001) *Chemistry - A European Journal* **7**: 2466–2480 © 2001 Wiley-VCH Verlag GmbH, Weinheim, Fed. Rep. of Germany]

References

ALSTER E. & SILVER B.L. (1986) *Molecular Physics*, **58**: 977-987.

ANDERSON P.W. (1963) "Exchange in insulators: superexchange, direct exchange and double exchange" in *Magnetism* vol II, Rado G.T. et Suhl H., eds., Academic Press, New York.

BARONE V. *et al.* (2011) *Journal of Chemical Theory and Computation* **7**: 699-706.

BARRA A.-L. *et al.* (2005) *Magnetic Resonance in Chemistry* **43**: S183-S191.

BELORIZKY E. & FRIES P.H. (1993) *Journal de Chimie Physique et de Physico-Chimie Biologique* **90**: 1077-1100.

BELORIZKY E. (1993) *Journal de Physique* **3**: 423-445.

BENCINI A. & GATTESCHI D. (1990) *EPR of Exchange Coupled Systems*, Springer-Verlag, Berlin.

BERSON J.A. (1988) in *The Chemistry of the Quinonoid Compounds*, vol. II, Patai S., Rappoport Z., eds, Wiley, New York.

BIEBER A., WAUTELET P., ANDRÉ J.-J. & TUREK P. (2003), unpublished results.

BLEANEY B. & BOWERS K.D. (1952) *Proceedings of the Royal Society of London. Series A* **214**: 451-465.

BOGANI L. & WERNSDORFER W. (2008) *Nature Materials* **7**: 179-186.

BONNET M. *et al.* (1995) *Molecular Crystals & Liquid Crystals* **271**: 35-53.

BORDEN W.T., IWAMURA H. & BERSON J.A. (1994) *Accounts of Chemical Research* **27**: 109-116.

BREITENKAMP R.B. & TEW G.N. (2004) *Macromolecules* **37**: 1163-1165.

BRIÈRE R. *et al.* (1965) *Bulletin de la Société Chimique de France* 3290-3297.

BRINKMANN M., TUREK P. & ANDRÉ J.-J. (1997) *Thin Solid Films* **303**: 107-116.

BRINKMANN M. *et al.* (2004) *Journal of Physical Chemistry A* **108**: 8170-8179.

CALVO R. (2000) *Journal of the American Chemical* Society, **122**: 7327-7341.

CATALA L. (1999) PhD thesis, *Nitronyl-nitroxide and imino-nitroxide oligoradicals: synthesis and study of the magnetic properties in the isolated state and in the crystalline phase*, Université Louis Pasteur, Strasbourg.

CATALA L. *et al.* (2005) *Chemistry - A European Journal* **11**: 2440-2454.

CATALA L. *et al.* (2001) *Chemistry - A European Journal* **7**: 2466-2480.

CATALA L. & TUREK P. (1999) *Journal de Chimie Physique et de Physico-Chimie Biologique* **96**: 1551-1558.

CHITTIPEDDI S. *et al.* (1987) *Physical Review Letters* **58**: 2695-2698.

CHOI S. H., KIM B. & FRISBIE C.D. (2008) *Science* **320**: 1482-1486.

CHU Q. & PANG Y. (2003) *Macromolecules* **36**: 4614-4618.

CIRUJEDA J. *et al.* (1995) *Journal of Materials Chemistry* **5**: 243-252.

CRAYSTON J.A., DEVINE J.N. & WALTON J.C. (2000) *Tetrahedron* **56**: 7829-7857.

DEI A. *et al.* (2004) *Journal of Magnetism and Magnetic Materials* **272-276**: 1083-1084.

DVOLAITZKY M., CHIARELLI R. & RASSAT A. (1992) *Angewandte Chemie International Edition* **31**: 180-181.

FANG S. *et al.* (1995) *Journal of the American Chemical Society* **117**: 6727-6731.

FITTIPALDI M. *et al.* (2009) *Physical Chemistry Chemical Physics* **11**: 6555-6568.

FRISCH M. J. *et al.* (1998) *Gaussian*, Inc., Pittsburgh PA.

FUJITA J. *et al.* (1996) *Journal of the American Chemical Society* **118**: 9347-9351.

GALLANI J.-L. *et al.* (2001) *Langmuir* **17**: 1104-1109.

GATTESCHI D. *et al.* (2006) *Coordination Chemistry Review* **250**: 1514-1529.

GLARUM S.H. & MARSHALL J.H. (1967) *Journal of Chemical Physics* **47**: 1374-1378.

HERNANDEZ-GASIÒ E. *et al.* (1994) *Chemistry of Materials* **6**: 2398-2411.

HIGASHIGUCHI K., YUMOTO K. & MATSUDA K. (2010) *Organic Letters* **12**: 5284-5286.

HIREL C. *et al.* (2005) *European Journal of Organic Chemistry* 348-359.

ITOH K. & KINOSHITA M. (2000) *Molecular Magnetism: New Magnetic Materials*, Kodansha et Gordon and Breach Science Publishers, Tokyo et Amsterdam.

KAHN O. & BRIAT B. (1976) *Journal of the Chemical Society Faraday Transactions II* **72**: 268-281.

KAHN O. (1993) *Molecular Magnetism*, VCH Publishers, New York.

KANNO F. *et al.* (1993) *Journal of the American Chemical Society* **115**: 847-850.

KINOSHITA M. *et al.*(1991) *Chemistry Letters* 1225-1228.

KYATSKAYA S. *et al.* (2009) *Journal of the American Chemical Society* **131**: 15143-15151.

LAHTI P. M. (1999) *Magnetic Properties of Organic Materials*, Marcel Dekker, New York.

LIU K. *et al.* (2008) *Journal of Physical Chemistry C* **112**: 4342-4349.

LU Q. *et al.* (2009) *ACS Nano* **3**: 3861-3868.

LUCKHURST G.R. & PEDULLI G. F. (1971) *Molecular Physics* **20**: 1043-1055.

MATSUDA K. & IWAMURA H. (1998) *Journal of the Chemical Society Perkin Transactions* **2**: 1023-1026.

METZNER K.E., LIBERTINI L.J. & CALVIN M. (1977) *Journal of the American Chemical Society* **99**: 4500-4502.

MILLER J.S. & DRILLON M. (2001-2005) *Magnetism: Molecules to Materials*, volumes I-V, Wiley-VCH-Weinheim.

MILLER J.S. & EPSTEIN A.J. (1994) *Angewandte Chemie International Edition* **33**: 385-415.

NOODLEMAN L. & CASE D.A. (1992) *Advances in Inorganic Chemistry* **38**: 423-470.

NOODLEMAN L. & DAVIDSON E. (1986) *Chemical Physics* **109**: 131-143.

NOODLEMAN L. (1981) *Journal of Chemical Physics* **74**: 5737-5743.

ONOFRIO N. & MOUESCA J.-M. (2010) *Journal of Physical Chemistry A* **114**: 6149-6156.

RAJADURAI C. *et al.*(2003) *Journal of Organic Chemistry* **68**: 9907-9915.

RAJCA A. (1994) *Chemical Reviews* **94**: 871-893.

SCHULTZ D., BOAL A.K. & FARMER G.T. (1997) *Journal of the American Chemical Society* **119**: 3846-3847.

SESSOLI R. *et al.* (1993) (a) *Nature* **365**: 141-143.

SESSOLI R. *et al.* (1993) (b) *Journal of the American Chemical Society* **115**: 1804-1816

SILVERMAN S.K. & DOUGHERTY D. (1993) *Journal of Physical Chemistry* **97**: 13273-13283.

SORACE L. *et al.* (2011) *Chemical Society Review* **40**: 3092-3104.

STEWART J.J.P. (1990) *MOPAC* 6.0 QCPE 455.

STEWART J.J.P. (1993) *MOPAC 93* Fujitsu Lim., Tokyo.

STROH C. (2002) PhD thesis, *Magnetic interactions in molecular systems containing unpaired electrons*, Université Louis Pasteur, Strasbourg.

STROH C. *et al.* (2004) *Chemical Communications* 2050-2051.

STROH C. *et al.* (2005) *Journal of Materials Chemistry* **15**: 850-858.

TAKAHASHI M. *et al.* (1991) *Physical Review Letters* **67**: 746-748.

TAKUI T. *et al.* (1995) *Molecular Crystals & Liquid* Crystals **271**: 55-66.

TAMURA M. *et al.* (1991) *Chemical Physics Letters* **186**: 401-404.

THOMAS L. *et al.* (1996) *Nature* **383**: 145-147.

TROIANI F. *et al.* (2010) *Nanotechnology* **21**: 274009-18.

TUREK P. *et al.* (1991) *Chemical Physics Letters* **180**: 327-331.

ULLMAN E.F. *et al.* (1972) *Journal of the American Chemical Society* **94**: 7049-7059.

Volume 1: BERTRAND P. (2020) *Electron Paramagnetic Resonance Spectroscopy - Fundamentals*, Springer, Heidelberg.

WAUTELET P. (1996) PhD thesis, *Synthesis and characterisation of phenylene ethynylene-type oligomers bearing stable organic radicals. Study of the long-distance intramolecular magnetic coupling*, Université Louis Pasteur, Strasbourg.

WAUTELET P. *et al.* (2001) *Polyhedron* **20**: 1571-1576.

WAUTELET P. *et al.* (2003) *Journal of Organic Chemistry* **68**: 8025-8036.

WEISSMAN S.I. & KOTHE G. (1975) *Journal of the American Chemical Society* **97**: 2537-2538.

YOON H. P. *et al.* (2010) *Nano Letters* **10**: 2897-2902.

YOSHIOKA N. *et al.* (1994) *Journal of Organic Chemistry* **59**: 4272-4280.

ZHELUDEV A. *et al.* (1994) *Journal of the American Chemical Society* **116**: 2019-2027.

ZIESSEL R. *et al.* (2004) *Journal of the American Chemical Society* **126**: 12604-12613.

Chapter 10

EPR of short-lived magnetic species

Maurel V. and Gambarelli S.

Laboratory of Magnetic Resonance, DSM/INAC/SCIB/LRM, CEA, Grenoble.

10.1 - Introduction

Acquisition of an EPR spectrum on a classical spectrometer generally lasts between a few seconds and a few hours, depending on the nature of the system studied, its concentration, and the desired signal-to-noise ratio. For this type of experiment to make sense, it is of course essential that the state of the sample remains relatively stable during this period of time. However, some paramagnetic centres – such as free radicals in solution, intermediates of chemical or photochemical reactions, photoexcited states – on which EPR can reveal interesting information, have a much shorter lifetime. For example, the lifetime for carbonated radicals in solution at room temperature is often less than a millisecond; molecules in a triplet state often persist for just 100 ms. The spin trapping method described in chapter 3 can be used to *detect* and *identify* free radicals using a conventional EPR spectrometer. With specific setups, the *spectrum* for paramagnetic centres with short lifetimes can be determined and their evolution can be monitored over time.

Determining an EPR spectrum as a function of time is equivalent to acquiring a signal as a function of two parameters: time t and magnetic field B. This 2D signal can be obtained in two ways (figure 10.1):

▷ "iso-t" mode: the EPR spectrum is recorded for a set value of t and the 2D signal is reconstituted by repeating the experiment with different values of t,

P. Bertrand, *Electron Paramagnetic Resonance Spectroscopy*,
https://doi.org/10.1007/978-3-030-39668-8_10

▷ "iso-B" mode: the evolution over time of the EPR spectrum is recorded for a set value of B and the 2D signal is reconstituted by repeating the experiment for different values of B.

Experimentally, two very different strategies can be applied. In the first, described in section 10.2, the evolution of the reactive system is blocked by freezing it or creating a steady-state, allowing its EPR spectrum to be recorded on a standard spectrometer. The other possibility consists in accelerating the acquisition of the EPR spectrum. This is known as *time-resolved EPR*, and is presented in section 10.3. The applications of this technique are then described for systems in solution (section 10.4) and in the solid state (section 10.5).

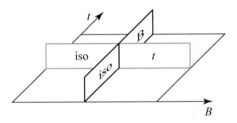

Figure 10.1 - The 2 possible methods by which a signal in the (B, t) plane can be acquired. A series of "iso-B" experiments or a series of "iso-t" experiments.

10.2 - Freeze quench and flow methods

In this section, we describe two classical approaches used to track the kinetics of some reactions in solution with a conventional EPR spectrometer. Their time-resolution is in the range 0.1 to 1 millisecond.

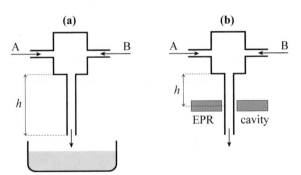

Figure 10.2 - Principle of the freeze quench and flow methods. **(a)** Freeze quench. Reagents A and B are mixed rapidly and react for a duration t, determined by the length h and the flow-rate. The mixture is snap-frozen and its spectrum recorded. **(b)** Flow method. The mixture circulates in a capillary which passes through the EPR cavity. See text.

10.2.1 - Freeze quench

The *freeze quench* technique was developed in the 1960s to monitor kinetics involving biological molecules containing metal centres [Bray, 1961; Ballou and Palmer, 1974]. It consists in rapidly mixing two reagents, leaving them to react for a pre-determined duration, t, then quenching the state of the mixture by freezing. In practice, two pistons controlled by compressed gas simultaneously project the reagents into a container; turbulent flow ensures rapid mixing. After traversing a capillary of length h, the mixture is ejected into an isopentane solution or onto a liquid-nitrogen-cooled metal block (figure 10.2a). The frozen particles are collected in an EPR tube and their spectrum is recorded at low temperature. The spectrum gives the composition of the mixture at time $t = h/v$ where v is the flow velocity in the tube. Freeze quench is therefore an "iso-t"-type experiment. To monitor the kinetics of the reaction, the experiment must be repeated for different values of t, by varying the length h or the flow-rate. The mixing and cooling times set the minimum time resolution at 0.1 ms. Its applications mainly relate to paramagnetic centres with an EPR spectrum only observable at low temperature, for which the flow methods described hereafter are not applicable.

This method was used to determine the kinetics of the electron-transfer chain for a membrane-bound protein complex (see appendix 4) [Verkhovskaya *et al.*, 2008]. This chain contains 8 iron-sulfur centres spaced around 12 Å apart. These centres are diamagnetic in the oxidised state, but *paramagnetic in the reduced state*, with $S = \frac{1}{2}$, and their spectra were determined by potentiometric titrations monitored by EPR[1]. Electron transfer was initiated by mixing a solution of complexes and the physiological reductant, NADH. The signals represented in figure 10.3 were obtained by optimising the time resolution.

---

[1] The EPR properties of iron-sulfur centres are presented in complement 2 to chapter 6, and section 6.3.2 describes an example of titration.

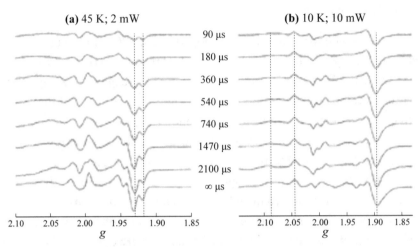

Figure 10.3 - EPR spectra obtained in an ultra-fast freeze quench experiment performed on complex 1 from the bacterium *E. coli*. A very precise experimental setup was used to achieve a time resolution of a few tens of microseconds. [From Verkhovskaya M.L. *et al.* (2008) *PNAS* 105: 3763–3767 © 2008 National Academy of Sciences, USA]

These signals are the result of superposition of the spectra for centres which are *reduced* at the times considered, and their simulation can be used to determine the reduction kinetics for each iron-sulfur centre. To simplify signal interpretation, the authors recorded the spectrum at several temperatures and several powers to make it possible to exploit the differences in relaxation properties for the centres: at 10 K the spectra for some centres are completely saturated at a power of 10 mW; at 45 K other spectra are no longer visible due to relaxation broadening [Volume 1, section 5.4.3].

10.2.2 - Continuous flow and stopped flow methods

Flow methods were developed in the 1920s to monitor the kinetics of chemical reactions [Hartridge and Roughton, 1923]. These methods were initially used with UV-visible absorption spectrometry, but they are currently applied with a wide variety of spectroscopies. EPR applications of these methods first emerged in the 1960s [Borg, 1964]. They can be used to study paramagnetic species for which a spectrum is observable in solution at room temperature, such as many radicals. In this approach, the two reagents are very rapidly mixed and the mixture flows through a capillary which passes through the EPR cavity (figure 10.2b). Two modes are possible:

◇ Continuous flow

The conditions are adapted to produce a *steady state* flow in the tube. The composition of the mixture present in the cavity is time-independent, and its spectrum can thus be recorded. As with the freeze quench method, the reagents remain in contact for $t = h/v$, and the experiment must be repeated for different values of t by varying the length of the capillary, h, or the flow-rate. This method gives good time resolution, but relatively large amounts of reagents are required to obtain steady state flow.

◇ Stopped flow

The completion of filling of the volume located in the cavity triggers the end to the flow. The composition of the mixture located in the cavity continues to evolve, and the amplitude of the EPR spectrum is recorded over time at a fixed B value ("iso-B" mode). To reconstitute the evolution of the EPR spectrum over time, the experiment must be repeated for a set of B values. Because of the blocking time, the time resolution with this method is not as good as with the continuous flow method.

Replacing the cavities by very small-volume resonators (*Loop Gap Resonators*, *Dielectric Resonators*) has made it possible to miniaturise these experiments. As a result, their time resolution has improved and the volumes of reagents required both in continuous flow [Grigoryants *et al.*, 2000; DeWeerd *et al.*, 2001] and stopped flow [Hubbell *et al.*, 1987; Sienkiewicz *et al.*, 1994, 1999; Lassman *et al.*, 2005] have been considerably reduced. With all these methods, the experimental setup must be calibrated based on reactions which are kinetically well-characterised.

10.3 - How can acquisition of an EPR spectrum be accelerated?

10.3.1 - Theoretical time resolution of an EPR experiment

Recording the spectrum for a paramagnetic species with a very short lifetime imposes severe technical constraints, and it also complicates the interpretation of the signals observed. Interaction of an electromagnetic wave with a paramagnetic centre generates a transient regime which evolves rapidly towards a steady-state [Volume 1, section 5.3.1]. When a continuous wave EPR spectrum is recorded by scanning the magnetic field, this steady-state is constantly attained unless a very high modulation frequency is used [Volume 1, section 9.2.1]. However, to detect species with very short lifetimes, the EPR signal must be observed as soon as possible after creation of the radicals, often during the transient regime where the lines are *broadened*. The magnitude of the broadening can be assessed using the time-energy uncertainty relation. If the signal is observed at a time τ after the creation of the radicals, the uncertainty for the transition energy is approximately given by:

$$\Delta E\, \tau \approx h \qquad\qquad [10.1]$$

This results in broadening $\Delta B = \Delta E / g\beta$ of the resonance line. For $g = 2.00$, this relation gives, for example, $\Delta B = 0.36$ mT for $\tau = 100$ ns.

In reality, the values for the relaxation times T_1, T_2 and the B_1 component of the electromagnetic field are also involved in the transient regime. In complement 1 we show how this regime can be modelled by determining the evolution of the EPR line over time by numerically solving the Bloch equations.

10.3.2 - Practical time resolution

In practice, the time resolution of conventional spectrometers is often limited by factors other than those mentioned above. The most significant are the time necessary to introduce the sample into the cavity and to adjust the coupling, the speed sweep for the magnetic field and the use of synchronous detection associated with field modulation. We will rapidly examine the constraints imposed by these various factors.

 ▷ The introduction of a sample into the cavity and adjustment of its coupling are often limiting factors in an EPR experiment. Indeed, it is generally impossible to perform these operations in less than a few seconds.

 ▷ Because of inductive phenomena, it is difficult to rapidly change the value of the magnetic field produced by an electromagnet. Even with *rapid scan* coils, sweep speeds are generally limited to a maximum of 100 mT s^{-1} with commercial EPR spectrometers. As a result, an EPR spectrum covering a 10 mT range cannot be recorded in less than 100 ms.

 ▷ To increase the signal-to-noise ratio, most commercial spectrometers use magnetic field modulation associated with synchronous detection. A weak sinusoidal magnetic field with a frequency between 100 Hz and 100 kHz is superposed on the magnetic field produced by the electromagnet. The EPR signal detected represents the variation of the total signal which oscillates at the same frequency and in phase with this variable field [Volume 1, chapter 1, complement 3]. A setup known as "synchronous detection" can be used to extract the derivative of the absorption signal from this signal [Volume 1, section 9.2]. For synchronous detection to function correctly, several oscillations of the signal must be observed, which further limits the time resolution. For example, with a modulation frequency of 100 kHz, it takes 100 µs to observe 10 oscillations.

To markedly increase the time resolution of a spectrometer, new acquisition techniques must be developed in which these constraints do not apply.

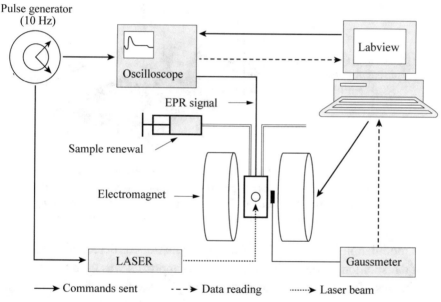

Figure 10.4 - Principle of a time-resolved continuous wave EPR experiment. The species studied are generated in situ by laser pulses. If the sample is a liquid, it must be regularly renewed during the experiment. [From Maurel V. (2004)]

We will review these points (figure 10.4):

▷ The reactive species must be generated *in situ* in the EPR cavity. Several methods exist, but the most frequently used consists in illuminating photo-active molecules with a laser pulse at a visible or UV wavelength. Lasers delivering 5-ns pulses can deliver sufficient power to generate an adequate number of paramagnetic species, despite modest quantum yields.

▷ The evolution of the amplitude of the EPR signal over time must be recorded *at constant field* (iso-*B* experiment). To monitor evolution of the spectrum, the experiment is repeated for a set of values for the magnetic field (figure 10.5).

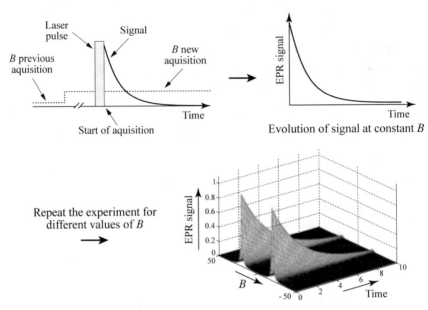

Figure 10.5 - Acquisition of a TR-EPR spectrum with a continuous wave spectrometer. For each field value, the species are generated *in situ* in the cavity and the signal is recorded as a function of time. The spectrum is constructed by repeating the experiment for different B values. [From Maurel V. (2004)]

▷ A modulation frequency of 100 kHz and synchronous detection can be used if a resolution of 100 µs is sufficient. This approach was used to record the kinetics of disappearance of signal 2 for photosystem II for the first time [Blankenship *et al.*, 1975] (section 6.4.1 in chapter 6). To further increase the time resolution, frequency modulation must be increased, or modulation eliminated and direct detection performed. For technical reasons, it is difficult to use a modulation frequency greater than 1 MHz, which corresponds to a resolution of 10 µs. If a better resolution is required, the modulation must be eliminated and the detection made direct. The EPR spectrometer will be much less sensitive, but the speed of acquisition is only limited by the time constants for the electronic circuits (a few ns).

By generating the paramagnetic species *in situ* using laser pulses, and monitoring the kinetics at constant field with direct detection, the theoretical time resolution discussed in section 10.3.1 can be achieved. This corresponds to *time-resolved EPR* (TR-EPR) per se.

TR-EPR experiments can also be performed with a pulsed spectrometer, by proceeding as indicated in figure 10.6. Pulsed EPR is presented in appendix to this book.

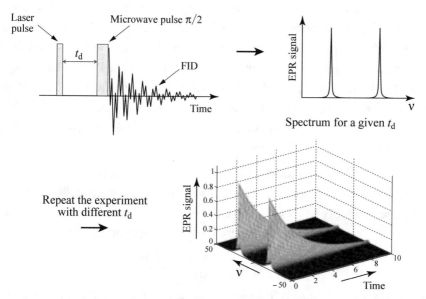

Figure 10.6 - Acquisition of a TR-EPR spectrum with a pulsed spectrometer. For each time value t_d after generation of the paramagnetic species, a series of microwave pulses is applied and the resulting signal is acquired. In the simplest case, a free induction decay (FID) signal is obtained (see appendix 1 to this book), the Fourier transform of which supplies the spectrum as a function of frequency or, as a function of the magnetic field (iso-t spectrum) (which is equivalent in this case). [From Maurel V. (2004)]

10.4 - Time-resolved EPR for radicals in solution

10.4.1 - Electron spin polarisation: the CIDEP effect

The spectra obtained by TR-EPR in the first microseconds following the creation of radicals by a photochemical process (typically by nanosecond laser pulses) are very different to those obtained in traditional EPR experiments. Whether recorded by continuous wave or pulsed TR-EPR, these spectra correspond to the *absorption signal* as modulation of the magnetic field or synchronous detection are generally not used at these time scales. The *positions* of the lines are determined by the hyperfine interactions in the radicals studied, but their *intensities* do not follow the usual rules. For example, the spectra can contain alternating *emission* and *absorption* lines (figures 10.7 and 10.8) or simply *emission* lines (figure 10.10). The inhabitual appearance of these spectra is due to the fact that the electron spins of the photochemically-produced radicals are *not at thermal equilibrium* in the first instants following their formation. This effect is known as CIDEP for *Chemically Induced Dynamical Electron spin Polarization*. It was first observed on the EPR signals for hydrogen atoms generated during a pulsed radiolysis experiment [Fessenden and Schuler, 1963]. Since then, the CIDEP effect has mainly been studied in photochemistry and it is the subject of numerous articles and reviews [McLauchlan, 1988, 1989; van Willigen *et al.*, 1993; Turro *et al.*, 2000]. In this section, we will define the notion of *spin polarisation*. In subsequent sections, we will present two examples of TR-EPR experiments involving photoinduced free radicals and describe the mechanisms likely to produce a CIDEP effect.

Consider an EPR line produced by a collection of N radicals. Its intensity is proportional to the difference in populations N_β and N_α of the radicals, characterised respectively by $M_S = -\frac{1}{2}$ and $M_S = +\frac{1}{2}$. At thermal equilibrium and at resonance, this difference is given by [Volume 1, section 1.4.4]:

$$N_\beta^0 - N_\alpha^0 = N\frac{\exp(h\nu/k_B T) - 1}{\exp(h\nu/k_B T) + 1}$$

where ν is the frequency of the spectrometer. For $T = 298$ K and $\nu = 9.5$ GHz, we obtain for example:

$$N_\beta^0 - N_\alpha^0 = 7.64 \times 10^{-4} N$$

When the radicals are subjected to the CIDEP effect, the N_β and N_α populations differ from those present at thermal equilibrium, and the intensity of the EPR line is multiplied by a *polarisation* factor defined by:

$$P = \frac{N_\beta - N_\alpha}{N_\beta^0 - N_\alpha^0}$$

If P is positive, an *absorption* line is obtained (in "increased" absorption if $P > 1$), if it is negative it is an *emission* line. The P factor can be very large, up to around 100 in favourable cases [Muus *et al.*, 1978; Forbes, 1997]. Polarisation due to the CIDEP effect is therefore often determinant for the detection of a TR-EPR signal at microsecond scale. Without this amplification of the signal, most of the radicals described in the literature could not be observed by these techniques. The shape of the spectrum evolves towards the spectrum at thermal equilibrium; the kinetics of this evolution is characterised by the spin-lattice relaxation time, T_1, which is typically around a microsecond for free radicals in solution at room temperature. The evolution over time of time-resolved EPR signals which benefit from a CIDEP effect therefore depends not only on the lifetime of the paramagnetic species observed, but also on their spin dynamics.

10.4.2 - Examples of spectra for photoinduced radicals in microsecond TR-EPR

◇ Photolysis of acetone in ethylene glycol

The photochemistry of ketones in alcohols has been studied in detail, and the radical intermediates created have frequently been observed by TR-EPR [Levstein and van Willigen, 1991]. Figure 10.7 represents the TR-EPR spectrum recorded 1 µs after a laser pulse, during photoreduction of acetone in ethylene glycol.

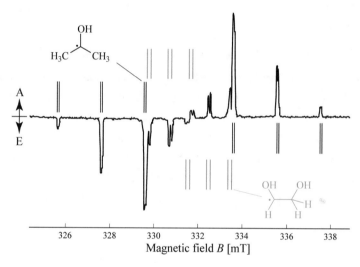

Figure 10.7 - Continuous wave TR-EPR spectrum recorded during the photolysis of acetone in ethylene glycol (2 mL in 100 mL) 1 µs after the laser pulse. The grey lines correspond to the signal for the **s1** radical, the black lines correspond to the **1** radical. The arrows A and E indicate absorption and emission lines. Laser power: 20 mJ/pulse, integration window: 50 ns. Frequency: 9.6 GHz, microwave power: 1 mW. [From Maurel V. (2004)]

The reaction observed is the following:

The spectrum presents emission signals at low-field and absorption signals at high-field, and it has an almost perfectly centro-symmetric appearance. Its shape is typical of a radical pair CIDEP mechanism, as we will see in section 10.4.3.

▷ The **s1** radical, derived from ethylene glycol, produces two triplets of doublets. The corresponding hyperfine couplings are clearly identifiable, with $a_{H, CH} = 1.74$ mT, $a_{2H, CH_2} = 0.99$ mT and $a_{H, OH} = 0.11$ mT. However, the relative intensities of the lines do not follow the expected $(1:1:2:2:1:1:1:1:2:2:1:1)$ pattern.

▷ The **1** radical, derived from acetone, contains 6 equivalent protons (2 CH_3 groups, $a_{6H, CH_3} = 1.99$ mT) and the more weakly-coupled OH proton $(a_{H, OH} = 0.05$ mT). Once again, the relative intensities for the lines do not correspond to those expected from Pascal's triangle, i.e., $(1:1:6:6:15:15:20:20:15:15:6:6:1:1)$ [Volume 1, chapter 2, complement 2]. In addition, the central doublet in **1** is missing as it corresponds to the spectrum's symmetry centre.

The spectrum's evolution over time is represented in figure 10.8.

We note that the spectra recorded at 200 ns, 500 ns and 1 μs are increasingly well-resolved. In particular, the coupling due to the protons in the OH groups only become visible from 500 ns after the laser pulse. Relation [10.1] accounts for the broadening observed at shorter times. Indeed, it predicts $\Delta B \approx 0.2$ mT for $\tau = 200$ ns, which is in quite good agreement with what is observed in figure 10.8. After 19 μs, the spectrum is no longer centro-symmetric and becomes more absorption-polarised as the populations approach thermal equilibrium.

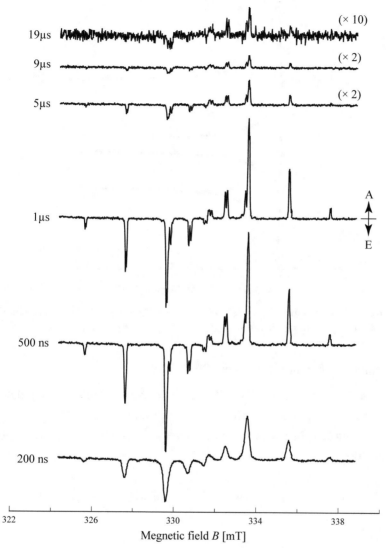

Figure 10.8 - TR-EPR spectra recorded at different times after the laser pulse during photolysis of acetone in ethylene glycol. For experimental conditions, see figure 10.7.
[From Maurel V. (2004)]

Figure 10.9 shows the evolution over time for 4 lines in the spectrum. The lines at 333.61 and 327.61 mT are due to radical **1**, the line at 330.70 mT to radical **s1**. At 333.49 mT, the EPR signals for the two radicals superpose.

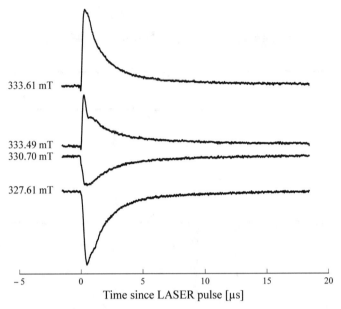

Figure 10.9 - Evolution of the intensity of 4 lines of the TR-EPR spectrum over time, as observed during photolysis of acetone in ethylene glycol. [From Maurel V. (2004)]

◇ *Photolysis of benzoquinone in ethylene glycol*

The second example of a TR-EPR spectrum obtained with a direct detection system relates to photoreduction of benzoquinone in ethylene glycol as solvent. This reaction has been described several times in the literature [Jäger and Norris, 2001].

$$O=\!\!\!\langle\ \rangle\!\!\!=O \ + \ HO \diagup \diagdown OH \ \xrightarrow{h\nu} \ HO-\!\!\!\langle\ \rangle\!\!\!-O^{\bullet} \ + \ HO \diagup \diagdown OH$$
$$\underline{\mathbf{2}} \qquad\qquad\qquad \underline{\mathbf{s1}}$$

The spectrum shown in figure 10.10 is entirely emission-polarised, which is an indicator of a triplet-type CIDEP mechanism (see section 10.4.4). The signal for the semiquinone radical **2** dominates, but some lines corresponding to radical **s1** are identifiable. For radical **2**, the two equivalent protons ortho to the $O^{\bullet}$ and a third proton produce a triplet of doublets with $a_{2H, Hortho} = 0.43$ mT and $a_{H, OH} = 0.16$ mT. The couplings due to the protons meta to the $O^{\bullet}$ are not resolved on the spectrum. The relative intensities of the different lines vary according to the expected (1:1:2:2:1:1) pattern for a triplet of doublets. The fact that the relative intensities are conserved is characteristic of a CIDEP effect due to a triplet-type mechanism.

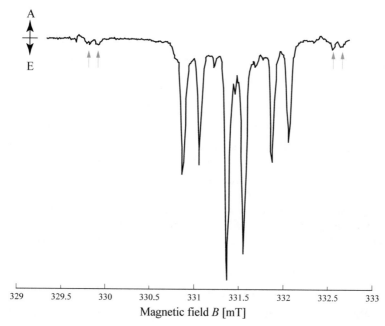

Figure 10.10 - TR-EPR spectrum recorded 5 microseconds after the laser pulse triggering photolysis of benzoquinone (6.4 mM) in ethylene glycol. The six main lines are due to the semiquinone radical **2**, the low-intensity groups of peaks indicated by the arrows are due to the radical **s1**. Laser power: 7 mJ/pulse (266 nm), microwave power: 0.2 mW, 16 accumulations per field value, flow-rate: 0.5 mL min^{-1}. [From Maurel V. (2004)]

The increase in intensity of the EPR signal due to the CIDEP effect was measured for this system [Muus *et al.*, 1978]. The P factor is equal to around -700, which is one of the largest values described in the literature. The magnification in figure 10.11a shows the experimental noise, with a signal-to-noise ratio of only around 30. Without the CIDEP effect, the TR-EPR signal would therefore be completely submerged in the experimental noise. Figure 10.11b shows the time-dependence of the line at 330.9 mT.

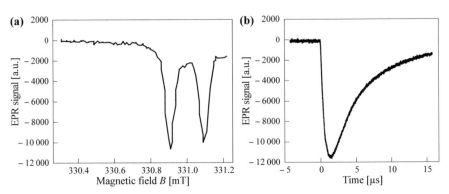

Figure 10.11 - TR-EPR signal recorded during photolysis of 1,4 benzoquinone in ethylene glycol. **(a)** spectrum obtained 2.6 µs after the pulse; **(b)** time-progression for the amplitude of the line at 330.9 mT. Laser power: 10 mJ/pulse. Microwave power: 2 mW, 32 accumulations. Benzoquinone concentration: 0.02 M. The flow-rate (0.5 mL min^{-1}) ensures complete renewal of the sample between laser pulses. The only treatment to which these signals are subjected is the subtraction of the baseline recorded outside resonance (no smoothing, or subtraction of the continuous level). [From Maurel V. (2004)]

The spectra presented in this section show that the CIDEP effect does not alter the *position* of the lines, but that it modifies the *populations* of the energy levels, causing absorption and emission lines to appear. In the following sections, we will briefly describe the main mechanisms likely to produce these effects.

10.4.3 - Radical pair CIDEP mechanism

The *Radical Pair Mechanism* (RPM) was proposed to explain the polarisation of NMR spectra for molecules generated in radical-based photochemical reactions [Closs, 1969] (*Chemically Induced Dynamic Nuclear Polarizatio* (CIDNP) effect). During a photochemical reaction, molecules in an excited singlet or triplet state can produce (A, B) radical pairs. The energy levels and spin states for the pair are determined by a Hamiltonian $\hat{H}_{RP}$ with the form [Volume 1, section 7.4.2]:

$$\hat{H}_{RP} = \hat{H}_{Zeeman, A} + \hat{H}_{hyperfine, A} + \hat{H}_{Zeeman, B} + \hat{H}_{hyperfine, B} - J\mathbf{S}_A \cdot \mathbf{S}_B$$

▷ Just after their formation, neighbouring radicals interact strongly and the exchange term, $\hat{H}_{RP}$, dominates. The possible spin states take the form $\{|S, M_S\rangle\}$, where $\mathbf{S} = \mathbf{S}_A + \mathbf{S}_B$ is the total spin for the pair [Volume 1, section 7.4]:

$$|1,+1\rangle = |\alpha(A)\alpha(B)\rangle \quad |1,0\rangle = \frac{1}{\sqrt{2}}(|\alpha(A)\beta(B)\rangle + |\beta(A)\alpha(B)\rangle)$$

$$|1,-1\rangle = |\beta(A)\beta(B)\rangle \quad |0,0\rangle = \frac{1}{\sqrt{2}}(|\alpha(A)\beta(B)\rangle - |\beta(A)\alpha(B)\rangle)$$

The pair remains in the spin state corresponding to that of the excited precursor, $|0, 0\rangle$ or $|1, 0\rangle$.

▷ When the radicals separate, the exchange interaction diminishes very rapidly, approximately exponentially with distance. The Zeeman terms then become comparable to the exchange terms and the $\{|S, M_S\rangle\}$ states no longer correspond to the eigenstates for the system. The state of the pair of radicals evolves towards a mixture of states $|0, 0\rangle$ and $|1, 0\rangle$ and it is this mixture that creates polarisation. Diffusion models show that two cases must be distinguished (figure 10.12):
  - if radicals A and B separate and do not meet again, the exchange interaction drops too rapily for the mixture of states $|0, 0\rangle$ and $|1, 0\rangle$ to produce significant polarisation,
  - in constrast, if the radicals do meet after having started to separate, J increases once again and the mixture of $|0, 0\rangle$ and $|1, 0\rangle$ states lasts long enough to produce significant polarisation of the EPR signal. This interaction occurs in the first 30 nanoseconds following the creation of the pair [McLauchlan, 1989, 1997]. In this case, the radicals are said to form a "G-pair" (for "*Geminate pair*").

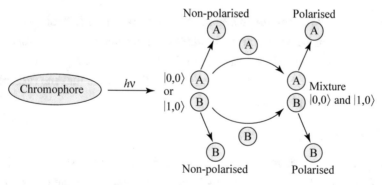

Figure 10.12 - Creation of a pair of polarised radicals (G-pair).

As the $|0, 0\rangle$ and $|1, 0\rangle$ states are composed of as many α as β spins, the RPM mechanism produces no *overall polarisation* for a radical. However, the mixture modifies the populations of the different hyperfine levels and therefore the relative intensities of the lines by creating as much emission as absorption polarisation. Depending on the nature of the excited precursor state ($|0, 0\rangle$ or $|1, 0\rangle$) and the sign of the exchange parameter J, an emission spectrum at low-field and an absorption spectrum at high-field (noted E/A), or an absorption spectrum at low-field and an emission spectrum at high-field (noted A/E) will be obtained (figure 10.13). The general appearance of the spectrum then provides information on the sign of J and on the excited precursor state of the radicals.

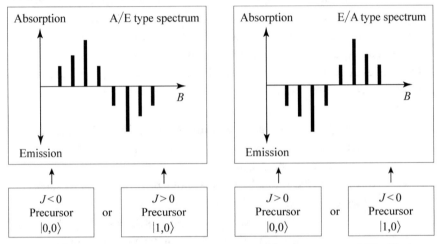

Figure 10.13 - Appearance of an EPR spectrum showing the RPM CIDEP effect's dependence on the sign of the exchange parameter and the nature of the excited precursor state of the pair of radicals.

In general, the J parameter is negative for neutral organic radicals, and the appearance of an RPM spectrum indicates the singlet or triplet nature of the excited precursor state. For example, in the case of acetone photolysis in ethylene glycol, the E/A-type spectrum (figure 10.7) shows that the radicals are produced from a triplet state.

10.4.4 - The triplet CIDEP mechanism

The *Triplet Mechanism* (TM) was proposed very early [Atkins and Evans, 1974; Chow *et al.*, 1992]. This mechanism only occurs if the chromophore molecule passes through an excited triplet state. Just after absorption of a photon, the chromophore molecule is necessarily in an excited *singlet* state $|S^1\rangle$, but transition towards a triplet state can occur due to spin-orbit coupling, through a phenomenon known as *Inter-System Crossing* (ISC). In the absence of magnetic field, the triplet states are the eigenstates of the zero-field splitting matrix, which take the form:

$$|T_x\rangle = \frac{1}{\sqrt{2}}(|(\beta(A)\beta(B)\rangle - |\alpha(A)\alpha(B)\rangle))$$

$$|T_y\rangle = \frac{1}{\sqrt{2}}(|(\beta(A)\beta(B)\rangle + |\alpha(A)\alpha(B)\rangle))$$

$$|T_z\rangle = \frac{1}{\sqrt{2}}(|(\beta(A)\alpha(B)\rangle + |\alpha(A)\beta(B)\rangle))$$

where $\{x, y, z\}$ are the principal axes of this matrix [Volume 1, section 6.2]. When the matrix is axially symmetric, the $|T_x\rangle$ and $|T_y\rangle$ states have the same energy and are separated from the $|T_z\rangle$ state by the energy D (figure 10.14a). The ISC is anisotropic: in the absence of magnetic field it favours transitions towards a particular state of the triplet. In figure 10.14a, this state was assumed to be $|T_z\rangle$.

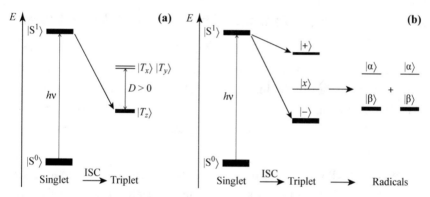

Figure 10.14 - Triplet CIDEP mechanism.
(a) Populations of the energy levels during ISC in the absence of magnetic field. The zero-field splitting term is assumed to be axially symmetric with $D > 0$ and the ISC is assumed to favour transitions towards the $|T_z\rangle$ state. The thickness of the line indicates the population size. **(b)** Populations of the energy levels of the triplet and the radicals under magnetic field for the same molecule, oriented with x parallel to the magnetic field. The $\{|+\rangle, |-\rangle, |x\rangle\}$ states are the states of the triplet when subjected to the zero-field splitting term and the Zeeman term.

In a TR-EPR experiment, the photochemical reaction takes place in a magnetic field. The anisotropy of the ISC then causes polarisation of the triplet state which is transmitted to the two radicals formed (figure 10.14b) [Buckley and McLauchlan, 1985; McLauchlan, 1987; Turro *et al.*, 2000]. This mechanism produces polarisation even when the molecules change orientation during the process, such as in solution.

For the polarisation to be observable on the spectrum, the passage from the polarised triplet to radicals must be rapid relative to the triplet's spin-lattice relaxation time T_1. This relaxation time is generally around 1 to 10 ns. The TM mechanism is therefore only effective for *very rapid* reactions, such as unimolecular reactions involving homolytic bond rupture or bimolecular reactions involving electron or hydrogen atom transfer between substrates available at high concentrations, e.g. when one of the substrates is the solvent. According to this theory, polarisation is created during ISC through the intermediate of spin-orbit coupling which does not involve nuclear spins. This explains why the *relative intensities* of the hyperfine lines are the same as on a conventional EPR spectrum (figure 10.10). The population difference created by this mechanism is 10 to 100 times greater than that at thermal equilibrium. Therefore, the emission or absorption signals are much more intense than those corresponding to thermal equilibrium.

10.4.5 - Superposition of the RPM and TM mechanisms. Other mechanisms

In numerous cases, the RPM and TM CIDEP mechanisms combine to produce the spectrum for photochemically-produced radicals. As a result, spectra where the relative intensities of the lines do not follow the rules for Pascal's triangle and for which the total polarisation is non-null are observed. For example, this is the case for the spectrum recorded during acetone photolysis in isopropanol, shown in figure 10.15.

A variant of the RPM mechanism is often encountered. Rather than involving a pair of radicals newly created by a photochemical reaction (G-pairs), the CIDEP effect occurs when radicals diffusing freely in solution for durations largely exceeding the 30-ns limit mentioned above re-form pairs (F-pairs for *Free-encounter pairs*). F-pairs create similar polarisation to G-pairs, except that the F-pairs observed in TR-EPR are necessarily pairs of radicals *in a triplet state*. Indeed, when the radicals meet freely and form a pair in the singlet state, they react to form a covalent bond, and thus a diamagnetic compound.

Other polarisation mechanisms are less frequently encountered. For example, an excited molecule in a triplet state can transmit polarisation to an already-formed radical. This phenomenon is known as the RTPM (*Radical Triplet Pair Mechanism*) [Kobori *et al.*, 1998]. Another mechanism is observed with photochemical systems containing heavy atoms characterised by a large spin-orbit coupling constant, where the radicals created are polarised by the spin-orbit coupling [Tero-Kubota *et al.*, 2001].

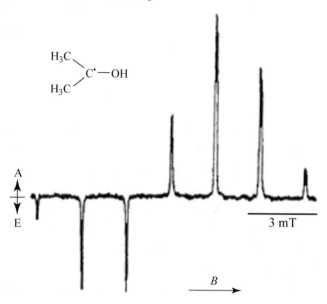

Figure 10.15 - TR-EPR spectra for photolysis of acetone in isopropanol. The CIDEP effect is the result of superposition of an RPM mechanism producing an E/A-type polarisation and a TM mechanism producing total absorption polarisation. [From *McLauchlan K.A.* (1988) *Accounts of Chemical Reasearch* 21: 54–59 © 1988 American Chemical Society]

10.5 - Time-resolved EPR for excited states in the solid phase

Time-resolved EPR can be used to study the paramagnetic intermediates produced by photochemical reactions in the solid phase. For example, a photoactive compound in the $|S^0\rangle$ ground state passes through an excited singlet state $|S^1\rangle$ upon absorption of a photon, and this state is de-excited by inter-system crossing (ISC, figure 10.14a), producing a triplet state which can be studied by TR-EPR. The EPR spectrum for these excited states present the same characteristics as those of radicals in solution: their observation is limited by their lifetime (here by de-excitation), and a CIDEP effect modifies the relative intensities of the different lines (but not their positions) for a duration approximately equal to the spin-lattice relaxation time T_1.

These experiments were very useful to identify the photoinduced species in natural photosynthetic systems, and more recently in artificial charge-separation systems. Derivatives of fluorenone provide good examples of photoinduced organic triplet states. Upon absorption of a photon in the visible part of the spectrum, these molecules produce a relatively long-lived (100 µs) excited triplet state with very good quantum yield (30 %) that can be readily studied by TR-EPR. The shape of the spectrum results from the combined effects of the zero-field splitting term in the Hamiltonian due to dipolar electron-electron interaction and CIDEP. This Hamiltonian is written [Volume 1, section 6.3]:

$$\hat{H} = \beta \mathbf{B} \cdot \tilde{\mathbf{g}} \cdot \mathbf{S} + \mathbf{S} \cdot \tilde{\mathbf{D}} \cdot \mathbf{S} = \beta \mathbf{B} \cdot \tilde{\mathbf{g}} \cdot \mathbf{S} + D(\hat{S}_z^2 - \tfrac{2}{3}) + E(\hat{S}_x^2 - \hat{S}_y^2)$$

As a simplification, we assume that the $\tilde{\mathbf{g}}$ matrix is isotropic and that $E = 0$. At thermal equilibrium, all the transitions produce absorption lines and the powder absorption EPR spectrum takes the characteristic shape of a Pake doublet (figure 10.16a) [Volume 1, section 6.5.2]. The CIDEP effect does not alter the position of the lines, but it modifies the *populations* of the energy levels, causing absorption and emission lines to appear (figure 10.16b). This effect varies with the orientation of the molecules in the magnetic field, resulting in a spectrum with a characteristic appearance which is well illustrated in figure 10.17. This figure shows the spectrum produced by the triplet excited state of a fluorenone-derivative.

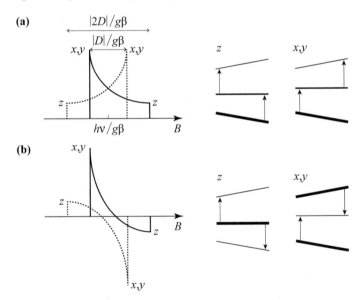

Figure 10.16 - **(a)** EPR transitions and appearance of the absorption spectrum for an axial triplet state when the populations are at *thermal equilibrium*. The thickness of the lines reflects the relative populations of the energy levels, **(b)** modification of the population of energy levels and appearance of the TR-EPR spectrum under the influence of a CIDEP effect.

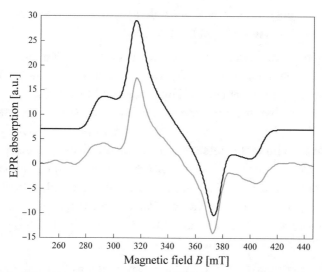

Figure 10.17 - TR-EPR spectrum for an excited triplet state of a fluorenone derivative (grey). The spectrum was generated by a 6-nJ laser pulse at 500 nm, lasting 5 ns. It was recorded on a pulsed spectrometer with the sequence $\pi/2 - \tau - \pi$ (pulse $\pi/2$ lasting 10 ns) at frequency 9.72 GHz, and $T = 70$ K. The simulation (black) was produced using in-house software.

10.6 - Conclusion

In this chapter we presented the different methods which can be used to record the EPR spectrum of short-lived paramagnetic species. We saw that in some cases the shape and intensity of the signal differ from those habitually observed. In these conditions, the reader may question whether EPR can be used to quantitatively study the *kinetics of disappearance* of these species. We will now re-examine the different methods from this point of view.

▷ The kinetics of appearance and disappearance of paramagnetic intermediates produced during "slow" bimolecular reactions can be monitored by the freeze quench (metal centres) or continuous flow (radicals) methods with a resolution in the 0.1 to 1 ms range. With these methods, a conventional spectrometer records a series of spectra at different moments after the start of the reaction (section 10.2). The variation in spectrum amplitude can also be monitored over time at fixed field for a sample placed in the cavity, whether prepared by the stopped flow method (section 10.2) or generated by a light pulse (section 10.3.2). With modulation at 100 kHz and synchronous detection, the spectrometer's time resolution is around

0.1 ms. To determine how the whole spectrum changes, the experiment must be repeated at different field values.

▷ To study more rapid kinetics, TR-EPR can be used with direct detection and recording the signal at fixed field. We saw that the spectra for the radicals generated by photochemical reactions are amplified by the CIDEP effect at times shorter than the spin-lattice relaxation time, T_1. The disappearance of the radicals leads to a reduction in their lifetime and thus *broadening* of the resonance line if the rate constant k characterising this disappearance is greater than $1/T_2$. Quantitative study of this broadening can therefore be used to determine k [Gatlik *et al.*, 1999]. A more general method consists in completely modelling how the resonance line changes taking the decrease in radical concentrations into account [Jäger and Norris, 2001]. This method is described in complement 1.

At times greater than T_1, the signal corresponding to thermal equilibrium is detected. Evolution of this signal over time can be monitored to the extent that its observation is compatible with the low-sensitivity of direct detection.

Complement 1 - Numerical solutions to the Bloch equations in the transient regime

In continuous wave EPR, we are generally interested in the *steady state regime* for which the Bloch equations can be solved in closed form. However, to interpret TR-EPR experiments, the *transient regime* preceding the steady state regime must be considered.

Suppose we use a continuous wave spectrometer to record the EPR signal produced by radicals created at time $t = 0$. The electromagnetic wave is characterised by its angular frequency ω and its amplitude B_1 is constant. We assume that the radicals are *chemically stable* and that the magnetisation they produce at $t = 0$ is $M_{eq} \, \mathbf{u}_z$, which corresponds to thermal equilibrium, where $\mathbf{u}_z$ is the unit vector in the direction z of the static field $\mathbf{B}_0$.

Changes to the components of the magnetisation $\mathbf{M}(t)$ in the rotating reference frame $(R) = (x', y', z' = z)$ are determined by equations [14] to [17] in appendix 1 to this book:

$$\frac{dM_{x'}}{dt} = (\omega_0 - \omega) M_{y'} - \frac{M_{x'}}{T_2}$$

$$\frac{dM_{y'}}{dt} = -(\omega_0 - \omega) M_{x'} - \frac{M_{y'}}{T_2} + \gamma B_1 M_z \qquad [1]$$

$$\frac{dM_z}{dt} = \frac{M_{eq} - M_z}{T_1} - \gamma B_1 M_{y'}$$

We set $\gamma = g\beta / \hbar$ and $\omega_0 = \gamma B_0$. T_1 and T_2 are the spin-lattice and spin-spin relaxation times for the radicals. By resolving these equations with the initial condition $\mathbf{M}(0) = M_{eq} \, \mathbf{u}_z$ it is possible to determine $M_{x'}(t)$, $M_{y'}(t)$, $M_z(t)$ for given values of $(\omega_0, \omega, T_1, T_2, B_1)$. The instantaneous power absorbed by the microwave field, with an angular frequency ω, is written:

$$P_\omega(t) = -\mathbf{B}_1 \cdot \frac{d}{dt}\mathbf{M} \quad \text{where } \mathbf{B}_1 = B_1 \, \mathbf{u}_{x'}.$$

The derivative is defined here in the laboratory's reference frame. It can be expressed as a function of the components of $\mathbf{M}$ in the rotating reference frame:

$$\frac{d}{dt}\mathbf{M} = \left(\frac{dM_{x'}}{dt} - \omega M_{y'}\right)\mathbf{u}_{x'} + \left(\frac{dM_{y'}}{dt} + \omega M_{x'}\right)\mathbf{u}_{y'} + \frac{dM_z}{dt}\mathbf{u}_z \qquad [2]$$

$\mathbf{u}_{x'}$, $\mathbf{u}_{y'}$, and $\mathbf{u}_z$ are the unit vectors of the reference frame (R). Only the first term of the second side is involved in the expression for $P_\omega(t)$. By replacing $\frac{dM_{x'}}{dt}$ by its expression (equations [1]), we obtain:

$$P_\omega(t) = \left[(2\omega - \omega_0) M_{y'} + \frac{M_{x'}}{T_2} \right] B_1 \tag{3}$$

▷ We will first examine what these expressions become in the *steady state regime* where the magnetisation is constant in the rotating reference frame (R). The first sides of equations [1] are null, and by resolving a 3-equation system with 3 unknowns the expressions of $M_{x'}$, $M_{y'}$, and M_z are found. The fact that $\dfrac{dM_{x'}}{dt} = 0$ in equation [2] leads to a constant absorbed power equal to $P_\omega = \omega B_1 M_{y'}$. Resolving the system [1] produces:

$$P_\omega = \frac{\gamma \omega B_1^2 M_{eq}(1/T_2)}{(\omega - \omega_0)^2 + (1/T_2)^2 + \gamma^2 B_1^2 (T_1/T_2)} \tag{4}$$

The line is Lorentzian with a full width at half maximum of

$$\Delta\omega = \frac{2}{T_2}\sqrt{1 + \gamma^2 B_1^2 T_1 T_2} \tag{5}$$

▷ During the *transient regime*, the equations [1] can be *numerically* resolved and the values of $M_{x'}(t)$ and $M_{y'}(t)$ used to calculate $P_\omega(t)$ (equation [3]) for values of ω close to the resonance angular frequency ω_0. An example is presented in figure 10.18 for $\nu_0 = \omega_0/2\pi = 9.400\,\text{GHz}, g = 2.00, T_1 = 2\,\mu s, T_2 = 1\,\mu s$, $B_1 = 0.0036$ mT (which corresponds to a microwave power of around 10 μW) and ν = 9.400 GHz; 9.401 GHz; 9.405 GHz. The signal oscillates with increasing frequency as we move away from resonance, and is more rapidly damped closer to resonance.

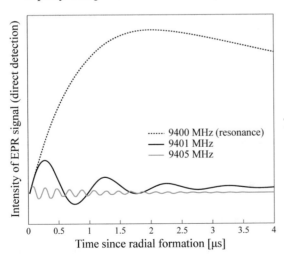

Figure 10.18 - Evolution over time of the EPR signal for three different microwave frequencies. The signal represented is the power absorbed by the system of radicals (equation [3]). Equations [1] were numerically resolved using a Labview routine based on the Runge and Kuta method, with $\nu_0 = 9.400$ GHz, $T_1 = 2$ μs, $T_2 = 1$ μs, $B_1 = 0.0036$ mT. [From Maurel V. (2004)]

By performing this calculation for frequencies between 9.390 GHz and 9.410 GHz in 0.05 MHz steps, it becomes possible to determine how the resonance line evolves (figure 10.19). The oscillations in figure 10.18 cause lateral bands to appear; these bands have already been experimentally observed [Hore *et al.*, 1981]. We observe that the lines calculated very soon after the formation of the radicals are very broad. For $t = 100$ ns, the width is equal to 6.1 MHz (0.22 mT), which is comparable to the width deduced from relation [10.1] (0.36 mT). At longer times, the line evolves towards the Lorentzian defined by equation [4], with a full-width at half-maximum $L = \Delta\omega / 2\pi$ where $\Delta\omega$ is given by equation [5]. With the values of (T_1, T_2, B_1) used in the calculation, we obtain $L = 0.42$ MHz which is effectively the linewidth observed at $t = 4$ µs (figure 10.19).

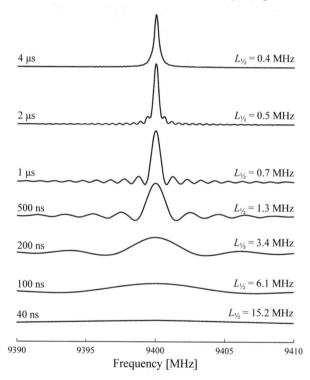

Figure 10.19 - Changes to the resonance line during the first microseconds following the creation of the radicals. The amplitude was calculated as indicated in the legend to figure 10.18. It is represented as a function of the frequency for different values of *t*. All the lines are traced at the same scale. The full width at half maximum is indicated for the central peak. [From Maurel V. (2004)]

▷ Up to now, we have assumed that the radicals created at $t = 0$ were chemically stable. If their concentration decreases with a rate constant k, the system of equations [1] must be modified as follows [Jäger and Norris, 2001]:

$$\frac{dM_{x'}}{dt} = (\omega_0 - \omega) M_{y'} - \frac{M_{x'}}{T_2}$$

$$\frac{dM_{y'}}{dt} = -(\omega_0 - \omega) M_{x'} - \frac{M_{y'}}{T_2} + \gamma B_1 M_z - kM_{y'}$$

$$\frac{dM_z}{dt} = \frac{M_{eq} - M_z}{T_1} - \gamma B_1 M_{y'} - kM_z$$

$$M_{eq}(t) = M_{eq}(t = 0)\exp(-kt)$$

When T_1 and T_2 are known, these equations can be used to numerically simulate the evolution of the resonance line and to deduce the rate constant k [Gatlik *et al.*, 1999].

References

ATKINS P.W. & EVANS G.T. (1974) *Molecular Physics* **27**: 1633-1644.

BALLOU D.P. & PALMER G.A. (1974) *Analytical Chemistry* **46**: 1248-1253.

BORG D.C. (1964) *Nature* **201**: 1087-1090.

BRAY R.C. (1961) *Biochemical Journal* **81**: 189-195.

BUCKLEY C.D. & MCLAUCHLAN K.A. (1985) *Molecular Physics* **54**: 1-22.

CHOW Y.L. *et al.* (1992) *Tetrahedron Letters* **33**: 3315-3318.

CLOSS G.L. (1969) *Journal of the American Chemical Society* **91**: 4552-4554.

DEWEERD K. *et al.* (2001) *Biochemistry* **40**: 15846-15855.

FESSENDEN R.W. & SCHULER R.H. (1963) *Journal of Chemical Physics* **39**: 2147-2195.

FORBES M.D.E. (1997) *Photochemistry and Photobiology* **65**: 73-81.

GATLIK I. *et al.* (1999) *Journal of the American Chemical Society* **121**: 8332-8336.

GRIGORYANTS V.M., VESELOV A.V & SCHOLES C.P. (2000) *Biophysical Journal*
78: 2702-2708.

HORE P.J. *et al.* (1981) *Chemical Physics Letters* **77**: 127-130.

HUBBELL W.L., FRONCISZ W. & HYDE J.S. (1987) *Review of Scientific Instruments*
58: 1879-1886.

HARTRIDGE H. & ROUGHTON F.J.W. (1923) *Proceeding of the Royal Society* (London)
94: 336-367.

JÄGER M. & NORRIS J.R. (2001) *Journal of Magnetic Resonance* **150**: 26-34.

KOBORI Y. *et al.* (1998) *Journal of Physical Chemistry* A **102**: 5160-5170.

LASSMAN G., SCHMIDT P.P. & LUBITZ W. (2005) *Journal of Magnetic Resonance*
172: 312-323.

LEVSTEIN P.R. & VAN WILLIGEN H. (1991) *Journal of Chemical Physics* **95**: 900-908.

MAUREL V. (2004) *Experimental characterization of some radical mechanisms in the
photochemistry of pyrimidines and of aromatic intro compounds: a time resolved
EPR study completed by mass spectroscopy-detected spin trapping experiments.*,
PhD thesis, Université Joseph Fourier - Grenoble I.

MCLAUCHLAN K.A. (1997) *Journal of the Chemical Society, Perkin Transactions*
2: 2465-2472.

MCLAUCHLAN K.A. (1989) "Time-resolved EPR" in *Advanced EPR, Applications
in Biology and Biochemistry*, Hoff A.J. ed., Elsevier, Amsterdam: 345-369.

MCLAUCHLAN K.A. & STEVENS D.G.(1988) *Accounts of Chemical Research*
21: 54-59.

MUUS L.T., FRYDKJAER S. & BONDRUP NIELSEN K. (1978) *Chemical Physics* **30**: 163-168.

SIENKIEWICZ A., QU K. & SCHOLES C.P. (1994) *Review of Scientific Instruments*
65: 68-74.

SIENKIEWICZ A. *et al.* (1999) *Journal of Magnetic Resonance* **136**: 137-142.

TERO-KUBOTA S., KATSUI A. & KOBORI Y. (2001) *Journal of Photochemistry and Photobiology C: Photochemistry Reviews* **2**: 17-33.

TURRO N.J., KLEINMAN M.H. & KARATEKIN E. (2000) *Angewandte Chemie, International Edition* **39**: 4436-4461.

VAN WILLIGEN H., LEVSTEIN P.R. & EBERSOLE M.H. (1993) *Chemical Reviews* **93**: 173-197.

VERKHOVSKAYA M.L. *et al.* (2008) *Proceeding of the National Academy of Sciences of the USA* **105**: 3763-3767.

Volume 1: BERTRAND P. (2020) *Electron Paramagnetic Resonance Spectroscopy - Fundamentals*, Springer, Heidelberg.

Chapter 11

Characterising contrast agents for magnetic resonance imaging

Belorizky E. [a] and Fries P.H. [b]

[a] *Interdisciplinary Laboratory of Physics, Grenoble Alpes University, Grenoble.*
[b] *Ionic Recognition and Coordination Chemistry, Inorganic and Biological Chemistry, UMR-E3, Grenoble Alpes University, Grenoble.*

11.1 - Introduction

Techniques for medical diagnosis, and more generally, preclinical research – used to non-invasively observe the inside of living organisms – have developed considerably over the last few decades. The most commonly used are based on X-rays [Ledley *et al.*, 1974], ultrasound [Correas *et al.*, 2009] or nuclear magnetic resonance of water protons from tissues, i.e., observation of how these nuclei, placed in a fixed external magnetic field, respond to excitation created by an oscillating magnetic field [Canet *et al.*, 2002]. This phenomenon constitutes the basis of Magnetic Resonance Imaging (MRI) [Mansfield and Pykett, 1978; Merbach and Toth, 2001; Callaghan, 2003]. The submillimetre resolution of this method is remarkable, but it lacks the requisite *sensitivity* for some examinations. For example, it cannot detect small anatomical anomalies such as burgeoning tumours or low-concentration molecular species characteristic of diseases such as those forming fatty deposits (atheromatous plaques) in the arteries. The contrast of images of tissues or anomalies can be improved by injecting biocompatible paramagnetic species, known as *contrast agents* [Bertini *et al.*, 2001]. These agents are often gadolinium Gd^{3+} complexes. The physical mechanisms explaining how they improve contrast are difficult to model as they involve multiple phenomena, such as translational Brownian motion of the water molecules, rotational Brownian motion of the complexes, and random evolution of the quantum states of the electronic spin of the Gd^{3+} causing its relaxation. EPR of Gd^{3+} complexes can effectively contribute to characterising these phenomena.

In this chapter, we briefly present the principle by which MRI images are constructed, in particular those based on spatial variations of local relaxation times for water protons. The role of contrast agents is explained by underlining the specific advantages of Gd^{3+} complexes. Then, the mechanisms leading to relaxation of the protons contained in the tissue water are analysed and classed as a function of the properties through which the water molecules associate with complexes in solution. We show how the electronic relaxation of Gd^{3+} affects the performance of contrast agents, and the key contribution of EPR to quantifying this relaxation. We propose a theoretical formalism through which to simulate EPR spectra and thus determine the electronic relaxation. This formalism is applied to a few representative Gd^{3+} complexes in aqueous solution and we assess their usefulness as contrast agents. Finally, we present a few images obtained using some of these complexes.

11.2 - MRI methods

11.2.1 - General principle

As indicated, biomedical MRI is a non-invasive technique based on observation of the Nuclear Magnetic Resonance (NMR) signal produced by the protons from the hydrogen atoms in water [Canet, 1996; Abragam, 1983]. Indeed, as water makes up around 70 % of the human body, ^1H protons are highly abundant and supply a good NMR signal. The intensity of the signal observed depends on the water concentration which is relatively constant in soft tissues, but is mostly influenced by the relaxation times for nuclear spins, which characterise the speed of return to equilibrium of the magnetic moments of the protons following excitation. More precisely, by applying a series of radiofrequency radiation pulses and magnetic field gradients in the three spatial directions, a region of the organism can be divided into juxtaposed elementary volumes (voxels) producing distinct NMR signals which are a function of the density ρ of protons and the longitudinal T_1 and transverse T_2 relaxation times specific to each voxel [Merbach and Toth, 2001; Callaghan, 2003]. The type of measurement performed reveals differences in the values of ρ, T_1 or T_2 between different voxels. These differences can be coloured in greyscale, to produce images that are weighted in proton density, T_1 or T_2. For example, lesions in tissues or tumours can be observed thanks to differences in the relaxation times for water between the healthy and affected areas.

In general, the T_1 and T_2 values vary as follows [Abragam, 1983]:

▷ T_1 is shorter in common solvents like water (a few seconds) than in viscous liquids and solids (several minutes or more). Indeed, rapid, large-amplitude molecular movement in non-viscous liquids promote transitions between Zeeman levels of protons causing faster T_1 relaxation than their slow, weak-amplitude equivalents in solids.

▷ T_2 is longer in common solvents than in viscous liquids as variations in the local field, which determine the T_2 relaxation, are caused by faster molecular movements and are thus more rapidly averaged to zero. This phenomenon is known as motional narrowing.

▷ T_1 is always greater than or equal to T_2. T_1 and T_2 have similar values in pure water and non-viscous liquids, but the differences between them can be considerable in solids (T_1 a few hours, T_2 a few ms).

In biological tissues, T_1 increases as the field B increases, whereas T_2 is relatively independent of the field [Abragam, 1983; Canet, 1996]. These relaxation times depend on the physico-chemical organisation of the water in the region of the organism observed. Radiologists exploit spatial variations in the values of T_1 or T_2 to produce anatomical images and use abnormal variations of these times to detect lesions and diseased areas in the tissues. In Table 11.1, we have indicated the values of T_1 and T_2 for some healthy tissues *in vivo* at 37 °C and for distilled water at 20 °C, in a 1.5-T field.

Table 11.1 - Relaxation times for protons in some healthy tissues *in vivo* and for water in a 1.5-tesla magnetic field.

Tissue	T_1 [ms]	T_2 [ms]
Adipose tissue [a]	282	111
Muscle [a]	749	43
Liver [a]	594	61
Cortex [b]	1304	93
White matter [b]	660	76
Distilled water [c]	2780	1400

[a] [Tadamura *et al.*, 1997], [b] [Vymazal *et al.*, 1999], [c] [Kiricuta and Simplaceanu, 1975].

In general, at the magnetic fields commonly used in MRI ($B \leq 1.5$ T), the relaxation times for protons in well-hydrated tissues, which are thus of low viscosity, increase with the water content as water is the most mobile species [Kiricuta and Simplaceanu, 1975] (also see sections 11.3.2 and 11.3.3). Schematically, in the case of an acute disorder, the proton density and T_1 and T_2 times vary in the same direction for a given tissue. Indeed, in most cases lesions are associated with inflammation and oedema characterised by infiltration of water into the affected tissue. The opposite occurs in scar tissue.

For many MRI sequences with appropriately selected parameters, the signal intensity for a voxel increases with the longitudinal relaxation rate $R_1 = 1/T_1$, but decreases with the transversal relaxation rate $R_2 = 1/T_2$ [Merbach and Toth, 2001; Callaghan, 2003].

11.2.2 - Role of contrast agents in MRI

The contrast agents considered here are chemical complexes which modify the NMR signal for the voxels where they are located, by increasing the relaxation rates $R_1 = 1/T_1$ and $R_2 = 1/T_2$ for water protons. The increases are generally proportional to the local concentration of relaxing agents, which depends on the properties of the different tissues. Thus, differences in the proton relaxation enhancement reflect biological differences such as the presence of tumour tissue. More significant differences generally produce better image contrast enhancement. The most frequently used contrast agents are complexes or chelates of paramagnetic metal ions, i.e., constructs composed of organic molecules (ligands) sequestering one or sometimes several of these ions. Less frequently, super-paramagnetic nanoparticles are also used, for example iron oxide particles composed of several thousand magnetic ions, with their individual magnetic moments aligned to produce a very high resultant paramagnetic moment [Merbach and Toth, 2001].

Here, we will only discuss complexes of a single metal ion bearing a high magnetic moment, capable of creating a large-amplitude fluctuating dipolar field in its surroundings, which notably accelerates proton relaxation. The appropriate ions belong either to the first transition series (manganese, iron), or to the lanthanide series (gadolinium, dysprosium), which have a $3d$ and a $4f$ incomplete atomic subshell, respectively. The contrast agents currently commercialised generally contain Gd^{3+} as it provides the best contrast enhancement for T_1-weighted images.

11.3 - Effect of a Gd^{3+} complex on relaxation of water protons

11.3.1 - General notions

Consider an aqueous solution of GdL complexes, where Gd is the paramagnetic ion Gd^{3+} and L is a multidentate ligand. Remember that the Gd^{3+} ion has a half-filled $4f^7$ subshell and that its ground term $^8S_{7/2}$ is characterised by a total orbital momentum $L = 0$, a total spin $S = 7/2$ and a Landé g factor close to that of the electron [Volume 1, section 8.1.1]. This ion has two isotopes with nuclear spin $I = 3/2$, but the hyperfine structure is not involved here due to the broad width of the EPR lines [Borel *et al.*, 2006]. The longitudinal relaxation rate $R_1 = 1/T_1$ for water protons in this solution is the sum of the diamagnetic contribution $R_{1,d}$ obtained in the absence of complexes and of the $R_{1,p}$ contribution due to paramagnetic complexes:

$$R_1 = R_{1,d} + R_{1,p} \qquad [11.1]$$

This relation only applies with dilute solutions such as those monitored by MRI. Experiments demonstrate, and the theory confirms (sections 11.3.2 and 11.3.3), that $R_{1,p}$, which represents the increase in the relaxation rate due to GdL complexes, is generally proportional to the molar concentration [GdL]. The *relaxivity* r_1 of these complexes is defined by the relation

$$R_1 = R_{1,d} + r_1 [GdL] \qquad [11.2]$$

i.e., by the slope of the line giving R_1 as a function of [GdL] expressed in mM (mmol L^{-1}).

Relaxivity is therefore expressed in s^{-1} mM^{-1}. Hereafter, we describe the different elements contributing to it.

Paramagnetic relaxation of the spin I of a water proton is due to fluctuations of its dipolar interactions with the magnetic moments of Gd^{3+} ions. In the point dipole approximation, in which the magnetic moment for each Gd^{3+} ion is considered to be localised at the centre of the ion [Yazyev and Helm, 2008], this interaction is described by the following Hamiltonian [Volume 1, section 7.2.4]:

$$\hat{H}_{dip} = \frac{\mu_0}{4\pi}\gamma_I\gamma_S\hbar^2\left[\mathbf{I}\cdot\mathbf{S} - 3(\mathbf{I}\cdot\mathbf{r})(\mathbf{S}\cdot\mathbf{r})\big/r^2\right]\Big/r^3 \qquad [11.3]$$

where μ_0 is the permeability of a vacuum, γ_I, γ_S are, respectively, the magnetogyric ratios of the proton and Gd^{3+}, and $\mathbf{r}$ is the vector linking the Gd^{3+} ion to the proton. Two relaxation mechanisms are generally distinguished: the inner sphere and the outer sphere mechanisms.

Inner sphere mechanism. As $\hat{H}_{dip}$ increases as $1/r^3$ when the distance r decreases, the protons in the water molecules close to the Gd^{3+} ions have a very large instantaneous relaxation rate. Thus, the molecular movements resulting in transient coordination of the water molecules to the metal ions and thus bringing the water protons close to these ions, play an important role in transmitting the effect of paramagnetic relaxation to the whole of the solvent. The water molecules bound (coordinated) to the metal form the first coordination sphere or *Inner-Sphere* (IS). These inner sphere water molecules, in which the protons have an extremely high *intrinsic intramolecular relaxation rate* $R_{1m} \equiv 1/T_{1m}$ ($T_{1m} \cong$ a few μs), then exchange with solvent protons such that, in turn, all the solvent molecules, when they come into contact with Gd^{3+} undergo the intrinsic inner sphere paramagnetic relaxation. This overall mechanism is described as the inner sphere contribution R_1^{IS} to $R_{1,p}$ (figure 11.1).

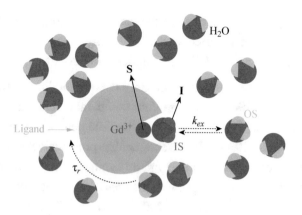

Figure 11.1 - Schematic representation of a Gd^{3+} complex (ion green, ligand red) to which an inner sphere (IS) water molecule is bound. This water molecule is surrounded by so-called outer sphere (OS) free water molecules from the solvent. k_{ex} is the rate constant for exchange between inner sphere water molecules and free molecules. It is linked to their lifetime τ_m by $k_{ex} = 1/\tau_m$ (see the text). τ_r is the rotational correlation time for the complex which is particularly involved in the expression for T_{1m} (equation [11.8]).

Outer sphere mechanism. For durations approximately equal to the lifetime τ_m of an inner sphere water molecule, no chemical exchange occurs but protons in the free water molecules experience another paramagnetic relaxation effect due to their relative *intermolecular translational diffusion* motion with respect to the Gd^{3+} ions. This motion is known as *Outer Sphere* OS motion and this effect produces the outer sphere contribution R_1^{OS} to $R_{1,p}$. We have thus defined the inner and outer sphere contributions to $R_{1,p}$ as a function of the intra- and intermolecular nature of the fluctuations in the dipolar interaction over times with an order of magnitude of around τ_m. The total increase in the relaxation rate for the protons due to GdL complexes can therefore be written

$$R_{1,p} = R_1^{IS} + R_1^{OS} \qquad [11.4]$$

and the total relaxivity is given by

$$r_1 = r_1^{IS} + r_1^{OS} \qquad [11.5]$$

With some complexes, another type of fluctuation of the dipolar interaction – intramolecular fluctuation – gives a "second sphere" contribution. This contribution is due to labile water molecules located close to the metal ion, which are not directly coordinated to the complex, as in the case of the inner sphere mechanism, but by *hydrogen bonding with the ligand*. The second sphere contribution $R_{1,p}^{2S}$ can be described as the inner sphere contribution. However, it is often negligible, or its effects are (incorrectly) taken into account in the outer sphere term [Aimé *et al.*, 2005; Bonnet *et al.*, 2010].

The Gd^{3+}-based contrast agents commonly used in medical MRI have been known since the mid-1980s. They produce comparable inner and outer sphere relaxivities, of around 1 to 2 s^{-1} mM^{-1} at the field generally used in imaging, 1 to 3 T [Powell *et al.* 1996]. However, chemists progressively obtained much larger inner sphere relaxivity values r_1^{IS} by synthesising a new generation of contrast agents developed according to the prescriptions of theoretical models (sections 11.3.2 and 11.9) [Aimé *et al.* 2005; Caravan, 2009; De Leon-Rodriguez *et al.* 2009]. Optimisation of the outer sphere contribution r_1^{OS}, in contrast, is much more difficult to conceive. We will now analyse the factors which determine the two main types of relaxivity and produce simplified theoretical expressions for R_1^{IS} and R_1^{OS}.

11.3.2 - Inner sphere proton relaxivity

The inner sphere contribution to proton relaxation results from fluctuations in the intramolecular dipolar interaction of the protons of the water molecules coordinating Gd^{3+} and exchanges between these molecules and the water molecules in the solution. This is a two-site exchange problem; the first site, represented by the water molecules *bound* to the metal ion, is much less populated than the second, represented by the *free* water molecules. The experimental NMR signal is that of protons in free water. In complement 1 we show that the inner sphere relaxation rate is given by

$$R_1^{IS} = P_m \frac{1}{T_{1m} + \tau_m} \qquad [11.6]$$

where P_m is the molar fraction of bound protons, τ_m is the lifetime of an inner sphere water molecule, linked to the exchange rate constant k_{ex} (figure 11.1) by $\tau_m = 1 / k_{ex}$ and T_{1m} is the intrinsic longitudinal relaxation time for bound protons in the absence of exchange, i.e., in the situation where the water molecule would remain indefinitely coordinated to Gd^{3+}. The molar fraction P_m is linked to [GdL], the molar concentration of complexes, and to q, the number of water molecules bound to a Gd^{3+} ion (solvation number), by the relation

$$P_m = \frac{q[GdL]}{55.5} \qquad [11.7]$$

For the Gd^{3+} complexes used as contrast agents, τ_m is typically around 10^{-7} to 10^{-6} s. In general, the main contribution to the relaxation rate $R_{1m} \equiv 1/T_{1m}$ is due to the modulation of the dipole-dipole interaction $\hat{H}_{dip}$ (equation [11.3]) by the reorientational Brownian motion of the $\mathbf{r}_{GdH}$ vector with constant magnitude, linking the centre of the Gd^{3+} ion to the bound water proton considered. In the absence of electronic relaxation of the paramagnetic ion, this relaxation rate is given by the simplified Solomon-Bloembergen-Morgan (SBM) equation, which is valid with the fields used in MRI [Merbach and Toth, 2001]:

$$R_{1m} = \frac{1}{T_{1m}} = \left(\frac{\mu_0}{4\pi}\right)^2 \frac{2}{5}\left(\frac{\gamma_I^2 g^2 \beta^2}{r_{GdH}^6}\right)S(S+1)\frac{\tau_r}{1 + \omega_0^2 \tau_r^2} \qquad [11.8]$$

where g is the Landé factor for the Gd^{3+} ion of spin $S = \frac{1}{2}$, β is the Bohr magneton ($\gamma_S \hbar S = -g\beta S$), ω_0 is the resonance angular frequency for the proton, and τ_r is the rotational correlation time for the complex. For a spherical complex of radius a, this time is approximately given by the Stokes rotational diffusion formula

$$\tau_r = \frac{4\pi a^3 \eta_{mic}^r}{3k_B T} \qquad [11.9]$$

where T is the temperature. The rotational microviscosity η_{mic}^r of the solution [Gierer and Wirtz, 1953] can differ significantly from the experimentally measured macroscopic viscosity because of the granular structure (finite size) of the molecules. Qualitatively, τ_r represents the mean time required for the complex to perform a rotation of 1 radian due to the effect of collisions with solvent molecules. For the complexes considered, τ_r is around 10^{-10} s at room temperature, but it increases to 10^{-8} s for bulkier complexes.

Equation [11.8] shows that a first advantage of the Gd^{3+} ion is the high value $S = \frac{1}{2}$ of its electronic spin, which can result in high values of R_{1m} thanks to the $S(S + 1)$ factor. The distance r_{GdH} varies little between complexes, being consistently around 0.31 ± 0.01 nm [Caravan, 2009]. For typical values $B \leq 1.5$ T, $\tau_r = 10^{-9}$ s, equation [11.8] gives a very high value, practically independent of B, $R_{1m} \cong 1.5 \times 10^6 \, s^{-1}$, or $T_{1m} \cong 0.7 \, \mu s$. In an imager with a given magnetic field B where the resonance frequency ω_0 of the protons is fixed, R_1^{IS} can be increased by optimising three parameters through chemical synthesis (equation [11.6]): the exchange rate $k_{ex} = 1/\tau_m$, the solvation number q (equation [11.7]), and the correlation time τ_r (equation [11.8]).

11.3.3 - Outer sphere proton relaxivity

The contribution of outer sphere proton relaxivity is the result of fluctuations in the magnetic dipolar interaction $\hat{H}_{dip}$ between the magnetic moment $\gamma_S \hbar S$ for each complexed Gd^{3+} ion and $\gamma_I \hbar I$, the magnetic dipole for each proton in the water molecules which are free relative to the complex, i.e., undergoing random relative translational diffusion motion. The corresponding relaxation rate R_1^{OS} can be determined by applying the Torrey [Torrey, 1953] and Solomon theory [Abragam, 1983; Solomon, 1955]:

$$R_1^{OS} = C[3j_2(\omega_I) + 7j_2(\omega_S)] \qquad [11.10]$$

where

$$C = \frac{32\pi}{405}\left(\frac{\mu_0}{4\pi}\right)^2 \gamma_I^2 g^2 \beta^2 S(S+1)\frac{N_A[GdL]}{bD} \qquad [11.11]$$

In these expressions, ω_I and ω_S are the resonance angular frequencies of the spins I and S. If we note $r(r, \theta, \varphi)$ the vector linking the proton to the Gd^{3+} ion, the spectral density $j_2(\omega)$ is the Fourier cosine transform of the temporal dipolar correlation function $g_2(t)$ of the random function $r^{-3}Y_{20}(\theta, \varphi)$, where $Y_{20}(\theta,\varphi) \equiv \sqrt{5/(4\pi)}(3\cos^2\theta - 1)/2$ is a second order spherical harmonic. In equation [11.11], $N_A[GdL]$ is the number of complexes per unit of volume, where N_A is Avogadro's number and $[GdL]$ is the concentration of Gd^{3+} ions in mM; b is the minimal approach distance between the interacting magnetic moments, and $D = D_I + D_S$ is the relative diffusion coefficient of a water molecule and a complex, equal to the sum of the coefficients for water self-diffusion D_I and diffusion of the complex D_S. The expression for $j_2(\omega)$ is difficult to calculate for the most general intermolecular motion which results from translational and rotational movements of the water and the complex as well as any changes in complex conformation. However, $j_2(\omega)$ has a simple closed form expression in the following case:

▷ the water molecules and the complex are hard impenetrable spheres with centred magnetic moments,

▷ the free water molecules are uniformly distributed around the complex. This is a reasonable approximation [Ayant *et al.* 1975; Hwang and Freed, 1975]. In this case, we obtain:

$$j_2(\omega) = \text{Re}\frac{1 + z/4}{1 + z + 4z^2/9 + z^3/9} \qquad [11.12]$$

Re designates the real part and $z = \sqrt{i\omega\tau_D}$, τ_D is a translational correlation time given by $\tau_D = b^2/D$, where b is the sum of the molecular radii for the two species. By order of magnitude, τ_D is the mean time necessary for a water molecule to cover the distance $b/2$. Typically, τ_D is around 10^{-10} s at room temperature. Equations [11.10] and [11.11] indicate that R_1^{OS} is proportional to $S(S + 1)$ as R_{1m} (equation [11.8]), so that the high spin $S = \frac{7}{2}$ of Gd^{3+} also favours high R_1^{OS} values. If we set $u \equiv \sqrt{2\omega\tau_D}$, expression [11.12] can be written in the form of a rational fraction with a real argument

$$j_2(\omega) = \frac{1 + 5u/8 + u^2/8}{1 + u + u^2/2 + u^3/6 + 4u^4/81 + u^5/81 + u^6/648} \quad [11.13]$$

The spectral density $j_2(\omega)$ is a decreasing function of ω which varies as $1-(\frac{3}{8})\sqrt{2\omega\tau_D}$ at low frequencies ($\omega\tau_D \ll 1$) and as $(\frac{81}{4})/(\omega\tau_D)^2$ at high frequencies ($\omega\tau_D \gg 1$).

It should be noted that the distinction between inner and outer sphere mechanisms implies $\tau_r, \tau_D \ll \tau_m$. This inequality is generally verified.

11.3.4 - Influence of the electronic spin relaxation of Gd^{3+}

Expressions [11.6], [11.8], [11.10], [11.12], which give the paramagnetic relaxation rates, were obtained by assuming that the electronic relaxation times for the Gd^{3+} complex are infinitely long. In fact, electronic relaxation of the complex affects the inner and outer sphere relaxivity of the water protons [Bertini et al., 2001; Kowalewski and Maler, 2006]. At the magnetic fields used in imaging ($B \geq 1$ T), the effects of electronic relaxation only depend on the longitudinal electronic relaxation time T_{1e} [Fries and Belorizky, 2007; Belorizky et al., 2008; Bonnet et al., 2008]:

▷ in the case of the inner sphere mechanism, expression [11.8] becomes:

$$R_{1m} = \frac{1}{T_{1m}} = \left(\frac{\mu_0}{4\pi}\right)^2 \frac{2}{5}\left(\frac{\gamma_I^2 g^2 \beta^2}{r_{GdH}^6}\right)S(S+1)\frac{\tau_1}{1 + \omega_0^2\tau_1^2} \quad [11.14]$$

the rotational correlation time τ_r is thus replaced by the *effective correlation time* τ_1 defined by:

$$\frac{1}{\tau_1} = \frac{1}{\tau_r} + \frac{1}{T_{1e}} \quad [11.15]$$

▷ in the case of the outer sphere mechanism, the z variable in equation [11.12] becomes $z = \sqrt{i\omega\tau_D + \tau_D/T_{1e}}$ [Bonnet et al., 2008].

At this stage, it is already obvious that electronic relaxation of Gd^{3+} complexes affects their relaxivity. This direct effect is sufficient to justify devotion of a whole chapter to contrast agents in a book on EPR applications. However, we will see in sections 11.4.1 and 11.4.2 that interpretation of the electronic relaxation can also be used to determine the correlation time τ_r involved in the inner sphere process. Finally, its role is important when analysing the relaxation of ^{17}O nuclei of water, a process which can provide information on the exchange rate k_{ex} involved in equation [11.6] [Merbach and Toth, 2001]. It is therefore essential to determine the relaxation time T_{1e} for the Gd^{3+} complex.

With current technology, T_{1e} can be measured at X-band by longitudinal detection of EPR (LODEPR) technique based on observation of the response of the longitudinal magnetisation when the microwave power is modulated at a frequency of around $1/T_{1e}$ [Borel *et al.*, 2002]. The T_{1e} values measured for Gd^{3+} chelates in aqueous solution at room temperature are around 2 to 4 ns. But this very promising approach is currently limited to X-band magnetic fields (0.34 T) which are considerably lower than those used in imaging. For MRI applications, T_{1e} must be *calculated* using an appropriate model relying on analysis of conventional EPR spectra (see section 11.6).

Examination of equations [11.14] and [11.15] indicates why Gd^{3+} complexes are considerably better contrast agents than complexes of other lanthanides with a higher angular momentum J:

▷ Lanthanides Ln^{3+} with an atomic number greater than that of Gd^{3+} have a very short electronic relaxation time T_{1e}, of around 10^{-13} s [Bertini *et al.*, 2001; Vigouroux *et al.*, 1999; Fries and Belorizky, 2012]. When T_{1e} is much shorter than τ_r which is around 10^{-9} to 10^{-10} s, equation [11.15] shows that $\tau_1 \cong T_{1e} \ll \tau_r$. In these conditions, at the frequencies used in MRI, the following inequality holds $\omega_0\tau_1 \ll \omega_0\tau_r \leq 1$. The value of R_{1m} given by [11.14] is therefore proportional to T_{1e}, and is thus much smaller than that obtained when electronic relaxation is neglected, where it is essentially proportional to τ_r (equation [11.8]). The strong reduction in R_{1m} considerably increases T_{1m} which, according to equation [11.6], causes R_1^{IS} to be significantly reduced since contrast agents must satisfy the $\tau_m < T_{1m}$ criterion promoting an efficient inner sphere mechanism.

Similarly, for the outer sphere contribution, a very short T_{1e} time such that $T_{1e} \ll \tau_D$ (τ_D is the translational correlation time) produces a much larger module of the variable z than in the absence of electronic relaxation. This effect translates into a considerable decrease in the spectral density $j_2(\omega)$ and in the relaxation rate R_1^{OS}.

▷ In contrast, for Gd^{3+}, T_{1e} is around 10^{-8} s at the fields generally used in imaging. Therefore $T_{1e} \geq \tau_r$ et $T_{1e} \gg \tau_D$, such that attenuation of the relaxivity due to T_{1e}, although problematic, particularly for bulky complexes, is no longer insurmountable.

In summary, Ln^{3+} ions in which the $4f$ subshell is more than half full, with much shorter relaxation times T_{1e} than those for Gd^{3+}, produce a relaxivity much smaller than that provided by Gd^{3+} even if their angular momentum J is higher than the $S = \frac{7}{2}$ spin of Gd^{3+}, as is the case for Dy^{3+} and Er^{3+} ($J = \frac{15}{2}$), Tb^{3+} ($J = 6$) and Ho^{3+} ($J = 8$). The relative slowness of the electronic relaxation of Gd^{3+} is due to the fact that its ground term is characterised by $L = 0$. This term is therefore not affected to first order of perturbation theory by the random electrostatic field produced by the ligands which causes electronic relaxation [Volume 1, section 8.3.5]. In contrast, the ground states for the other Ln^{3+} ions, for example Tb^{3+}, Dy^{3+}, Ho^{3+} and Er^{3+} characterised by $L = 3, 5, 6, 6$, respectively, are directly affected by the dynamic effects of the ligand field [Merbach and Toth, 2001, chapter 8].

Below, we will model the relaxation mechanisms for Gd^{3+} complexes and provide simplified expressions for T_{1e} as a function of a restricted number of parameters, then we will show how these parameters can be determined from simulated EPR spectra.

11.4 - Modelling Gd^{3+} electronic relaxation

11.4.1 - Fluctuations of the zero-field splitting term

Fluctuation of the zero-field splitting term has been the subject of numerous studies [Rast et al. 2001; Kowalewski and Maler, 2006; Fries and Belorizky, 2007; Helm, 2006]. We previously mentioned that, for an ion like Gd^{3+}, the electrostatic potential of the ligands does not affect the ground term characterised by $L = 0$ to first order of perturbation theory. However, it produces second order effects thanks to the spin-orbit coupling $\lambda \mathbf{L} \cdot \mathbf{S}$, which removes the degeneracy of this term in the absence of an external magnetic field. This removal of the degeneracy which is equal to 8 for Gd^{3+} ($S = \frac{7}{2}$) is the *Zero-Field Splitting* (ZFS) [Volume 1, sections 6.2 and 8.3.3].

In a molecular reference frame (M) defined by a system of orthogonal axes $\{X, Y, Z\}$ linked to the complex and moving with it, this interaction can be described by a spin Hamiltonian $\hat{H}_{ZFS}^{(M)}(t)$ which depends on the time and generally includes terms of degree $n = 2$, 4 and 6 in $\hat{S}_X, \hat{S}_Y, \hat{S}_Z$. In perturbation theory, these terms involve spin-orbit coupling at increasing degrees and their magnitude decreases rapidly when n increases. We can often limit ourselves to second order terms, except in the case of complexes with cubic or octahedral symmetry for which the mean for this term is null, and for which higher order terms must therefore be considered [Rast *et al*, 2001].

The Hamiltonian $\hat{H}_{ZFS}^{(M)}(t)$ is the sum of its temporal mean $\hat{H}_{ZFS,S}^{(M)}$ and its time-dependent "transient" residual part $\hat{H}_{ZFS,T}^{(M)}(t)$ due to vibrations of the complex and to deformations produced by collisions with neighbouring molecules:

$$\hat{H}_{ZFS}^{(M)}(t) = \hat{H}_{ZFS,S}^{(M)} + \hat{H}_{ZFS,T}^{(M)}(t) \qquad [11.16]$$

The form of the "static" part $\hat{H}_{ZFS,S}^{(M)}$ must respect the mean symmetry of the complex. When the calculation is restricted to second order terms, there is a molecular reference frame $\{X, Y, Z\}$ where this Hamiltonian is written

$$\hat{H}_{ZFS,S}^{(M)} = D_X \hat{S}_X^2 + D_Y \hat{S}_Y^2 + D_Z \hat{S}_Z^2 \qquad [11.17]$$

with $D_X + D_Y + D_Z = 0$. It is convenient to write this term in the following form [Volume 1, section 6.2.2]:

$$\hat{H}_{ZFS,S}^{(M)} = D_S \left[\hat{S}_Z^2 - S(S+1)/3 \right] + E_S (\hat{S}_X^2 - \hat{S}_Y^2) \qquad [11.18]$$

where $D_S = 3 D_Z/2$, $E_S = (D_X - D_Y)/2$. To characterise the magnitude of $\hat{H}_{ZFS,S}^{(M)}$, the following parameter can be used

$$\Delta_S = \sqrt{D_X^2 + D_Y^2 + D_Z^2} = \sqrt{2D_S^2/3 + 2E_S^2} \qquad [11.19]$$

For Gd^{3+} complexes, Δ_S is around 0.05 cm^{-1}, which corresponds to $\Delta_S/\hbar \approx 10^{10}$ rad s^{-1}.

In the laboratory's reference frame (L) identified by the system of orthogonal axes $\{x, y, z\}$, where z is the direction of the field **B**, the total spin Hamiltonian acting on the Gd^{3+} ion is written

$$\hat{H}_{spin}^{(L)}(t) = \hat{H}_{Zeeman} + \hat{H}_{ZFS}^{(L)}(t) \qquad [11.20]$$

The Zeeman term is given by

$$\hat{H}_{Zeeman} = g\beta B\hat{S}_z = \hbar \omega_S \hat{S}_z \qquad [11.21]$$

where ω_S is the resonance angular frequency in the absence of zero-field splitting. $\hat{H}_{ZFS}^{(L)}(t)$ is obtained from $\hat{H}_{ZFS}^{(M)}(t)$ by expressing the components of the spin **S** in the reference frame (M) as a function of its components in the reference frame (L). Due to rotational Brownian motion of the complex, $\hat{H}_{ZFS,S}^{(L)}(t)$ is a random time perturbation, characterised by the rotational correlation time τ_r introduced in section 11.3.2 (equation [11.9]). This perturbation induces transitions between the Zeeman sublevels of the complex, thus contributing to its spin-lattice relaxation. In complement 2 we detail the form of $\hat{H}_{ZFS,S}^{(L)}$ deduced from expression [11.18] through the action of a *rotation R* which transforms the laboratory's reference frame into the molecular reference frame.

It is obviously very complex to rigorously express the transient part of $\hat{H}_{ZFS,T}^{(M)}(t)$ (equation [11.16]). Indeed, deformations of the complex due to collisions, even if they are restricted to the vibrational modes, are difficult to describe mathematically. In addition, this interaction must be expressed in (L) if we wish to determine the random perturbation $\hat{H}_{ZFS,T}^{(L)}(t)$ contributing to the electronic relaxation. We therefore take a very simplified model of $\hat{H}_{ZFS,T}^{(L)}(t)$, assuming that it is produced by a second order Hamiltonian $\hat{H}_{ZFS,T}^{(M)}$ that is *independent of time*, and displays *axial* symmetry in an $\{X', Y', Z'\}$ reference frame linked to the complex:

$$\hat{H}_{ZFS,T}^{(M)} = D_T[\hat{S}_{Z'}^2 - S(S+1)/3] \qquad [11.22]$$

This reference frame is assumed to undergo *pseudo-rotational Brownian motion*, characterised by a correlation time τ_v with the same order of magnitude as the characteristic times of the vibrations and deformations of the complex, i.e., 10^{-12} to 10^{-11} s. τ_v is therefore much shorter than τ_r which is around 10^{-10} to 10^{-8} s (section 11.3.2). It should be noted that the reference frame $\{X', Y', Z'\}$ generally differs from $\{X, Y, Z\}$ in which the static contribution $\hat{H}_{ZFS,S}^{(M)}$ takes the simplified form given in [11.18]. The magnitude of $\hat{H}_{ZFS,T}^{(M)}(t)$, given by the expression $\Delta_T = \sqrt{2D_T^2/3}$ analogous to [11.19], is of the same order as Δ_S. The form of $\hat{H}_{ZFS,T}^{(L)}(t)$ is similar to that of $\hat{H}_{ZFS,S}^{(L)}(t)$ (complement 2).

11.4.2 - Expression for the electronic relaxation time T_{1e} for Gd^{3+}

The electronic relaxation time T_{1e} is the result of random fluctuations of the Hamiltonian

$$\hat{H}_{ZFS}^{(L)}(t) = \hat{H}_{ZFS,S}^{(L)}(t) + \hat{H}_{ZFS,T}^{(L)}(t) \qquad [11.23]$$

where fluctuations of $\hat{H}_{ZFS,S}^{(L)}(t)$ and $\hat{H}_{ZFS,T}^{(L)}(t)$ are characterised by the correlation time τ_r and τ_v, respectively, with $\tau_v \ll \tau_r$. It is difficult to obtain a closed form expression of the relaxation time T_{1e} for the Gd^{3+} ion due to the effect of the Hamiltonian [11.23]. Redfield's theory, based on second order perturbation theory, is much too complicated to be presented here. To be valid, on principle, the two conditions: (1) $\Delta_S \tau_r / \hbar \ll 1$ and (2) $\Delta_T \tau_v / \hbar \ll 1$ must be simultaneously satisfied, which is rarely the case. Indeed, most Gd^{3+} complexes – particularly the bulkiest ones – are characterised by slow reorientation movements such that $\tau_r > 0.1$ ns. Thus, condition (1) is not satisfied, even if condition (2) generally is. The values of T_{1e} calculated by applying Redfield's theory are therefore expected to be false, and indeed this has been demonstrated at low-field [Fries and Belorizky, 2007; Bonnet *et al.*, 2008]. However, unexpectedly, we have demonstrated theoretically and verified by numerical simulation (see section 11.6.2) that this approximation provides a good description of longitudinal electronic relaxation of the Gd^{3+} ion at intermediate MRI fields and higher magnetic fields: the electronic magnetisation M_z tends towards its equilibrium value M_{eq} according to a mono-exponentiel law characterised by a time T_{1e}, with

$$\frac{1}{T_{1e}} = \frac{1}{T_{1e,S}} + \frac{1}{T_{1e,T}} \qquad [11.24]$$

where the contributions from $\hat{H}_{ZFS,S}^{(L)}(t)$ and $\hat{H}_{ZFS,T}^{(L)}(t)$ are given by the expressions obtained by McLachlan from the Redfield approximation [Fries and Belorizky, 2007; McLachlan, 1964]:

$$\frac{1}{T_{1e,S}} = \frac{1}{25}[4S(S+1)-3]\left(\frac{\Delta_S}{\hbar}\right)^2 \tau_r \left(\frac{1}{1+\omega_S^2\tau_r^2} + \frac{4}{1+4\omega_S^2\tau_r^2}\right) \qquad [11.25]$$

$$\frac{1}{T_{1e,T}} = \frac{1}{25}[4S(S+1)-3]\left(\frac{\Delta_T}{\hbar}\right)^2 \tau_v \left(\frac{1}{1+\omega_S^2\tau_v^2} + \frac{4}{1+4\omega_S^2\tau_v^2}\right) \qquad [11.26]$$

The angular frequency ω_S is defined by equation [11.21]. Since $\tau_r \gg \tau_v$, $1/T_{1e,S}$ is much greater than $1/T_{1e,T}$ for $\omega_S = 0$, thus at zero field. But when B and thus ω_S increase, $1/T_{1e,S}$ decreases much faster than $1/T_{1e,T}$. The simulations described in section 11.6.2 show that at low-field ($B < 0.1$ T), electronic relaxation is mainly due to $\hat{H}_{ZFS,S}^{(L)}(t)$, even when the Redfield approximation does not apply, such that [11.25] does not apply. In contrast, $1/T_{1e,T}$ by far exceeds $1/T_{1e,S}$ at intermediate magnetic fields used in imaging ($0.5 \le B \le 1.5$ T) [Bonnet *et al.*, 2010].

11.5 - Estimation of the parameters determining the paramagnetic relaxation of protons

11.5.1 - Survey of parameters

To determine the paramagnetic relaxation rate $R_{1,p}$ (equation [11.4]) or in an equivalent manner, the relaxivity r_1 (equation [11.5]), we should know the following:

▷ from equations [11.6] to [11.8]: the solvation number q, the Gd-proton r_{GdH} distance, the residence time τ_m for a water molecule in the inner sphere of the complex, and the rotational correlation time τ_r for the complex,

▷ according to equations [11.10] to [11.13]: the minimal approach distance b between a proton from a water molecule and the Gd^{3+} ion in the complex, and the translational correlation time $\tau_D = b^2/D$.

▷ finally, T_{1e} must be calculated, which requires Δ_S, Δ_T and τ_v in addition to τ_r to be determined according equations [11.24] to [11.26]

Adjustable parameters are often used to interpret relaxivity experiments. This easy solution should be avoided, and the maximum number of parameters should be independently determined using methods that we will present briefly.

▷ Several methods can be used to determine the number q of water molecules coordinating Gd^{3+}. For example, it is possible to use variations in the NMR frequency for ^{17}O nuclei in water molecules induced by LnL complexes (L = ligand), where Ln^{3+} is a cation close to Gd^{3+} but for which the electronic relaxation is very fast, e.g. Dy^{3+}. These variations, known as induced paramagnetic shifts, which are independent of the nature of the ligands, are proportional to the solvation number q and the complex concentration. They can therefore be used to measure q [Alpoim et al., 1992]. EuL or TbL are similar to GdL and the τ_{H_2O} and τ_{D_2O} lifetimes for laser-induced luminescence of one of these complexes can be measured in H_2O and D_2O, then the proportionality relation $q \propto \left(\tau_{H_2O}^{-1} - \tau_{D_2O}^{-1}\right)$ [Horrocks and Sudnick, 1979; Parker and Williams, 1996] can be used. The values of q provided by these methods vary between 0 for the Gdttha complex (ttha = triethylenetetraaminehexaacetate) [Chang et al., 1990] and 8 for the aqua-ion $[Gd(H_2O)_8]^{3+}$ [Powell et al., 1996]. However, it should be noted that a Gd^{3+} complex must satisfy $q \leq 2$, for its thermodynamic stability in water to be sufficient to allow its safe use in clinical studies.

▷ The value of r_{GdH} is very important because of the $1/r^6{}_{GdH}$ dependence of R_{1m} in equation [11.8]. However, it is often only estimated. With Gd^{3+} poly(aminocarboxylate) complexes, a generally accepted reasonable value is $r_{GdH} = 0.31 \pm 0.1$ nm [Caravan, 2009].

▷ To determine τ_m, the transverse relaxation rate for ^{17}O nuclei of water molecules is measured as a function of temperature, together with variations in their resonance frequency in the presence of Gd^{3+} complexes [Merbach and Toth, 2001].

▷ The rotational correlation time τ_r for the complex is an essential component of inner sphere relaxivity, but its value is difficult to determine independently. Use of Stokes's rotational diffusion formula [11.9] is limited by the fact that we neither know the effective radius a of the complex, nor the effective rotational viscosity or microviscosity η^r_{mic} of the solution [Rast *et al.*, 2000]. However, the Stokes formula can be used within a family of relatively globular complexes, which have similar chemical structures and hydration properties, and are thus assumed to be subject to the same microviscosity. According to equation [11.9], τ_r is proportional to the volume of the complex. τ_r can be assumed to be proportional to its molecular mass given the relatively constant density of organic matter, such that the values of τ_r for complexes in the family considered vary in proportion to their masses [Aimé *et al.*, 2005]. τ_r can also be estimated by measuring the longitudinal relaxation rate for ^{17}O nuclei from water molecules as a function of temperature [Merbach and Toth, 2001]. Finally, τ_r can be deduced from the longitudinal relaxation times T_1 of deuterium substituted for hydrogen in the L ligand in diamagnetic complexes LaL, YL or LuL, which are analogous to GdL [Merbach and Toth, 2001; Bonnet *et al.*, 2008].

▷ The *relative diffusion* coefficient D can be estimated by independently measuring the coefficients of self-diffusion of water and of a diamagnetic analogue of GdL: LaL, YL or LuL. These coefficients of self-diffusion can readily be measured by pulsed-magnetic field gradient NMR techniques producing spin echos (of the signal) for protons from diffusing species [Canet *et al.*, 2002; Bonnet *et al.*, 2008].

▷ Four very important parameters remain: Δ_S, Δ_T and τ_v and b. They can be determined by three methods: (1) measurement of the relaxation time T_1 for water protons as a function of the applied field, and in some cases of the temperature, designated by NMRD (*Nuclear Magnetic Resonance Dispersion*) which involves these four parameters (2) the NMRD for protons of probe solutes [Bonnet *et al.*, 2010] and (3) analysis of simulated EPR spectra. The two latter techniques only involve Δ_S, Δ_T and τ_ν. These three techniques are complementary and they must all be used to verify the coherence of the results. NMRD methods, which have been the subject of numerous studies [Bertini *et al.*, 2001; Kowalevski and Maler, 2006; Helm, 2006; Korb and Bryant, 2005], are beyond the scope of this book.

It should be noted that the correlation times τ_r and τ_ν vary strongly with temperature, according to Arrhénius-type laws, with activation energies E_A^r and E_A^ν of around 10 kJ mol^{-1} [Merbach and Toth, 2001; Powell *et al.*, 1996].

11.5.2 - What EPR provides

We have seen that the longitudinal electronic relaxation time T_{1e} for the Gd^{3+} ion plays a key role in proton relaxivity, but that it is difficult to measure directly. In contrast, the Δ_S, τ_r, Δ_T, τ_ν parameters can be extracted from *simulations of the EPR spectrum* for a liquid solution of complexes, recorded at several temperatures and several frequencies. These parameters can be used to calculate T_{1e}, either by applying equations [11.24] to [11.26], or by the numerical procedure described below in section 11.6. The simulation is made difficult by the complex evolution of the transverse magnetisation of Gd^{3+}, which generally cannot be described by the Redfield approximation based on second order perturbation theory in $\hat{H}_{ZFS}^{(L)}(t)$ and for which the validity conditions detailed in section 11.4.2 must be satisfied [Fries and Belorizky, 2007]. Even when applying this approximation, the evolution is given by a combination of four decreasing exponentials which are tedious to calculate [Rast *et al.*, 2001]. For these reasons, in the following section we present a rigorous method which can be used to simulate both the EPR spectrum and the longitudinal relaxation. This method is conceptually simpler than the Redfield approximation, and it can be readily numerically implemented given the power of modern computers.

EPR can also be used to determine the Landé g factor for the complexed Gd^{3+} ion, which is involved in the expressions [11.8] of R_{1m} and [11.10], [11.11] of R_1^{OS}. However, g is always very close to $g_e = 2.0023$, and this value can be used given the uncertainty of the measurements and the precision of the relaxivity models (at most a few percent).

Experimentally, the EPR spectrum must be recorded over a broad range of frequencies (a few GHz to more than 500 GHz) and temperatures so as to verify the validity of the structural and dynamic models used in the simulations. The electronic relaxation times for Gd^{3+} complexes in aqueous solution are too short (e.g. $T_{2e} < T_{1e} \cong 10^{-9}$ s for $B = 0.34$ T) to allow free induction decay or echo signals to be observed. We therefore cannot use pulsed techniques, and EPR spectra are therefore recorded using continuous wave spectrometers. This is not without difficulties as the high values of the static and transient ze-ro-field splitting terms, around 0.03 cm^{-1}, and their associated correlation times $\tau_r > 100$ ps and $\tau_v \cong$ a few ps, produce linewidths of several tens of mT at X-band. High concentrations of complexes ($1 - 10$ mM) are therefore required to obtain a satisfactory signal/noise ratio. The appropriate spectrometers are described in the reference work on the chemistry of contrast agents [Merbach and Toth, 2001].

11.6 - Simulating the EPR spectrum and longitudinal relaxation of gadolinium complexes

11.6.1 - General theory

When a paramagnetic centre of spin S is subjected to a random Hamiltonian $\hat{H}(t)$, the absorption spectrum of the energy of a field $\mathbf{B}_1$ orthogonal to $\mathbf{B} // z$, which oscillates at an angular frequency ω, is given by [Abragam, 1983]:

$$f(\omega) = A \int_0^\infty \cos \omega t \, G_x(t) \, dt \qquad [11.27]$$

where $G_x(t)$ is the correlation function for the transverse component S_x of the spin:

$$G_x(t) = \langle S_x(t) S_x(0) \rangle = \frac{1}{2S+1} \overline{\text{Tr}[S_x(t) S_x(0)]} \qquad [11.28]$$

In this expression, $S_x(t)$ is defined by

$$S_x(t) = U(t)^\dagger S_x U(t) \qquad [11.29]$$

where $U(t)$ is the evolution operator of the spin states. It is determined by the spin Hamiltonian $\hat{H}(t)$ through Schrödinger's time-dependent equation

$$i\hbar \frac{dU(t)}{dt} = \hat{H}(t) U(t) \qquad [11.30]$$

with the initial condition $U(0) = 0$. In equation [11.28] the trace Tr is taken for the space of spin states of dimension $2S + 1$ and the bar indicates an *ensemble average* calculated over the different realisations of the random Hamiltonian $\hat{H}(t)$. At the initial time, we can write

$$G_x(0) = \frac{1}{2S+1} \operatorname{Tr} S_x^2 = \frac{1}{3} S(S+1) \tag{11.31}$$

In this general formalism, which is valid whatever the speed of fluctuations of $\hat{H}(t)$, the *transverse* electronic relaxation is described by the time dependence of the correlation function $G_x(t)$. Similarly, the *longitudinal* electronic relaxation is described by the time-dependence of the correlation function $G_z(t)$ of the S_z component defined by

$$G_z(t) = \langle S_z(t) S_z(0) \rangle = \frac{1}{2S+1} \overline{\operatorname{Tr}[S_z(t) S_z(0)]} \tag{11.32}$$

where
$$S_z(t) = U(t)^\dagger S_z U(t) \tag{11.33}$$

and
$$G_z(0) = \tfrac{1}{3} S(S+1) \tag{11.34}$$

Numerical simulation of $G_z(t)$ can be used to *test the validity of equations* [11.25] and [11.26].

For small-sized complexes which perform rapid Brownian rotation, such as the hydrated Gd^{3+} ion (see section 11.7.1), Redfield's theory can be applied to calculate the $G_x(t)$ and $G_z(t)$ functions [Rast *et al.*, 2000, 2001]. But generally, it is necessary to resort to numerical calculations, which are indeed conceptually simpler.

11.6.2 - Numerical simulation of the electronic relaxation of Gd³⁺ by the Monte Carlo method

Consider a solution of Gd^{3+} complexes. In the laboratory's reference frame (L), the random Hamiltonian $\hat{H}(t)$ for the general theory (equation [11.30]) is $\hat{H}_{spin}^{(L)}(t)$, given by equation [11.20], which is the sum of the Zeeman term (equation [11.21]) and the static and transient zero-field splitting terms (equation [11.23]). Fluctuations of these terms are the result of random and pseudo-rotational reorientations (deformations and vibrations) of the complex. We call *spin system* a specific Gd^{3+} complex with its Hamiltonian $\hat{H}_{spin}^{(L)}(t)$. To numerically simulate the evolution of the correlation functions $G_x(t)$ and $G_z(t)$ defined by equations [11.28] and [11.32], a large number of N_{sys} ($500 \le N_{sys} \le 3000$) realisations of the spin system must be generated to represent the various possible dynamic situations with good statistical precision. For each realisation, the molecular reference frames $\{X, Y, Z\}$ and $\{X', Y', Z'\}$ associated with the static and transient zero-field splitting terms will be randomly and independently oriented in (L). The dynamics of the spin state for a realisation can be calculated for a given duration $[0, t_{max}]$ by decomposing this duration into a large number N of very short steps Δt ($t_{max} = N \Delta t$) such that $\hat{H}^{(L)}_{spin}(t)$ can be considered constant and equal to $\hat{H}^{(L)}_{spin}(n\Delta t)$ during the interval $n\Delta t \leq t \leq (n + 1)\Delta t$ ($0 \leq n \leq N - 1$). During this interval, equation [11.30] can be immediately integrated:

$$U[(n + 1)\Delta t] = \exp\left[-\frac{i\hat{H}^{(L)}_{spin}(n\Delta t)\Delta t}{\hbar}\right]U(n\Delta t) \qquad [11.35]$$

The exponential function is calculated by numerical diagonalisation of $\hat{H}^{(L)}_{spin}(n\Delta t)$, which depends on the adjustable parameters Δ_S, τ_r, Δ_T, τ_v. By successive application of equation [11.35] with increasing n, the evolution operator $U(n\Delta t)$ is calculated step-by-step for each realisation. Practical calculation of $\hat{H}^{(L)}_{spin}(t)$, as defined by [11.23], and the method used to generate the random rotational trajectory of the complex are explained in complement 2. The correlation function $G_x(t)$, which is the arithmetic mean of the traces calculated for the different realisations of the spin system (equation [11.28]), can be used to reproduce the EPR spectrum for any field B with a single set of adjusted parameters (equation [11.27]). The correlation function $G_z(t)$, which is obtained in the same way from the $\text{Tr}[S_z(t)S_z(0)]$ traces, describes the longitudinal relaxation of the complex. At high fields, this function is observed to decrease in a mono-exponential manner, with a characteristic time T_{1e} given with good precision by equations [11.25] and [11.26] obtained in the framework of the Redfield approximation.

11.7 - Examples of simulations of EPR spectra for Gd^{3+} complexes

11.7.1 - The hydrated Gd^{3+} complex

Although not compatible with MRI because of its toxicity, the hydrated complex $[\text{Gd}(H_2O)_8]^{3+}$ has been used in several studies because it is simple to prepare and its symmetric structure is well known. The Gd^{3+} ion is at the centre of a square-based anti-prism the corners of which are occupied by H_2O ligands (figure 11.2). This figure can be obtained by deformation of a rectangular prism with two opposing square faces: we simply symmetrically pivot these faces by an angle of $\pm\pi/8$ around the axis of the prism passing through the centres of the faces. This complex has D_{4h} group symmetry. EPR spectra were recorded for this complex for various resonance frequencies and temperatures. These spectra were interpreted by assuming the implication of the static and transient zero-field splitting terms.

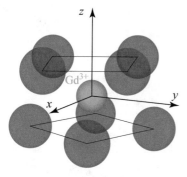

Figure 11.2 - The $[Gd (H_2O)_8]^{3+}$ complex. The apexes of the parallel squares are occupied by water molecules.

Figure 11.3 shows representative experimental spectra and their simulations. These were calculated by the Redfield method and validated by Monte Carlo numerical simulation, with the parameters indicated in table 11.2 [Rast *et al.*, 2001]. The parameters were adjusted so as to reproduce a large number of spectra recorded at different frequencies and temperatures, allowing variation of the relative contributions of the static and transient terms to obtain a single set of reasonable values. In particular, we verified that the τ_r value determined complies with Stokes's law [11.9] for a complex with the same size as $[Gd (H_2O)_8]^{3+}$ in water.

Table 11.2 - Zero-field splitting parameters and correlation times used to simulate EPR spectra for $[Gd(H_2O)_8]^{3+}$, $[Gd(DOTA)(H_2O)]^-$ and GdACX complexes at various frequencies and temperatures. The g factors are indicated, as are the activation energies when they were determined.

	$[Gd (H_2O)_8]^{3+ a}$	$[Gd (DOTA)(H_2O)]^-$	GdACX
$\Delta_S/\hbar \ [10^{10} \ \text{rad s}^{-1}]$	0.38	0.35	0.45
$\tau_r(T_0) \ [10^{-12} \ \text{s}]^{\ b}$	140	491	260
$E^r_A \ [\text{kJ mol}^{-1}]$	18.9	16.4	
$\Delta_T/\hbar \ [10^{10} \ \text{rad s}^{-1}]$	0.65	0.43	0.34
$\tau_v(T_0) \ [10^{-12} \ \text{s}]^{\ b}$	0.63	0.54	8.0
$E^v_A \ [\text{kJ mol}^{-1}]$	9.2	6.0	
g	1.99273	1.99252	1.985

[a] For simplicity, the amplitudes for static zero-field splitting terms of order 4 and 6, which contribute slightly to electronic relaxation, are not presented [Rast *et al.*, 2001]. [b] $T_0 = 298.15$ K.

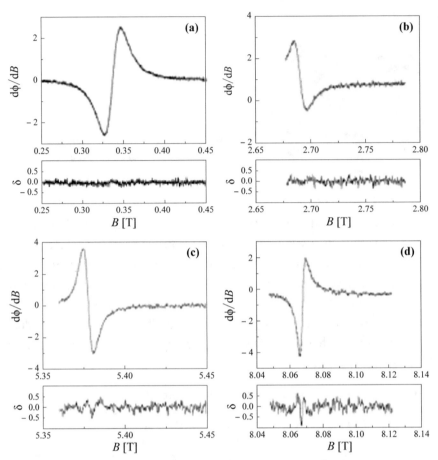

Figure 11.3 - EPR spectra (continuous lines) for the $[Gd(H_2O)_8]^{3+}$ complex recorded at various frequencies and temperatures. **(a)** 9.435 GHz; 354.0 K, **(b)** 75 GHz; 315.1 K, **(c)** 150 GHz; 365.0 K, **(d)** 225 GHZ; 320.1 K. Simulations (dashed lines) were obtained using the parameters listed in table 11.2. The difference Δ between the experimental and calculated spectra is indicated under each spectrum. Note that the experimental spectrum is remarkably well reproduced by the theory. [From Rast S. *et al.* (2001) *J. Am. Chem. Phys.* 123: 2637–2644 © 2001 American Chemical Society, reproduced with permission]

11.7.2 - The GdDOTA complex

To exploit the magnetic properties of the Gd^{3+} ion – which is toxic in its free aqua form – in MRI, it must be sequestered in a multidentate ligand. A multi-dendate ligand has several electron-donor atoms coordinating the Gd^{3+} ion to form a thermodynamically stable and kinetically inert bio-compatible complex. Poly amino carboxylates of Gd^{3+} were found to be particularly appropriate, and several are now used as contrast agents in clinical practice. Here, we will focus on the [Gd (DOTA)(H$_2$O)]$^-$ complex, where DOTA = [1,4,7,10-tetrakis(carboxyme-thyl)-1,4,7,10-tetra-azacyclo dodecane]. Its structure is shown in figure 11.4.

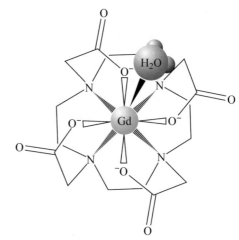

Figure 11.4 - Schematic representation of the structure of the [Gd (DOTA)(H$_2$O)]$^-$ complex.

Its EPR spectrum was simulated at various frequencies and temperatures by applying the Redfield theory and the parameters listed in table 11.2, in line with the Monte Carlo numerical simulation method [Rast *et al.*, 2001]. Figure 11.5 shows a typical X-band EPR spectrum and its simulation.

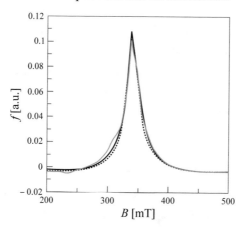

Figure 11.5 - Integrated EPR spectrum for the [Gd (DOTA)(H$_2$O)]$^-$ complex at X-band and $T = 274$ K. The black line represents the experimental spectrum, the grey line is the simulation given by Redfield's approximation. The dashed line was obtained by the Monte Carlo method. [From Rast S. *et al.* (2001) *J. Chem. Phys.* **115**: 7554–7563 © 2001 AIP Publishing LLC, reproduced with permission]

11.7.3 - The GdACX complex

The numerical simulation method described in section 11.6.2 was recently used to reproduce the X-band EPR spectrum for the Gd^{3+}– containing GdACX complex – in which ACX is the cyclodextrine derivative [Bonnet *et al.*, 2008] shown in figure 11.6.

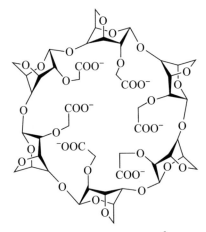

Figure 11.6 - Structure of the ACX^{6-} ligand.

This complex is not stable enough for human injection, but it can be used in animals where it displays acceptable toxicity. It has a high relaxivity due to a strong inner sphere contribution due to the $q = 4$ water molecules coordinating the metal. In animals, it can be used to study models of human brain cancer as, unlike GdDOTA, it remains intra-vascular in the presence of tumour-induced lesions of the hematoencephalic barrier [Lahrech *et al.*, 2008]. This property can be used to quantify the cerebral blood volume, which is an important parameter in tumour vascularisation. Figure 11.7 shows the X-band EPR spectrum for this complex at $T = 298$ K together with simulations by the Redfield method and the Monte Carlo numerical method using the parameters listed in table 11.2.

Note that for all the complexes studied, the values of T_{1e} at the fields used in imaging can be calculated by applying equations [11.24] to [11.26] and the parameters from table 11.2.

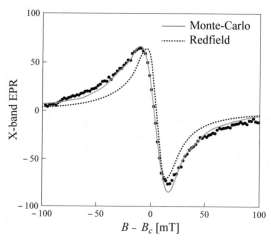

Figure 11.7 - X-band EPR spectrum for the Gd(ACX) complex in aqueous solution at $T = 298$ K. The continuous grey curve is the Monte Carlo simulation, the dashed black line corresponds to the Redfield approximation, which cannot be applied to this slow-rotating bulky complex. B is the applied field, B_c its central value. [From Bonnet C.S. *et al.* (2008) *J. Am. Chem. Phys.* **130**: 10401–10413 © 2008 American Chemical Society, reproduced with permission]

11.8 - Performance of Gd^{3+} complexes as contrast agents

In section 11.3, we described the various molecular parameters of Gd^{3+} complexes which determine the relaxivity of the water protons. Among these, the lifetime τ_m of the water molecules bound to the complex plays an essential role, as indicated by expression [11.6] for R_1^{IS}. A value of $\tau_m = 1/k_{ex}$ shorter than T_{1m} is required in order for exchanges with free water molecules to be as fast as possible and that the free water protons are effectively relaxed by the paramagnetic ions. Numerous derivatives of the basic complexing patterns have been synthesised by chemists, incorporating one or more hydrophobic side-chains likely to increase k_{ex} to more than 10^7 s^{-1}, resulting in $\tau_m \leq 100$ ns. The relaxivity r_1 obviously depends on the magnitude B of the applied field and on the temperature. The first GdDOTA contrast agents had relaxivities $r_1 \cong 4$ to 7 s^{-1} mM^{-1} at 25 °C in the $B \cong 0.5$ T field, which corresponds to an NMR frequency of 20 MHz. With some recent complexes, a relaxivity of around 50 s^{-1} mM^{-1} was obtained in the same conditions.

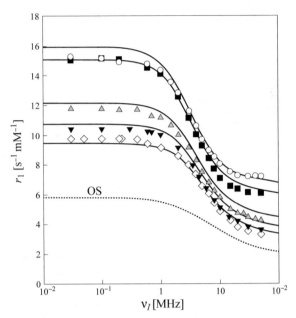

Figure 11.8 - Relaxivity curve $r_1(\nu_I)$ for water protons in a saline solution of [Gd(DOTA)
(H$_2$O)]$^-$ complexes at 277.2 K (white circles), 283.2 K (black squares), 298.2 K (grey
triangles), 305.2 K (black triangles), 298.2 K (white diamonds). The continuous theoretical curves were predicted by applying a simplified model limited to the effect of a
transient zero-field splitting term (equation [11.22]) to calculate the electronic relaxation.
The dashed curve at the bottom of the figure indicates the *outer sphere* contribution to
the relaxivity calculated at 298.2 K. [From Powell H., D. *et al.* (1996) *J. Am. Chem. Phys.* **118**:
9333–9346 © 1996 American Chemical Society, reproduced with permission]

In figure 11.8, the relaxivity values $r_1(\nu_I)$ *measured* for the [Gd(DOTA)(H$_2$O)]$^-$
complex in saline solution are presented as a function of the resonance frequency ν_I
of the water protons at various temperatures, along with predictions from a
theoretical model [Powell *et al.*, 1996]. Figure 11.9 shows the curve for water
proton relaxivity induced by the Gd(ACX) complex in a 0.1 M KCl solution
at 298 K [Bonnet *et al.*, 2008]. High relaxivity is obtained over a broad frequency range. The calculated curve was obtained using the parameters listed in
table 11.2, deduced from simulation of the EPR spectra and other independent
measurements. It reproduces the experimental data perfectly.

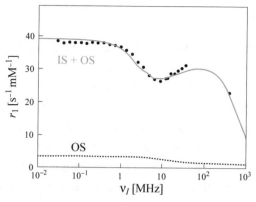

Figure 11.9 - Total relaxivity $r_1(v_I)$ (black dots) for water protons in an aqueous solution of Gd(ACX) containing 0.1 M KCl at 298 K. The grey curve is obtained by numerical simulation using the Δ_S, Δ_T, τ_v and g parameters from table 11.2. The value of τ_r was deduced by independent NMR measurements, and the solvation number $q = 4$ was deduced by measuring the lifetime of the luminescence using the method evoked in section 11.5.1. The dashed curve represents the calculated contribution for outer sphere relaxivity. [From Bonnet C.S. *et al.* (2008) *J. Am. Chem. Phys.* **130**: 10401–10413 © 2008 American Chemical Society, reproduced with permission]

Currently, in around 30 % of MRI examinations, radiologists inject Gd^{3+} complexes to improve image contrast and facilitate diagnosis. Figure 11.10 shows the T_1-weighted images of a patient's cerebral tumour before (left) and after (right) injection of the $[Gd(DOTA)(H_2O)]^-$ contrast agent. The contrast agent accelerates the speed of relaxation of the water protons in the tumour which appears as a lighter region, with much better definition, in the right-hand part of the image.

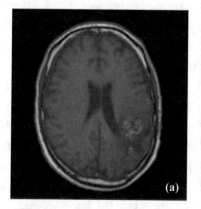
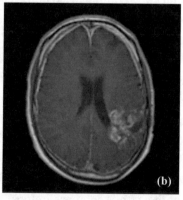

Figure 11.10 - T_1-weighted images acquired at Grenoble University Hospital at 1.5 T, showing a brain tumour **(a)** before and **(b)** after injection of the $[Gd(DOTA)(H_2O)]^-$ contrast agent. The signal for the tumour increases in the presence of the contrast agent – DOTAREM (Guerbet laboratories) – injected at a dose of 0.1 mmol kg^{-1} body weight.

11.9 - Perspectives

Several clues exist to further improve the contrast obtained with Gd^{3+} complexes while maintaining their non-toxicity, which is a pre-requisite for any medical application [Merbach and Toth, 2001; Aimé *et al.*, 2005; Caravan, 2009; De Leon-Rodriguez *et al.*, 2009; Vander Elst *et al.*, 2003; Caravan, 2006; Avedano *et al.*, 2007]. Ligands may be synthesised to produce the highest possible solvation number q compatible with good thermodynamic stability of the complex. Ligands could also be prepared which simultaneously complex several Gd^{3+} ions. Finally, the size of the macromolecular ligand could be increased to enhance the rotational correlation time τ_r, which shortens the longitudinal relaxation time T_{1m} for protons of water molecules coordinating the Gd^{3+} ions at intermediate imaging fields (equation [11.8]) and contributes to increasing R_1^{IS} (equations [11.6] and [11.8]). As mentioned above, the rate of exchange for the water molecules in the first solvation sphere can also be increased, which would reduce the duration of their coordination lifetime τ_m. In summary, the molecular parameters of inner sphere relaxivity must be optimised to satisfy the following conditions:

$$\tau_m < T_{1m} \text{ with } T_{1m} \text{ minimum} \qquad [11.36]$$

Currently, in normal MRI conditions, i.e., at room temperature and fields of 1.5 T, the highest values obtained for r_1 relaxivity are around $50 \text{ s}^{-1} \text{ mM}^{-1}$.

Several groups recently demonstrated that high relaxivities can be obtained by confining Gd^{3+} complexes in porous hollow nanostructures. Aimé, Sherry and their teams [Aimé *et al.*, 2002; Vasalatiy *et al.*, 2006] observed a considerable increase in the relaxivity of water protons when Gd^{3+} complexes were sequestered in apoferritin nanovesicles. Each nanovesicle has a cavity measuring approximately 7.5 nm in diameter, with pores of 0.3 to 0.4 nm, which are thus permeable to small water molecules but not to bulkier Gd^{3+} complexes. A nanovesicle can contain 6,000 to 8,000 water molecules and around 10 complexes, producing a high local concentration $c \cong 92$ mM. The large relaxivity measured $r_1 \cong 80 \text{ s}^{-1} \text{ mM}^{-1}$, can be explained by a rapid exchange between the water molecules inside the nano vesicles and the surrounding water molecules, combined with a strong increase in local viscosity inside the nanovesicles. These effects result in much longer local translational τ_D and rotational τ_r correlation times than in liquid water which is less viscous at the same temperature [Fries and Belorizky, 2010].

The theory predicts even larger relaxivities for optimised semi-permeable nanovesicles in which Gd^{3+} complexes are sequestered [Fries and Belorizky, 2010]. Finally, we note the very large values of r_1 – up to 100 to 200 s^{-1} mM^{-1} – which were experimentally obtained by sequestering Gd^{3+} complexes in small porous silica particles [Ananta *et al.*, 2010].

Complement 1 - Influence of the rate of exchange of inner sphere water molecules on proton relaxivity

Here, we present a simplified demonstration of expression [11.6] for R_1^{IS}, which was initially obtained by Luz and Meiboom [Luz and Meiboom, 1964]. A more rigorous demonstration, with an extension to three proton sites in chemical exchange, can be found in [Fries and Belorizky, 2010].

We are interested in the longitudinal relaxation rate for protons in water molecules of the solvent which exchange with water molecules coordinated to the complexed Gd^{3+} ions. Thus, the evolution of the component M_f (f for *free*) of their magnetisation along the field **B** towards its equilibrium value $M_{f,eq}$ must be studied. We note M_c (c for *coordinated*) the magnetisation of the protons in water molecules coordinated to the Gd^{3+} ions, and $M_{c,eq}$ its equilibrium value. The orders of magnitude for the two equilibrium magnetisations are very different. Indeed, as the number n_c of protons from the coordinated water molecules is much smaller than the number n_f of protons in free water molecules, we can write:

$$M_{c,eq}/M_{f,eq} = n_c/n_f = P_m \ll 1 \qquad [1]$$

where P_m is the molar fraction of coordinated water molecules.

The magnetisations M_c and M_f vary due to the effect of longitudinal relaxation which already occurs in the absence of chemical exchange. In the presence of exchange and in the absence of a radiofrequency field, the equations describing the evolution of these magnetisations are written:

$$\frac{dM_c}{dt} = -R_{1m}(M_c - M_{c,eq}) - W_{c \to f} M_c + W_{f \to c} M_f$$

$$\frac{dM_f}{dt} = -R_{1d}(M_f - M_{f,eq}) - W_{f \to c} M_f + W_{c \to f} M_c \qquad [2]$$

$R_{1m} = 1/T_{1m}$ is the longitudinal relaxation rate for "bound" protons in the absence of exchange, and R_{1d} is the intrinsic relaxation rate for "free" protons. $W_{c \to f}$ and $W_{f \to c}$ are, respectively, the transition probabilities per unit of time from c towards f and from f towards c. $W_{c \to f}$ is none other than the rate constant k_{ex} defined in figure 11.1:

$$W_{c \to f} = k_{ex} = \frac{1}{\tau_m} \qquad [3]$$

where τ_m is the lifetime for an inner sphere water molecule of the complex. The principle of detailed balance leads to the following relation

$$M_{c,eq} W_{c \to f} = M_{f,eq} W_{f \to c} \qquad [4]$$

Using equations [1], [3] and [4], we can therefore write

$$W_{f \to c} = \frac{P_m}{\tau_m} \qquad [5]$$

We now introduce the *differences* between the magnetisations and their equilibrium values

$$m_c \equiv M_c - M_{c,eq} ; \quad m_f \equiv M_f - M_{f,eq} \qquad [6]$$

Given [4] and [5], system [2] becomes

$$\frac{dm_c}{dt} = -\frac{1}{T_{1m}} m_c - \frac{1}{\tau_m} m_c + \frac{P_m}{\tau_m} m_f$$
$$\frac{dm_f}{dt} = -R_{1d} m_f - \frac{P_m}{\tau_m} m_f + \frac{1}{\tau_m} m_c \qquad [7]$$

From these coupled equations, we can show that the return of magnetisations M_c and M_f to their equilibrium values is monotonous except during a very short initial period. According to equation [1], we have $|m_c| \ll |m_f|$. Thus, in equations [7], we can neglect $|dm_c/dt|$ in comparison with $|dm_f/dt|$ after the transient period, i.e., throughout most of the relaxation. With this approximation, the first equation [7] produces the relation

$$m_c \left(\frac{1}{T_{1m}} + \frac{1}{\tau_m} \right) = \frac{P_m}{\tau_m} m_f \qquad [8]$$

By substituting m_c into the second equation [7], we obtain:

$$\frac{dm_f}{dt} = -R_{1d} m_f - \frac{P_m}{\tau_m} m_f + \frac{P_m}{\tau_m} \frac{T_{1m}}{T_{1m} + \tau_m} m_f = -\left(R_{1d} + \frac{P_m}{T_{1m} + \tau_m} \right) m_f \qquad [9]$$

Relaxation of M_f towards its equilibrium value is therefore described by the following equation

$$\frac{dM_f}{dt} = -(R_{1d} + R_1^{IS})(M_f - M_{f,eq}) \qquad [10]$$

where

$$R_1^{IS} = \frac{P_m}{T_{1m} + \tau_m} \qquad [11]$$

This is equation [11.6] in the main text. This expression shows how the exchange rate of water molecules affects the inner sphere paramagnetic relaxation of protons.

Complement 2 - Elements of the method used to simulate EPR spectra for Gd^{3+} complexes

◇ *Expressions for static and transient zero-field splitting (ZFS) Hamiltonians of Gd^{3+} in the laboratory's reference frame*

In section 11.4.1, we provided the expressions for the "static" $\hat{H}^{(M)}_{ZFS,S}$ and "transient" $\hat{H}^{(M)}_{ZFS,T}$ parts of the zero-field splitting Hamiltonian in a *molecular reference frame* (M); we now wish to obtain them in the *laboratory's {x, y, z} reference frame* when we shift from one reference frame to another by *rotation*.

▷ In the molecular reference frame {X, Y, Z}, the static part is written (equation [11.18]):

$$\hat{H}^{(M)}_{ZFS,S} = D_S\left[\hat{S}_Z^2 - S(S+1)/3\right] + E_S(\hat{S}_X^2 - \hat{S}_Y^2) \quad [1]$$

It is convenient to express $\hat{H}^{(M)}_{ZFS,S}$ as a function of the standard second order irreducible tensor operators $\hat{T}_2^q$ ($-2 \leq q \leq 2$) defined by [Messiah, 2014]:

$$\hat{T}_2^0 = \sqrt{\frac{3}{2}}\left[\hat{S}_Z^2 - S(S+1)/3\right], \quad \hat{T}_2^{\pm 1} = \mp\frac{1}{2}(\hat{S}_Z\hat{S}_\pm + \hat{S}_\pm\hat{S}_Z), \quad \hat{T}_2^{\pm 2} = \frac{1}{2}\hat{S}_\pm^2 \quad [2]$$

where $\hat{S}_\pm = \hat{S}_X \pm i\hat{S}_Y$. We therefore obtain:

$$\hat{H}^{(M)}_{ZFS,S} = \sqrt{\frac{2}{3}}D_S\hat{T}_2^0 + E_S(\hat{T}_2^2 + \hat{T}_2^{-2}) = \sum_{q=-2}^{2} C_2^q\hat{T}_2^q \quad [3]$$

where $C_2^0 = \sqrt{2/3}\,D_S$, $C_2^{\pm 1} = 0$, $C_2^{\pm 2} = E_S$.

Now consider a rotation $R(\mathbf{u}, \omega)$ by an angle ω around a unit vector $\mathbf{u}$, which transforms the laboratory's {x, y, z} reference frame into the {X, Y, Z} molecular reference frame. The advantage of irreducible operators is that in this rotation, each operator $\hat{T}_2^q$ becomes $\sum_{q'=-2}^{2} D_{q'q}^2(R)\hat{t}_2^{q'}$, where the $D_{q'q}^2(R)$ quantities are well known elements of the Wigner matrix $D^2(R)$ of dimension 5, and $\hat{t}_2^{q'}$ are the tensor operators [2] expressed using the components $\hat{S}_x, \hat{S}_y, \hat{S}_z$ of **S** in the {x, y, z} reference frame [Messiah, 2014]. By replacing the $\hat{T}_2^q$ by their expression as a function of the $\hat{t}_2^{q'}$ in equation [3] we obtain:

$$\hat{H}^{(L)}_{ZFS,S} = \sum_{q'=-2}^{2} x_{q'}\hat{t}_2^{q'}; \quad x_{q'} = \sum_{q=-2}^{2} D_{q'q}^2(R)C_2^q \quad [4]$$

▷ The same method can be used to obtain the expression $\hat{H}^{(L)}_{ZFS,T}$ for the transient part in the {x, y, z} reference frame: start with expression [11.22] of $\hat{H}^{(M)}_{ZFS,T}$ in the molecular reference frame {X', Y', Z'}, then introduce the pseudo-rotations $R_T(\mathbf{u}, \omega)$ in place of the overall spatial rotations $R(\mathbf{u}, \omega)$ of the complex.

◇ *Simulation of the random rotational trajectory of a complex*

The calculation of $\hat{H}_{ZFS}^{(L)}$ given by [11.23] requires knowledge of the $D_{q'q}^2(R_t)$ elements of the Wigner matrix, which are used to transform the laboratory's reference frame into the molecular reference frame by applying a random rotation R_t at time t. The rotations R_t defining the random rotational trajectory of the complex, which is assumed to be spherical, are generated by a succession of elementary rotations $R(\mathbf{u}, \Delta\omega)$, where $\Delta\omega$ is a random angle and the direction $\mathbf{u}$ is spatially distributed in an isotropic manner. The distribution of the angles $\Delta\omega$ can be taken to be uniform over a $[-\Delta\omega_{max}, +\Delta\omega_{max}]$ domain, where $\Delta\omega_{max}$ is small. Alternatively, $\Delta\omega$ can randomly take the values $\pm\Delta\omega_{max}$. As $R_{t+\Delta t} = R(\mathbf{u}, \Delta\omega) R_t$, $R_{n\Delta t}$ is obtained by applying n successive elementary rotations to the initial orientation defined by R_0. For each time step Δt, the angle θ of rotation of an axis linked to the molecular reference frame, for example Z for the real rotation modulating the static part of zero-field splitting, is given by the scalar product $\cos\theta = \mathbf{e}_{t+\Delta t} \cdot \mathbf{e}_t$, where $\mathbf{e}_t$, $\mathbf{e}_{t+\Delta t}$ are the unit vectors along the Z axis at times t and $t + \Delta t$, respectively. The extremity of the $\mathbf{e}$ vector moves with two-dimensional Brownian motion over the unit sphere. According to the Einstein relation, the mean quadratic value of the rotational angle θ is given by

$$\overline{\theta^2} = 4D^r\Delta t \qquad [5]$$

where D^r is the rotational diffusion constant which is linked to the correlation time τ_r introduced in equation [11.9] by the relation $D^r = 1/(6\tau_r)$. The angles $\Delta\omega$ of the random elementary rotations and pseudo-rotations of the complex are chosen such that $\overline{\theta^2}$ is small enough for Δt to be shorter than the relevant correlation time, i.e., τ_r and τ_v for the static and transient parts of the zero-field splitting, respectively. Equation [11.35] can thus be resolved step-by-step to obtain a numerical solution to the evolution equation [11.30] for each realisation of the spin system.

References

ABRAGAM A. (1983) *The Principles of Nuclear Magnetism*, Clarendon Press, Oxford.

AIMÉ S. *et al.* (2005) *Advances in Inorganic Chemistry* **57**: 173-237.

AIMÉ S. *et al.* (2002) *Angewandte Chemie International Edition* **41**: 1017-1019.

ALPOIM M.C. *et al.* (1992) *Journal of the Chemical Society, Dalton Transaction*: 463-467.

ANANTA J.S. *et al.* (2010) *Nature Nanotechnologies* **5**: 815-821.

AVENADO S. *et al.* (2007) *Chemical Communications* **45**: 4726-4728.

AYANT Y. *et al.* (1975) *Journal de Physique (Paris)* **36**: 991-1004.

BELORIZKY E. *et al.* (2008) *Journal of Chemical Physics* **128**: 052315, 1-17.

BERTINI I. *et al.* (2001) *Solution NMR of Paramagnetic Molecules*, Elsevier, Amsterdam.

BONNET C.S. *et al.* (2008) *Journal of the American Chemical Society* **130**: 10401-10413.

BONNET C.S. *et al.* (2010) *Journal of Physical Chemistry B* **114**: 8770-8781.

BOREL A. *et al.* (2002) *Journal of Physical Chemistry A* **106**: 6229-6231.

BOREL A. *et al.* (2006) *Journal of Physical Chemistry A* **110**: 12434-12438.

CALLAGHAN P.T. (2003) *Principles of Nuclear Magnetic Resonance Microscopy*, Oxford University Press, New York.

CANET D. (1996) *Nuclear Magnetic Resonance: concepts and methods*, Wiley, New-York.

CARAVAN P. (2009) *Account of Chemical Research* **42**: 851-862.

CARAVAN P. (2006) *Chemical Society Review* **35**: 512-523.

CHANG C.A. *et al.* (1990) *Inorganic Chemistry* **29**: 4468-4473.

CORREAS J.M. *et al.* (2009) *Nouvelles recommandations pour l'utilisation des agents de contraste ultrasonores : mise à jour 2008* ; Elsevier Masson SAS: Issy-les-Moulineaux, France.

DE LEON-RODRIGUEZ L.M. *et al.* (2009) *Account of Chemical Research* **42**: 948-957.

FRIES P.H. & BELORIZKY E. (2007) *Journal of Chemical Physics* **126**: 204503, 1-13.

FRIES P.H. & BELORIZKY E. (2010) *Journal of Chemical Physics* **133**: 0244504, 1-6.

FRIES P.H. & BELORIZKY E. (2012) *Journal of Chemical Physics* **136**: 074513, 1-10.

GIERER A. & WIRTZ K. (1953) *Zeitshrift für Naturforschung A: Physical Science* **8**: 532-538.

HELM L. (2006) *Progress in Nuclear Magnetic Resonance Spectroscopy* **49**: 45-64.

HORROCKS W.D. *et al.* (1979) *Journal of the American Chemical Society* **101**: 334-340.

HWANG L.P. & FREED J.H. (1975) *Journal of Chemical Physics* **63**: 4017-4025.

KIRICUTA I.C. & SIMPLACEANU V. (1975) *Cancer Research* **35**: 1164-1167.

KORB J.P. & BRYANt R.G. (2005) *Advances in Inorganic Chemistry* **57**: 293-326.

KOWALEWSKI J. & MALER L. (2006) *Nuclear Spin Relaxation in Liquids: Theory, Experiments, and Applications,* Taylor & Francis, Londres.

LAHRECH H. *et al.* (2008) *Journal of Cerebral Blood Flow and Metabolism* **28**: 1017-1029.

LEDLEY R.S. *et al.* (1974) *Science* **186**: 207-212.

LUZ Z. & MEIBOOM S. (1964) *Journal of Chemical Physics* **40**: 2686-2692.

MANSFIELD P. & PYKETT I.L. (1978) *Journal of Magnetic Resonance* **29**: 355-373.

MCLACHLAN A.D. (1964) *Proceedings of the Royal Society of London Series A* **280**: 271-288.

MERBACH A.E. & TOTH E. (2001) *The Chemistry of Contrast Agents in Medical Magnetic Resonance Imaging,* Wiley, New York.

MESSIAH A. (2014) *Quantum Mechanics ICS,* Dover Publications, New-York.

PARKER D. & WILLIAMS J.A.G. (1996) *Journal of the Chemical Society, Dalton Transaction* **18**: 3613-3628.

POWELL D.H. *et al.* (1996) *Journal of the American Chemical Society* **118**: 9333-9346.

RAST S. *et al.* (2000) *Journal of Chemical Physics* **113**: 8724-8735.

RAST S. *et al.* (2001) *Journal of the American Chemical Society* **123**: 2637-2644.

RAST S. *et al.* (2001) *Journal of Chemical Physics* **115**: 7554-7563.

SOLOMON I. (1955) *Physical Review* **99**: 559-565.

TADAMURA E. *et al.* (1997) *Journal of Magnetic Resonance Imaging* **7**: 220-225.

TORREY H.C. (1953) *Physical Review* **92**: 962-969.

VANDER ELST L. (2003) *European Journal of Inorganic Chemistry* **13**: 2495-2501.

VASALATIY O. *et al.* (2006) *Contrast Media Molecular Imaging* **1**: 10-14.

VIGOUROUX C. *et al.* (1999) *The European Physical Journal D* **5**: 243-255.

Volume 1: BERTRAND P. (2020) *Electron Paramagnetic Resonance Spectroscopy - Fundamentals,* Springer, Heidelberg.

VYMAZAL J. *et al.* (1999) *Radiology* **211**: 489-495.

YAZYEV O.V. & HELM L. (2008) *European Journal of Inorganic Chemistry* **2**: 201-211.

Chapter 12

Ferromagnetic resonance spectroscopy: basics and applications

von Bardeleben H.J., Cantin J.L., Gendron F.

Paris Institute of NanoSciences, Pierre and Marie Curie University, Paris VI.

12.1 - Introduction

A ferromagnetic resonance (FMR) experiment consists in applying a microwave field to excite a *single crystal* sample in which the spins are strongly coupled by *ferromagnetic exchange interactions*, in the presence of an external magnetic field. These interactions result from the combined effects of electrostatic interactions and the exclusion principle [Volume 1, section 7.2]. As in EPR, the frequency of the resonant mode depends both on the direction and the value of the applied magnetic field, in addition to the interactions internal to the system. In ferromagnetic systems, these internal interactions are the *energy of the demagnetising field and the magnetocrystalline anisotropy energy*. As in EPR, an FMR spectrum is recorded at fixed frequency while sweeping the magnetic field to achieve resonance. By studying how the resonance fields vary depending on the direction of the field in different crystallographic planes, the parameters characterising the sample's magnetic anisotropy can be determined. Analysis of the linewidths can also be used to determine the characteristic times for relaxation of the magnetisation. This description relates to the *uniform* resonance mode, in which precession of all the spins in the system are in-phase. Sometimes, non-uniform modes known as *spin waves* can be excited, their study provides other information, in particular on the value of the exchange parameter.

© Springer Nature Switzerland AG 2020
P. Bertrand, *Electron Paramagnetic Resonance Spectroscopy*,
https://doi.org/10.1007/978-3-030-39668-8_12

The very first FMR experiment was performed in 1946 by Griffiths at the *Clarendon Laboratory in Oxford* [Griffiths, 1946]. FMR then developed in parallel to EPR over the 1950s thanks to the availability of low-noise microwave sources, initially klystrons and more recently Gunn diodes. Its first applications helped us to better understand the properties of numerous crystals. While not exhaustive, we can cite studies of single crystals of iron [Kip and Arnold, 1949], of which the anisotropic effects are very well explained by Kittel's theory [Kittel, 1948]. Analyses of single crystals of magnetite Fe_3O_4 as a function of temperature were used to precisely study the behaviour of the anisotropy constant and the magnetogyric factor g [Bickford, 1950]. Bloembergen was more specifically interested in linewidth variations as a function of temperature for nickel and *Supermalloy* – an alloy of nickel (80 %), iron (15 %) and molybdenum (5 %) – which is used as a magnetic material in transformers and magnetic amplifiers [Bloembergen, 1950]. The experimental results agreed well with the theories, in particular that proposed by Van Vleck [Van Vleck, 1950]. The anisotropy constants and the magnetogyric factor for numerous ferrites were also measured, starting with ferrites of nickel $NiOFe_2O_3$ [Yager *et al.*, 1950; Healy, 1952]. In the case of ferrites of manganese in crystal form, experiments performed at several frequencies (24 GHz; 9.1 GHZ; 5.6 GHz and 2.8 GHz) revealed very narrow resonance lines (widths of around 5 mT) which allowed the width of these lines to be studied as a function of the orientation [Tannewald, 1955]. Finally, Artman [Artman, 1957a] produced the expressions describing angular variations in ferrite crystals and studied resonance behaviour as a function of the presence of magnetic domains. He also studied mono- and multi-domain metallic crystals [Artman, 1957b].

Today, many studies focus on the magnetic properties of nanostructured materials, thin or ultrathin films (down to monolayers), and nanoparticles. A specificity of these systems is that they have comparable numbers of atoms in volume and area, where the reduced symmetry creates surface magnetic anisotropy in addition to volume magnetocrystalline anisotropy. In the case of epitaxial films grown on a substrate with different characteristics, lattice mismatch causes deformations in the film, which create additional magnetic anisotropy. Below, we will first present the principles of FMR (section 12.2) and the experimental setups used (section 12.3). We then illustrate the advantages of using FMR to study nanostructured systems, e.g. magnetic metallic iron films, which are used to store information at high density (section 12.4), and semiconductor films rendered magnetic by doping (GaAs:Mn alloys) which could play a significant role in the future of spin electronics (spintronic) (section 12.5). Subsequently, we show that FMR can provide detailed information on the magnetic properties of suspensions of iron oxide nanoparticles which make up "ferrofluids" (section 12.6). An understanding of these properties and control of the reproducibility of the measured quantities are essential in biomedical applications of nanoparticles, which are currently being extensively developed.

12.2 - Principle of FMR

We saw above that FMR is similar in principle to EPR. However, the specific properties of the *ferro-* and *ferri-magnetic* bodies cause the parameters determining the positions and widths of the resonance lines to differ considerably between the two techniques. We will now briefly review these properties taking the example of ferromagnetic systems:

▷ In ferromagnetic materials, the strong exchange coupling between atoms bearing unpaired electrons tends to align their magnetic moments, resulting in much stronger magnetisation than that of paramagnetic substances.

▷ The system of magnetic dipoles creates an intense field $\mathbf{H}_d$ with a value which is very dependent on the shape of the sample. As this field has an opposite direction to the magnetisation $\mathbf{M}$ inside the material, it is known as the *demagnetising field*.

▷ In crystalline materials, the combined effects of spin-orbit coupling and Coulomb interaction with ions located at the nodes of the lattice produce a *magnetocrystalline* energy. This energy is minimal when the magnetisation $\mathbf{M}$ is directed along certain axes of the crystal known as the *easy axes of magnetisation*.

In the absence of external magnetic field, minimisation of the total energy resulting from the sum of the exchange, demagnetising field and magnetocrystalline anisotropy energies, results in spontaneous structuring of the ferromagnetic materials into *magnetic domains* with different directions of the magnetisation. This structure is such that the total magnetisation of the sample is generally weak or null. Application of an external field $\mathbf{H}$ modifies the organisation of the domains, and creates a total magnetisation which increases with H and reaches saturation when the sample becomes *monodomain*. The fields used in FMR to satisfy the resonance conditions are large enough for this limit to be reached in nanostructured materials.

Below, we will present the expressions for the energies of the demagnetising field and the magnetocrystalline anisotropy which will be used to obtain the resonance condition.

Readers interested in the properties of magnetic materials and in the methods used to study them can consult [du Trémollet de Lacheisserie, 2004]. Despite its age, Herpin's work [Herpin, 1968] has retained all its value. A recent publication by Alloul [Alloul, 2011] is particularly well written and documented.

12.2.1 - Expression of the energy of the demagnetising field

The demagnetising field $\mathbf{H}_d$ is the resultant of the magnetic fields created by all the magnetic dipoles in the sample, and its direction is opposite to that of the magnetisation $\mathbf{M}$. When the sample has a simple geometry, $\mathbf{H}_d$ can be expressed as a function of $\mathbf{M}$:

▷ In a uniformly magnetised cylindrical sample where $\mathbf{M}$ is parallel to the axis of the cylinder, the demagnetising field at the centre O of the cylinder is given by [du Trémollet de Lacheisserie, 2004]:

$$\mathbf{H}_d = -(1 - \cos\theta)\,\mathbf{M}$$

where θ designates half of the angle at which an extremity of the cylinder can be seen from point O. When the cylinder is very long, $\theta = 0$ and $\mathbf{H}_d = \mathbf{0}$ except near the extremities. For discs or cylindrical blades with low thickness relative to their other dimensions, $\theta = \pi/2$ and $\mathbf{H}_d = -\mathbf{M}$ far from the edges.

▷ In a uniformly magnetised sample with an *ellipsoid of revolution* shape, the demagnetising field is uniform and is written [du Trémollet de Lacheisserie, 2004]:

$$\mathbf{H}_d = -\tilde{\mathbf{N}} \cdot \mathbf{M}$$

where $\tilde{\mathbf{N}}$ is the tensor of the coefficients of the demagnetising field, also known as the "shape factors". The tensor's principle axes $\{x, y, z\}$ are the axes of the ellipsoid, and its trace is equal to 1. In this case, the energy per unit volume (also known as the energy density) associated with the dipolar interactions is given by:

$$F_d = \frac{1}{2}\mu_0 \mathbf{M} \cdot \tilde{\mathbf{N}} \cdot \mathbf{M} \qquad\qquad [12.1]$$

This energy is minimal when the magnetisation is oriented in the direction of the largest dimension of the sample. Because of this energy, we will see that the resonance condition for ferromagnetic materials depends on their shape.

12.2.2 - Expression for the magnetocrystalline anisotropy energy

In transition ion complexes, electrostatic interactions between the ligands and the orbital angular momentum of the electrons act on the spin angular momentum through spin-orbit coupling. This phenomenon is responsible for the existence of zero-field splitting terms in the spin Hamiltonian [Volume 1, section 6.2 and appendix 2]. Similarly, in ferromagnetic materials, ions in the crystal lattice interact with the spin angular momentum of the electrons thanks to spin-orbit coupling, which generates magnetocrystalline anisotropy energy. In monodomain crystalline materials, the magnetisation spontaneously orients in the "easy direction" of magnetisation, for which this energy is minimal. This energy is very difficult to calculate *ab initio* and it is generally simply described phenomenologically, assuming that it depends only on the direction of **M** relative to the crystal axes. This direction is either identified by angles (θ, ϕ) in a system of spherical coordinates, or by components $(\alpha_1, \alpha_2, \alpha_3)$ of the unit vector in the direction of **M**. In the latter case, the magnetocrystalline anisotropy energy is described by a polynomial in $\{\alpha_i\}$ only involving even powers. The terms of this polynomial depend on the crystal's symmetry as the energy must remain invariant under its symmetry operations. For a crystal of cubic symmetry, the magnetocrystalline anisotropy energy density is written:

$$F_K = K_4(\alpha_1^2\alpha_2^2 + \alpha_2^2\alpha_3^2 + \alpha_3^2\alpha_1^2) + K_4'\alpha_1^2\alpha_2^2\alpha_3^2 + ... \qquad [12.2]$$

If, in addition, the sample is subjected to uniaxial deformation in the Z direction, a term with the form $K_2 \sin^2\theta$ is added to F_K, where θ is the angle between **M** and Z.

12.2.3 - Expression for the resonance frequency

Consider a ferromagnetic monodomain sample and the "uniform" resonant mode in which all the magnetic moments rotate in-phase while conserving the amplitude of the magnetisation. We will first assume that the energy of the demagnetising field and of the magnetocrystalline anisotropy are negligible. If the sample is placed in a uniform external field **H**, provided neither energy dissipation nor relaxation occur, the rate of variation of the magnetisation **M** can simply be written

$$\frac{d\mathbf{M}}{dt} = -\gamma\mu_0 \mathbf{M} \times \mathbf{H} \qquad [12.3]$$

where $\gamma = g\beta/\hbar$. Due to the effect of the $\mu_0\mathbf{M} \times \mathbf{H}$ couple, the magnetisation precesses around $\mathbf{H}$ with an angular velocity $\omega_0 = \gamma\mu_0 H$. The angle between $\mathbf{M}$ and $\mathbf{H}$ can be increased by applying a second field $\mathbf{H}_1$ perpendicular to $\mathbf{H}$ and rotating with an angular velocity ω_0. In a ferromagnetic resonance experiment, polarisation of the microwave field $\mathbf{H}_1$ is, like in EPR, linear rather than circular. However, this field can be considered to be the sum of two circularly polarised components rotating in opposite directions, and only the component rotating in the same direction as $\mathbf{M}$ is efficient. Like in the case of a paramagnetic system, the resonance frequency is $\omega_0/2\pi$. If the frequency of the spectrometer is set to $\omega/2\pi$, resonance occurs for the resonance field

$$H_{res} = \frac{\omega}{\gamma\mu_0} \qquad\qquad [12.4]$$

In the presence of magnetocrystalline anisotropy and a demagnetising field, precession of the magnetisation takes place around its equilibrium direction, which is not necessarily the direction of $\mathbf{H}$, with an angular velocity ω that differs from ω_0. To determine ω, the sample's volumic energy density F must be considered. F depends on the directions of $\mathbf{H}$ and $\mathbf{M}$ relative to the axes of the crystal, which are identified by the angles (θ_H, ϕ_H) and (θ, ϕ), respectively (figure 12.1). To designate the crystallographic directions and planes, hereafter we will use the notations generally used in solid-state physics: if a vector has the components (l, m, n), we note $[l, m, n]$ a *direction parallel* to this vector and (l, m, n) a *plane perpendicular* to this vector. For further details, see [Artman, 1957a].

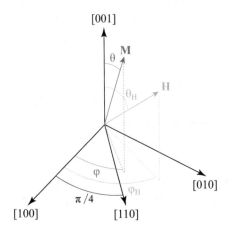

Figure 12.1 - Identification of the directions of the magnetisation and the magnetic field relative to the crystallographic axes.

The values (θ_{eq}, ϕ_{eq}) which define the direction of $\mathbf{M}_{eq}$ can be determined by writing that $F(\theta, \phi)$ is minimal, thus $\dfrac{\partial F}{\partial \theta} = \dfrac{\partial F}{\partial \phi} = 0$. For small precession angles around $\mathbf{M}_{eq}$, the angular precession frequency ω is given by the Smit-Beljers equation [Smit-Beljers, 1955]:

$$\left(\frac{\omega}{\gamma}\right)^2 = \frac{1}{M^2 \sin^2\theta_{eq}} \left[\frac{\partial^2 F}{\partial\theta^2}\frac{\partial^2 F}{\partial\phi^2} - \left(\frac{\partial^2 F}{\partial\theta\partial\phi}\right)^2\right] \qquad [12.5]$$

The partial derivatives are calculated for $\theta = \theta_{eq}$, $\phi = \phi_{eq}$.

To illustrate the preceding statements, consider a sample of a ferromagnetic crystal with cubic symmetry which has the shape of an ellipsoid of revolution, and which is subjected to a uniaxial constraint in the [001] direction. Its energy density can be written:

$$F = -\mu_0 \mathbf{M} \cdot \mathbf{H} + \frac{1}{2} \mu_0 \mathbf{M} \cdot \tilde{\mathbf{N}} \cdot \mathbf{M} + K_4 (\alpha_1^2\alpha_2^2 + \alpha_2^2\alpha_3^2 + \alpha_3^2\alpha_1^2) + K_2 \sin^2\theta \qquad [12.6]$$

The first term represents the energy of the interaction with the external field $\mathbf{H}$, the second represents the energy of the demagnetising field (equation [12.1]). The two other terms represent the anisotropy energies due, respectively, to the cubic symmetry of the crystal (equation [12.2]) and the uniaxial constraint. Components $(\alpha_1, \alpha_2, \alpha_3)$ of $\mathbf{M}$ can be expressed as a function of (θ, ϕ) by the usual relations (see for example [Volume 1, section 5.2.1]). Using equation [12.5], the expression for the resonance frequency can be calculated, and consequently the expression for the resonance field can be determined for a given direction of $\mathbf{H}$. In the specific case where the $\{x, y, z\}$ axes of the ellipsoid are parallel to the crystal axes, for $\mathbf{H}$ parallel to z, the following equation is obtained [Kittel, 1947, 1948]:

$$\left(\frac{\omega}{\gamma\mu_0}\right)^2 = \left[H_{res} + (N_x - N_z)M + \frac{2K_2}{M} + \rho\frac{K_4}{M}\right]\left[H_{res} + (N_y - N_z)M + \frac{2K_2}{M} + \rho\frac{K_4}{M}\right] \qquad [12.7]$$

Parameter ρ is equal to 2 when z is the crystallographic direction [100], and $-\frac{4}{3}$, when it is the direction [111]. Comparison of this expression to equation [12.4] clearly shows the effect of the demagnetising field and of magnetocrystalline anisotropy on the value of the resonance field. For spherical samples, $N_x = N_y = N_z = \frac{1}{3}$ and the demagnetising field is not involved. Note that Van Vleck [Van Vleck, 1950] obtained equation [12.7] from a purely quantum calculation. In the general case, where the ellipsoid and crystallographic axes do not coincide, and for any direction of the applied field, calculation of the resonance field is more complicated [Artmann, 1957b].

Equation [12.7] shows that, unlike in EPR, in FMR the resonance field depends on the magnitude M of the magnetisation. In a FMR experiment, the values of the spectrometer frequency and of the magnetisation are measured, and the variation in the resonance field is generally monitored as a function of the direction of **H** in certain crystallographic planes. By simulating these variations using expressions similar to equation [12.7], we can determine the directions of the easy and hard axes of magnetisation, the value of g and the anisotropy constants linked to the magnetocrystalline energy.

12.2.4 - Dissipation and relaxation phenomena: width of the resonance line

To consider the relaxation of the magnetisation **M** towards its equilibrium direction, terms must be added to equation [12.3]. These additional terms macroscopically account for the underlying phenomena of creation, annihilation and scattering of magnetic excitations: *the spin waves or magnons*. Several formulations have been proposed on purely phenomenological bases. The first was proposed in 1935 by Landau and Lifshitz well before the discovery of FMR [Landau and Lifshitz, 1935]. Currently, the Gilbert equation [Gilbert, 1956, 2004] is generally used, which qualitatively takes a broader range of damping phenomena into account:

$$\frac{d\mathbf{M}}{dt} = -\gamma\mu_0\mathbf{M} \times \mathbf{H} + \frac{\alpha}{M_S}\mathbf{M} \times \frac{d\mathbf{M}}{dt}$$

The dimensionless parameter α is known as the *damping factor*. This equation can be used to interpret the dissipation/relaxation phenomena by attributing them to an "effective field" proportional to $\frac{d\mathbf{M}}{dt}$. Note that another parameter G, known as the *Gilbert* coefficient is sometimes used in the literature. It is linked to α by:

$$\alpha = \frac{G}{\gamma\mu_0 M} \qquad [12.8]$$

where γ is the magnetogyric ratio and M is the value of the magnetisation. The Gilbert coefficient can be determined by studying how the width of the resonance line in the uniform mode varies as a function of the microwave frequency. Indeed, the peak-to-peak width ΔH_{pp} can be shown to be the sum of two terms [Heinrich and Bland, 2005]:

$$\Delta H_{pp}(\omega) = \Delta H_{inhom} + \Delta H_{hom} = \Delta H_{inhom} + \frac{2G}{\sqrt{3}\,\gamma^2 M}\omega \qquad [12.9]$$

The first term describes the broadening of the line linked to the magnetic inhomogeneities of the film. It is independent of frequency, but strongly influenced by the microstructure of the films. The second varies linearly with the microwave frequency and contains the Gilbert coefficient.

12.3 - Experimental aspects

An important aspect of FMR spectroscopy is homogeneous excitation of the sample throughout its volume. In the case of samples which have a metallic electrical conductivity, microwaves do not penetrate beyond a certain depth (skin effect). But as this depth is around a μm for microwaves at a frequency of 10 GHz, this phenomenon is not involved in the materials currently studied which are generally sub-micrometric films.

The basic elements of an FMR spectrometer are the same as for an EPR spectrometer [Volume 1, section 1.5.2]. In the standard configuration, the sample, which is a few nm thick and has an area of 1 to 10 mm^2, is placed in the centre of a resonant cavity operating in TE or TM mode. The quality factor for the cavity varies between 2,000 and 20,000. This cavity is linked to a microwave bridge composed of a microwave source (Gunn diode), an attenuator, a detector composed of a Schottky diode, a circulator and a reference arm. The most frequently used microwave frequency ranges are L- (4 GHz), X- (9 GHz) and Q- (35 GHz) bands. Electromagnets typically supply fields up to 2 teslas. The quantity presented on the abscissa for the spectra is $\mu_0 H$ [T]. The microwave magnetic field is maximal at the centre of the cavity and is generally vertically oriented. The Gunn diode delivers microwave radiation with a microwave power of up to 200 mW. Its frequency is measured using a frequency meter with precision better than 10^{-6}, and the magnetic field applied to the sample is measured using Hall probes at 10^{-5}. Coils create a sinusoidal magnetic field with a frequency of 100 kHz and an amplitude of around 1 mT, which superposes on the quasi-static magnetic field to allow synchronous detection of the signal. Consequently, the resonance spectrum takes the shape of a Lorentzian derivative [Volume 1, complement 3 to chapter 1]. Cryostats allow spectra to be recorded from 2 K to 300 K and from 300 K to 700 K.

To calculate the quantities that appear in the expressions given in the previous section, the magnetisation M must be known. Two types of particularly sensitive devices can be used to perform magnetic measurements over a range of temperatures from 2 K to 400 K. These are *Superconducting Quantum Interference Devices* (SQUID) and *Vibrating Sample Magnetometers* (VSM). Magnetic and ferromagnetic resonance measurements are often complementary, each providing information required by the other [Kopec *et al.*, 2009].

In addition to "classical" spectrometers with resonant cavities, such as those used for EPR, inductive measurements can be used when the system produces high-intensity FMR signals. These cavity-free systems, in which absorption of the microwaves at resonance is detected by spectral analysers, have the advantage of allowing continuous variation of the microwave frequency over a very broad range [Kuanr *et al.*, 2003].

12.4 - FMR of ultrathin metallic layers: epitaxial Fe films grown on (100) GaAs

Modern growth techniques such as molecular jet epitaxy allow thin monocrystalline layers of Fe of excellent crystallographic quality to be grown on GaAs surfaces [Wastlbauer *et al.*, 2005]. Layers of variable thickness, from the nm to the monolayer, can thus be prepared, allowing detailed study of their magnetic properties. However, this heteroepitaxy is characterised by lattice mismatch between the Fe, which has a cubic structure, and the GaAs, with a ZnS-type structure (zincblende). The mismatch is only around 8 %, despite significant differences between the crystallographic parameters. Epitaxy is achieved as two Fe cells can adapt to that of GaAs; indeed, if the parameters of the Fe and GaAs cells are noted a_{Fe} and a_{GaAs}, respectively, we observe that $2a_{Fe} = 5.73$ Å and $a_{GaAs} = 5.653$ Å. Nevertheless, the lattice mismatch causes tetragonal deformation of the Fe layer, which modifies its magnetic anisotropies.

As FMR measurements are generally performed in air after growth, a protective film is deposited on the Fe layer to prevent its oxidation. This protective layer modifies the magnetocrystalline anisotropy of ultrathin layers. To take this effect into account, surface/interface anisotropy is introduced into the energy expression, in addition to the usual volume anisotropies. For thicknesses exceeding 5 nm, volume properties dominate, and the properties are the same as full-mass Fe samples [Farle, 1998; Heinrich, 2005].

For a thin layer parallel to the crystallographic plane (001), the energy of the demagnetising field defined by equation [12.1] is $F_d = -\frac{1}{2}\mu_0 M^2 \sin^2\theta$, and the total energy density is written [Zakeri, 2007]:

$$F = -MH\left[\sin\theta\sin\theta_H\cos(\phi-\phi_H) + \cos\theta\cos\theta_H\right] - \left(\frac{1}{2}\mu_0\,M^2 - K_{2\perp}\right)\sin^2\theta$$

$$+ K_4\sin^2\theta - \frac{K_4}{8}(7+\cos 4\phi)\sin^4\theta + K_{2//}\sin^2\theta\cos^2\left(\phi - \frac{\pi}{4}\right)$$

The first term represents the interaction between the magnetisation and the applied field. K_4, $K_{2\perp}$, $K_{2//}$ are the anisotropy constants. K_4 is linked to the cubic structure of Fe (equation [12.6]), $K_{2\perp}$ and $K_{2//}$ are linked to out of plane and in plane uniaxial deformations, respectively. The resonance field is obtained using equation [12.5]. Its expression is simplified when the magnetic field is applied in a high-symmetry crystallographic plane:

▷ For a variation of **H** in the plane (110), between directions [001] and [1−10], it becomes:

$$\left(\frac{\omega}{\mu_0\gamma}\right)^2 = \left[H_{res}\frac{\cos\Delta\theta}{\sin\theta_{eq}} + \left(M_{eff} - \frac{2K_4}{M}\right)\frac{\cos 2\theta_{eq}}{\sin\theta_{eq}}\right.$$

$$\left. + \frac{K_4}{M}\left(3\sin^3\theta_{eq} - 9\sin\theta_{eq}\cos^2\theta_{eq}\right) - \frac{2K_{2//}}{M}\frac{\cos 2\theta_{eq}}{\sin\theta_{eq}}\right]$$

$$\left[H_{res}\sin\theta_H - \frac{2K_4}{M}\sin^3\theta_{eq} + \frac{2K_{2//}}{M}\sin\theta_{eq}\right] \qquad [12.10]$$

We have set $\Delta\theta = \theta_{eq} - \theta_H$ and $\Delta\phi = \phi_{eq} - \phi_H$. The quantity $M_{eff} = (M - 2K_{2\perp}/M)$ is generally named "*effective magnetisation*".

▷ For a variation of **H** in the plane (001) parallel to the film, we obtain:

$$\left(\frac{\omega}{\mu_0\gamma}\right)^2 = \left[H_{res}\cos\Delta\phi + \frac{2K_4}{M}\cos 4\phi_{eq} + \frac{2K_{2//}}{M}\cos 2\left(\phi_{eq} - \frac{\pi}{4}\right)\right]$$

$$\left[H_{res}\cos\Delta\phi - M_{eff} - \frac{2K_4}{M} + \frac{K_4}{2M}(7+\cos 4\phi_{eq}) + \frac{2K_{2//}}{M}\cos^2\left(\phi_{eq} - \frac{\pi}{4}\right)\right]$$

$$[12.11]$$

In some cases, two equilibrium directions exist for the magnetisation and the Smit-Beljers equation has two solutions. Thus, two resonance lines are observed, that of the uniform mode at high-field and an additional line at lower field known as the unsaturated mode line. This effect is particularly important in the case of thin layers of Fe, where the uniform mode is only observable at X-band for a few degrees near to [1−10].

To determine the surface anisotropy constants, a series of measurements must be performed on films of various thicknesses. Significant surface anisotropy exists for thicknesses of a few monolayers, and it can result in reorientation of the easy magnetisation axis: for thick layers, the demagnetising field dominates and the easy axis is oriented in the plane of the film, whereas for nanometric thicknesses the easy axis becomes normal to the plane of the film [Zakeri *et al.*, 2006]. The direction of the easy axis of magnetisation can be deduced from the sign of the M_{eff} quantity. In fact, the concept of demagnetising field corresponds to a macroscopic description of the system. This description loses its signification in the case of ultrathin layers composed of just a few atomic planes, for which the demagnetising field must be described at atomic scale [Heinrich, 2005].

Figures 12.2 and 12.3 represent FMR spectra for a 1.3-nm thick Fe layer protected by a thin layer (2 nm) of aluminium. The linewidths are small, around a few mT.

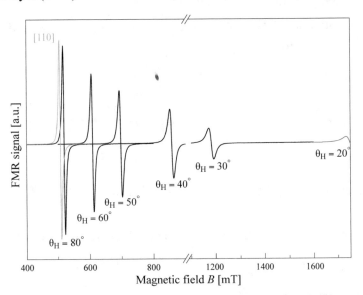

Figure 12.2 - Q-band FMR spectra for Al /Fe (1.3 nm) /(100)GaAs Rotational plane (110), T = 300 K.

This film presents uniaxial anisotropy in the plane due to the strong uniaxial constraints which dominate the cubic anisotropies. The resonance field for the uniform mode measured when the magnetic field is normal to the plane of the film ($\theta_H = 0$) is equal to 2.2 T at X-band, and 3.1 T at Q-band; these values exceed the magnetic fields that can be obtained with standard electromagnets, which are often limited to 2 T. Figure 12.4 shows the angular variation of the resonance field as a function of the direction of the magnetic field applied in various planes, and the simulations calculated using equations [12.10] and [12.11].

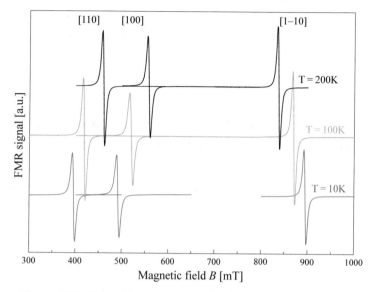

Figure 12.3 - Q-band FMR spectra for Al /Fe (1.3 nm) /(100)GaAs for H // [110], [100], [1−10].

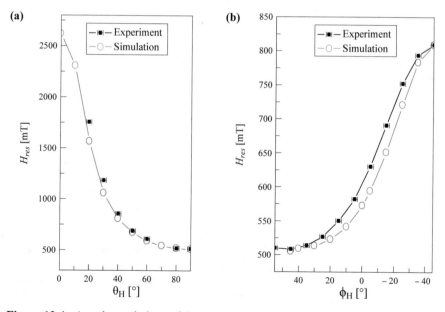

Figure 12.4 - Angular variations of the resonance field for the uniform mode measured at Q-band at 300 K for a sample of Al /Fe(1.3 nm) /(100)GaAs, in the **(a)** (110), **(b)** (001) planes.

12.4.1 - Surface/interface anisotropies

When surface and/or interfacial anisotropies exist, they must be explicitly accounted for. The K_i constants introduced previously are decomposed as the sum of a component K_i^V related to the volume of the system and one or more components K_i^S linked to the surfaces/interfaces such that:

$$K_i = K_i^V + \frac{K_i^{S,\,eff}}{d}$$

where d is the thickness of the film. Figure 12.5 shows how the surface anisotropy constants vary with the inverse of the film thickness in the case of an Fe film.

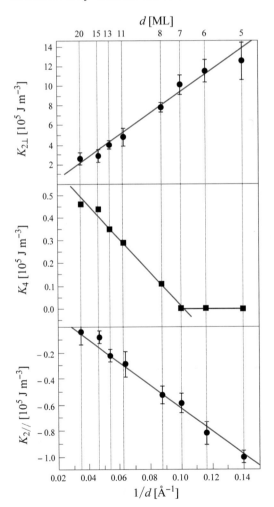

Figure 12.5 - Anisotropy constants $K_{2\perp}$, K_4, $K_{2//}$ as a function of the thickness of the Fe film on (100)GaAs measured by FMR at X-band at 300 K. At the top of the figure, d is expressed as a number of monolayers (ML).
[From Zakeri Kh. *et al.* (2006) *Phase transition* **79**(9): 793–813, Magnetic anisotropy in nanoscaled materials probed by ferromagnetic resonance: figure 6 © Taylor & Francis Ltd (http://www.tandf.co.uk/journals), reproduced with permission]

In table 12.1, we list typical values for the parameters measured by FMR at 300 K for an Fe film deposited on a (100) GaAs substrate. Analysis of the magnetic properties of the system requires components describing the anisotropy of the free surface of the film to be introduced [Zakeri, 2007]; the Fe/(100) GaAs interface effects are generally negligible.

Table 12.1 - Anisotropy constants characterising the volume and surface of an Fe film.

K_4^V	$K_{2\perp}^V$	$K_{2/\!/}^V$	$K_4^{s,eff}$	$K_{2\perp}^{s,eff}$	$K_{2/\!/}^{s,eff}$
$0.66\ 10^5$ $J\,m^{-3}$	$-1.7\ 10^5$ $J\,m^{-3}$	$0.18\ 10^5$ $J\,m^{-3}$	$-6.1\ 10^5$ $J\,m^{-2}$	$1.17\ 10^{-3}$ $J\,m^{-2}$	$-8.9\ 10^{-5}$ $J\,m^{-2}$

The surface anisotropies are modified when the surface of the film is no longer free but corresponds to an interface with another material. These interface anisotropies depend on the chemical nature of the material in contact with the Fe film. The interfaces which have been most studied are Au/Fe, Al/Fe, Cu/Fe, Ag/Fe or MgO/Fe. The anisotropy coefficients generally display significant temperature-dependence which is mainly linked to temperature-induced variations in the magnetisation. A detailed discussion of this phenomenon can be found in [Callen and Callen, 1996] and [Dzwiatkowski *et al.*, 2004].

12.4.2 - Relaxation mechanisms

Relaxation mechanisms can also be studied by FMR spectroscopy. They are not involved in the value of the resonance field, but result in broadening of the resonance line for the uniform mode. This broadening can be analysed as a function of the microwave frequency used (section 12.2.4). In the case of Fe, the broadening of the lines in the uniform mode is well described by equation [12.9]. In thin metallic films, the homogeneous term generally dominates for frequencies greater than 9 GHz. From G we can deduce the value of the damping factor α (equation [12.8]); with Fe, typical values of α are around 0.001.

12.5 - III–V ferromagnetic semiconductors: $Ga_{1-x}Mn_xAs/(100)GaAs$

The most commonly used semi-conductors from groups IV or III–V, such as Si or GaAs, are diamagnetic. When they are doped with magnetic ions at concentrations not exceeding their solubility ($\sim$ 100 ppm), they become paramagnetic. It is possible to dope some III–V compounds with Mn^{2+} ions at much higher concentrations – several % [Ohno *et al.*, 1992]. This doping is performed thanks to a new growth technique – low temperature molecular jet epitaxy – which allows layers with thicknesses ranging from a few nm to a few 100 nm to grow in out of thermodynamic equilibrium conditions. With this technique, abrupt GaMnAs/GaAs interfaces are produced without point or extended defects. In the case of GaAs and III–V compounds in general, Mn doping was found to be of great interest to introduce a ferromagnetic state. In III–V compounds, manganese not only provides magnetic Mn^{2+} ions which substitute for Ga with a spin $S = \frac{5}{2}$, it also constitutes an electrical doping agent which introduces free carriers (holes) in the material. Antiferromagnetic interaction between the spin of the delocalised holes and that of the Mn^{2+} ions creates ferromagnetic interaction between the Mn^{2+} ions, which generates long-range order when the proportion x of Mn exceeds 2 %. The ferromagnetic state of GaMnAs can then be described by a "Zener kinetic" model [Jungwirth *et al.*, 2006]. Curie temperatures of around 200 K were reported for highly doped layers (10 %) [Chen *et al.*, 2009]. Because of the contribution of the free holes to the exchange interaction between Mn^{2+} ions, the magnetic anisotropy of these layers reflects the anisotropy of the valence band characterised by the presence of heavy and light holes.

The magnetic properties of III–V ferromagnetic semi-conductors doped with Mn^{2+} are very different from those of magnetic metals such as Fe or Co. Mn doping levels of 5 % to 10 % produce dilute magnetic compounds, with much lower saturation magnetisation than that of the metals. For thin layers of $Ga_{0.95}Mn_{0.05}As$, the saturation magnetisation is close to 5×10^4 A m^{-1} (50 emu cm^{-3}) [Khazen *et al.*, 2008a], whereas for example, it is equal to 1.75×10^6 A m^{-1} (1750 emu cm^{-3}) for Fe. Consequently, the demagnetising field remains weak relative to the anisotropy fields. As a result, it is possible to adjust the orientations of the easy axes of magnetisation in the plane or out of plane by adjusting the constraints, for example. Indeed, as in the case of the ultrathin Fe films described in the previous section, the epitaxy of GaMnAs on GaAs generates uniaxial deformation linked to weak lattice mismatch which is proportional to the concentration of Mn [Thevenard *et al.*, 2005].

The main magnetic parameters (saturation magnetisation, Curie temperature, magnetocrystalline anisotropy) also depend on the concentration of the holes [Khazen, 2008a]. It is possible to set the hole concentration independently to that of the Mn^{2+} ions by p or n co-doping, hydrogenation, forming an alloy with phosphorous, or by depletion in a diode structure. The latter aspect is of particular interest as it can be used to control the orientation of the magnetisation by applying a voltage.

12.5.1 - Constraints and magnetocrystalline anisotropy

High-resolution X-ray diffraction experiments can be used to determine the cell parameter $a_\perp$ for the layer, which differs from that $a_{//}$ in the plane due to lattice mismatch. The latter is considered to be equal to that for the GaAs substrate. The incorporation of Mn^{2+} ions increases the cell parameter, which induces a uniaxial constraint in the layer, perpendicular to the plane. It has been shown [Thevenard et al., 2005] that a different substrate with a greater cell parameter, like $Ga_{1-y}In_yAs$, can eliminate or even reverse the constraints. More recently, the idea of manipulating the magnetic properties of these systems by playing on the constraints led to the production of quaternary $Ga_{1-x}Mn_xAs_{1-y}P_y$ alloys in which the state of constraint can be imposed independently of the Mn^{2+} concentration by adjusting the phosphorus content [Lemaitre et al., 2009; Cubukcu et al., 2010]. In these systems, the effective fields associated with the constraints, which can reach several hundred of mT, dominate the effects of the demagnetising field which rarely exceed 30 mT.

12.5.2 - FMR studies

Thin $Ga_{1-x}Mn_xAs$ layers have been studied by FMR spectroscopy at X- and Q-band [Liu et al., 2003; Liu and Furdunya, 2006; Khazen et al., 2008b]. The growth temperature is around 250 °C, but the magnetic properties of the layers can be improved by subsequent annealing between 180 °C and 200 °C. This annealing, which is most effective with high doping levels (> 5 %), allows the concentration of free carriers in the material to be increased, to reach values between 5×10^{20} cm^{-3} and 1×10^{21} cm^{-3}. Obviously, with this concentration of holes, which is considerably higher than the Mott transition, the conductivity of the layers is of the metallic type [Mott, 1982]. It is around a few $10^2 \, \Omega^{-1}$ cm^{-1} and varies very little between 300 K and 4 K; it presents a minimum at the Curie temperature. Layers with lower doping (2 %) present a ferromagnetic state with a low Curie temperature, of around 50 K. For higher concentrations of Mn, around 5 %, the Curie temperatures are around 130 K, and with higher doping levels (> 10 %), they can be up to 195 K [Chen et al. 2009].

The results obtained with epitaxial GaMnAs films on (100) GaAs or on (100) GaInAs [Liu *et al.*, 2003] show that the FMR spectra can be parameterised by four anisotropy constants: $K_{4//}$ and $K_{4\perp}$ linked to the deformed zincblende structure, and $K_{2//}$ and $K_{2\perp}$ linked to the uniaxial constraints in the plane and out of plane. The g factor can also be determined [Khazen *et al.* 2008a]. Because of the coupling between the paramagnetic ions and the holes, it differs from that of the Mn^{2+} ions ($g = 2.002$). Its value varies between $g = 1.95$ and $g = 2.00$.

Analysis of the FMR spectra is based on the expression for the energy density and the Smit-Beljers equation. The energy density in this case is written [Liu *et al.*, 2003]:

$$F = -MH\left[\cos\theta\cos\theta_H + \sin\theta\sin\theta_H\cos(\phi - \phi_H)\right] - \frac{1}{2}\mu_0 M^2\sin^2\theta$$

$$-K_{2\perp}\cos^2\theta - \frac{1}{2}K_{4\perp}\cos^4\theta - \frac{1}{2}K_{4//}\frac{(3 + \cos 4\phi)}{4}\sin^4\theta - K_{2//}\sin^2\theta\sin^2\left(\phi - \frac{\pi}{4}\right)$$

By applying equation [12.5], we obtain the following expressions for the resonance field:

▷ For a variation of **H** in the plane (110):

$$\left(\frac{\omega}{\mu_0\gamma}\right)^2 = \left[H_{res}\cos\Delta\phi + \frac{2K_{4//}}{M}\cos 4\phi_{eq} - \frac{2K_{2//}}{M}\cos\left(2\phi_{eq} - \frac{\pi}{2}\right)\right]$$

$$\left[H_{res}\cos\Delta\phi + 4\pi M - \frac{2K_{2\perp}}{M} + \frac{K_{4//}}{2M}(3 + \cos 4\phi_{eq}) + \frac{2K_{2//}}{M}\sin^2\left(\phi_{eq} - \frac{\pi}{4}\right)\right]$$

▷ For a variation of **H** in the plane (001):

$$\left(\frac{\omega}{\mu_0\gamma}\right)^2 = \left[H_{res}\cos\Delta\theta + \left(-4\pi M + \frac{2K_{2\perp}}{M} + \frac{K_{4\perp}}{M} - \frac{K_{//}}{2M}\right)\cos 2\theta_{eq}\right.$$

$$\left. + \left(\frac{K_{4\perp}}{M} + \frac{K_{4//}}{2M}\right)\cos 4\theta_{eq}\right]\left[H_{res}\cos\Delta\theta + \left(-4\pi M + \frac{2K_{2\perp}}{M} + \frac{K_{4//}}{M}\right)\cos^2\theta_{eq}\right.$$

$$\left. + \left(\frac{2K_{4\perp}}{M} + \frac{K_{4//}}{M}\right)\cos^4\theta_{eq} - \frac{2K_{4//}}{M} - \frac{2K_{2//}}{M}\right]$$

We have set $\Delta\phi = \phi_{eq} - \phi_H$ and $\Delta\theta = \theta_{eq} - \theta_H$. The value of M is obtained by measuring the magnetisation, and the Landé g factor is assumed to be equal to 2.00. Measurement of the resonance field for the uniform mode in the four orientations [001], [110], [100], [1−10] of the applied field can then be used to determine the four anisotropy constants.

In figure 12.6, we present a few typical FMR spectra for a thin layer (50 nm) of epitaxial $Ga_{0.95}Mn_{0.05}As$ on (100) GaAs.

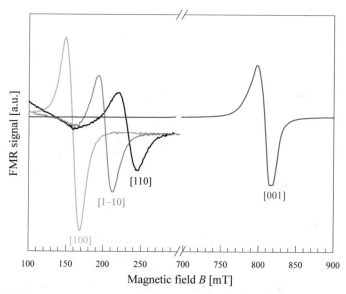

Figure 12.6 - FMR spectra for a film of $Ga_{0.95}Mn_{0.05}As$ recorded at X-band at 4 K.

The magnetic anisotropies are dominated by the constant $K_{2\perp}$, which is negative, with a value close to -5×10^3 J m^{-3} at 70 K (figure 12.7). This indicates, in agreement with the results obtained by X-ray diffraction, that the layers are under compressive constraints in the plane of growth. The ε_{zz} constraint is around 0.2 %.

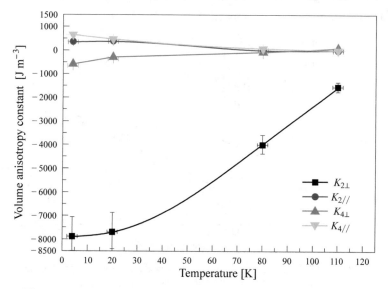

Figure 12.7 - Temperature variation of volume anisotropy constants for a film of $Ga_{0.95}Mn_{0.05}As$.

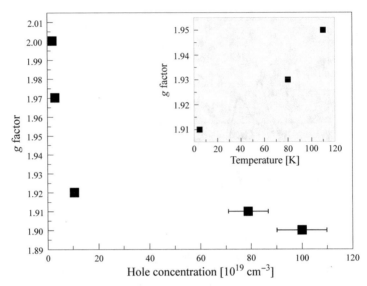

Figure 12.8 - Variation of the g value as a function of the hole concentration.

Further analysis of the angular variations of the FMR spectra and measurements at two microwave frequencies reveal the value of g to vary as a function of the concentration of holes (figure 12.8) [Khazen *et al.*, 2008a]. The damping factor α was also determined for various orientations of the field and various temperatures (figure 12.9) [Khazen *et al.*, 2008b] It is around 0.005 at 20 K and is anisotropic.

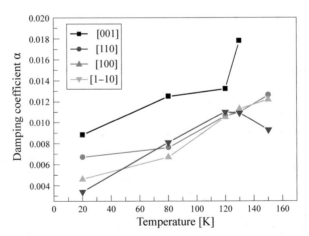

Figure 12.9 - Variation of the damping coefficient with temperature for various orientations of the applied magnetic field.

12.6 - Maghemite nanoparticle ferrofluids

Ferrofluids are composed of nanometric ferrimagnetic iron oxide particles suspended in a carrier liquid. The physicochemical parameters of synthesis and stabilisation of these colloids allow size-sorting and control of the interactions between particles. Technological interest in these magnetic nanostructures is considerable, particularly in the field of biomedical applications [Lévy *et al.*, 2010]. Indeed, iron oxide nanoparticles can penetrate cells and the "magnetic cells" produced by this labelling can be manipulated by magnetic fields, allowing them to be directed towards a target and localised by Magnetic Resonance Imaging (MRI) [Frasca *et al.*, 2011]. Magnetic labelling must however be sufficient for the magnetic force produced by the field to displace each cell without modifying the cell's function [Wilhelm *et al.*, 2007].

Ferrimagnetic nanoparticles have extensive surface effects, and due to their small size they display *superparamagnetic* behaviour: at zero-field, above a certain temperature lower than the Curie temperature, the thermal fluctuations of the direction of the magnetisation inside a nanoparticle become rapid enough for its mean value to be null. As a result, the system of nanoparticles behaves like a paramagnetic system composed of "giant" magnetic moments. The dynamic response of a ferrofluid is determined by the *internal dynamics* of the magnetic moment relative to the crystallographic lattice of the particle and by the *external dynamics* due to Brownian motion of the particles in the fluid. These two phenomena occur simultaneously and have different characteristic times:

▷ The characteristic time of the *internal dynamics* is that of fluctuations in the orientation of the magnetisation of a nanoparticle, which allows it to shift from one easy direction of magnetisation to another. This is called the "Néel time" and its temperature-dependence is that of an activated process:

$$\tau_N = \tau_0 \exp\left(\frac{E_a}{k_B T}\right)$$

τ_0 is around 10^{-9} s [Néel, 1949].

▷ The characteristic time for the *external dynamics* is the correlation time for the rotational Brownian motion for a spherical particle, which is given by

$$\tau_B = \frac{3\eta V_H}{k_B T}$$

where V_H is the hydrodynamic volume of the particle in the fluid with viscosity η.

When the anisotropy energy of the particles is weak, the activation energy E_a is small and $\tau_B \gg \tau_N$. The magnetisation of each particle can practically rotate independently of its motion in the fluid. When the anisotropy energy of the particles is strong, the activation energy E_a is large and $\tau_B \ll \tau_N$. The magnetisation of the nanoparticles is practically blocked along the easy axis of magnetisation. In this situation, it is the rotational Brownian motion of the nanoparticles which predominantly determines the magnetic relaxation of the system: its magnetic degrees of freedom are determined by the hydrodynamic degrees of freedom.

12.6.1 - Presentation of the nanoparticles

The nanoparticles studied are composed of maghemite γ-Fe_2O_3. The crystallographic structure is a lacunar inverse spinel with cubic symmetry; for particles, the cell parameter is between 8.35 Å and 8.36 Å. The spontaneous magnetisation is 3.3 Bohr magnetons per cell. This magnetisation is due to the Fe^{3+} cations which occupy 16 octahedral sites (B sites) and 8 tetrahedral sites (A sites) in each cell. Exchange interactions between cations are all antiferromagnetic, but the dominant one takes place, through oxygen ions, between two neighbouring cations, one of which is located on an A site and the other on a B site. As a result, all the spins belonging to the same lattice are parallel and the spins for the two lattices are antiparallel. The volume magnetocrystalline anisotropy is cubic, with [110] as the easy axis of magnetisation and [100] as the hard axis of magnetisation. The Curie temperature is 590 °C.

When their diameter is less than around a hundred nanometers, the particles are magnetic monodomains. This is the case of the nanoparticles used in the experiments presented below. They can thus be considered nano-magnets with a magnetic moment $\mu = V\mathbf{M}$, where V is the volume of the particle and $\mathbf{M}$ is its magnetisation. Electron microscopy images show that these nanoparticles are quasi-spherical. Their diameter d can be determined by measuring their magnetisation at room temperature. It is distributed according to a "log-normal" law:

$$P(d) = \frac{1}{\sqrt{2\pi}\, s d_{mp}} \exp\left[-\frac{1}{2s^2} \ln^2\left(\frac{d}{d_{mp}}\right)\right],$$

where d_{mp} is the most probable diameter and s is the standard deviation of $\ln d$. The mean of d^3 is given by $\langle d^3 \rangle = d_{mp}^3 \exp(7.5s^2)$. The characteristics of the distribution of d for the samples used in this study are described in table 12.2.

Table 12.2 - Characteristics of the nanoparticles in the samples studied.

Samples	d_{mp} [nm]	s [dimensionless]	$V = \frac{\pi}{6}\langle d^3 \rangle$ [nm³]
Initial synthesis 1	7	0.35	448
2	10	0.2	706
3	9.3	0.1	452
4	7.7	0.1	256
5	6.5	0.15	168
6	4.8	0.2	77

The initial synthesis produces nanoparticles with diameters which are too dispersed for our purposes. Indeed, with $0.35 < s < 0.4$, we obtain equal quantities of particles with a diameter of 3 nm and of 14 nm, and we will see that these two types of particles display very different dynamic behaviours. It is thus necessary to size-sort the particles. Size-sorting can be done by performing a succession of controlled phase-separations, which produces the different samples presented in table 12.2. It should be noted that the carrier liquid, glycerol, ethylene glycol or water, has no effect on the magnetic properties of the nanoparticles. We used glycerol, which has a solidification temperature of 210 K, as it allows resonance experiments to be performed between 3 K and 300 K without imposing mechanical constraints on the nanoparticles.

12.6.2 - FMR study of the magnetic anisotropy of the nanoparticles

Figure 12.10 shows X-band spectra recorded at 3.5 K for samples 2 and 6, containing the largest and smallest nanoparticles, respectively (table 12.2).

Two different procedures were used to cool the samples:

1- The sample was cooled from 300 K to 3.5 K in the absence of magnetic field. The particles (and their anisotropy axes) are thus randomly oriented in the frozen fluid.

2- The sample was subjected to an external magnetic field $\mathbf{H}_{FC}$ (*Field Cooling* FC) throughout the cooling phase. The particles are thus oriented with their anisotropy axes in the $\mathbf{H}_{FC}$ direction.

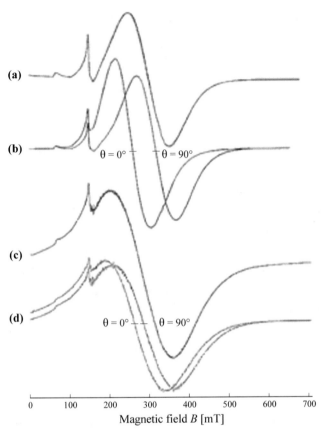

Figure 12.10 - Spectra of samples 2 and 6 measured at X-band (9.26 GHz) at 3.5 K.
(a) Sample 2 (d_{mp} = 10 nm) cooled from 300 K to 3.5 K in the absence of magnetic field.
(b) Sample 2 cooled under H_{FC} = 1 T. Spectra were obtained for θ = 0° and
θ = 90°, where θ is the angle between the direction of H_{FC} and that of the field
applied to induce resonance.
(c) Sample 6 (d_{mp} = 4.8 nm) cooled from 300 K to 3.5 K in the absence of magnetic field.
(d) Sample 6 cooled under H_{FC} = 1 T. Spectra were obtained for θ = 0° and θ = 90°.
The lines observed between 0 and 170 mT are due to residual impurities contained
in glass pipettes.

Resonance spectra were recorded for various values of the angle θ between
the direction of H_{FC} and that of the field **H** used to record the spectrum. As
expected, the resonance field for samples prepared according to procedure 1
was practically independent of the direction of **H**. In contrast, for samples in
which the nanoparticles were oriented before recording the spectrum, the resonance field passes through a minimum when **H** is parallel to H_{FC} and through a

maximum when **H** is perpendicular to **H**$_{FC}$ (figure 12.10b, d). This anisotropy is more noticeable for sample 2, which contains larger particles.

The minimal and maximal values increased with the temperature of the sample (figure 12.11). Above the freezing temperature for the carrier liquid, the particles are free to move, they orient themselves in the direction of the applied field **H** and the corresponding line is observed at $\theta = 0$ whatever the sample-preparation procedure. The position of the line produced by the sample cooled in the absence of magnetic field is also temperature-dependent. It tends towards the field $H_R = \dfrac{\omega}{\gamma\mu_0} = 330$ mT which would be observed if the sample was magnetically isotropic.

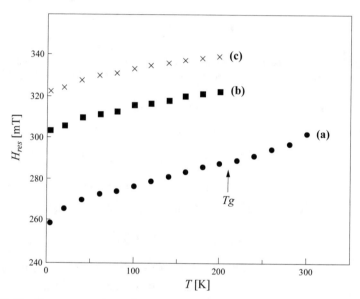

Figure 12.11 - Temperature-dependence of the resonance fields of sample 2. **(a)** and **(c)** sample 2 cooled in a H$_{FC}$ = 1 T field with **(a)** $\theta = 0°$, **(c)** $\theta = 90°$, **(b)** sample cooled in the absence of magnetic field. The arrow indicates the freezing temperature for the carrier liquid.

Figure 12.12 shows the variation of the resonance field as a function of θ for sample 2, at 3.5 K. It is well described by the following equation

$$H_{res}(\theta) = H_{res}(0) + [H_{res}(90) - H_{res}(0)]\sin^2\theta \qquad [12.12]$$

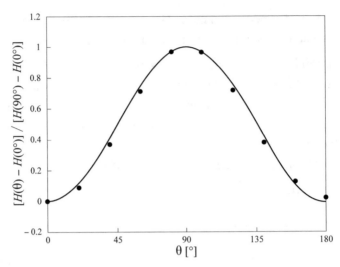

Figure 12.12 - Variation of the resonance field as a function of θ for sample 2 cooled under $H_{FC} = 1$ T, at 3.5 K. The curve is calculated using equation [12.12].

This equation has the same form as that giving the resonance field when the particle displays uniaxial anisotropy [de Biasi and Devesas, 1978]:

$$H_{res}(\theta) = \frac{\omega}{\gamma\mu_0} - H_a(1 - \frac{3}{2}\sin^2\theta)$$

where H_a is the anisotropy field for the particles. If all the particles were perfectly aligned in the field $\mathbf{H}_{FC}$, we could deduce by comparison

$$H_{res}(90) - H_{res}(0) = \frac{3}{2}H_a \qquad [12.13]$$

In fact, the orientation of the anisotropy axes relative to the field is not identical for all the particles in the sample. This distribution is the result of combined effects of the interaction between the magnetic moment and the field $\mathbf{H}_{FC}$, the magnetocrystalline anisotropy energy, which tends to align the magnetic moment along the anisotropy axes (section 12.2.2), and thermal motion. By modelling these different contributions, an expression describing the distribution of the orientation of the anisotropy axes throughout the sample can be obtained [Raikher and Stepanov, 1994]. A simple means of taking this distribution into account consists in replacing equation [12.13] by:

$$H_{res}(90) - H_{res}(0) = \frac{3}{2}H_a\langle\cos^2\theta\rangle \qquad [12.14]$$

The quantity $\langle\cos^2\theta\rangle$ can be calculated from the expression for the distribution function [Gazeau *et al.*, 1998]. The difference $H_{res}(90) - H_{res}(0)$ is deduced from the spectra, and equation [12.14] can be used to determine H_a. The calculation shows that the factor $\langle \cos^2 \theta \rangle$ decreases rapidly with the particle size [Gazeau et al., 1998], which explains why the difference $H_{res}(90) - H_{res}(0)$ is smaller for sample 6 than for sample 2 (figure 12.10). Indeed, the results show that H_a is proportional to the inverse of the particle diameter (figure 12.13a).

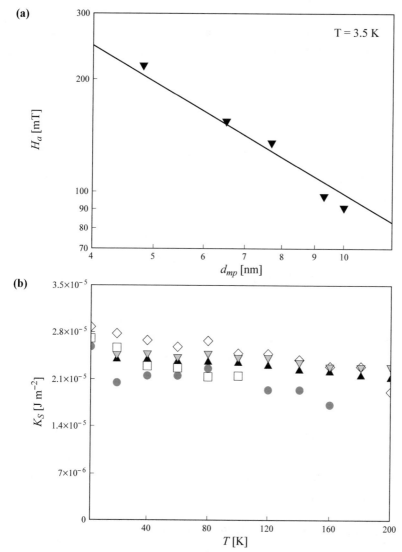

Figure 12.13 - **(a)** Variation of H_a as a function of the particle diameter d_{mp} (double log scale). The values of H_a are deduced from angular variations measured at 3.5 K. The slope of the curve is –1. **(b)** Variation of the K_S parameter defined by equation [12.15] as a function of temperature for samples 6 to 2. (White squares) $d_{mp} = 4.8$ nm; (black dots) $d_{mp} = 6.5$ nm; (white diamonds) $d_{mp} = 7.7$ nm; (grey triangles) $d_{mp} = 9.3$ nm; (black triangles) $d_{mp} = 10$ nm.

This variation is typical of *uniaxial surface* anisotropy (section 12.4.1). Interpretation of the data with a model where the particles display volume anisotropy, and thus an anisotropy field $H_a = \frac{2K}{M}$, results in a contradiction. Indeed, the smaller the diameter of the particles, the greater the value of K, whereas this quantity should remain constant if the origin of the anisotropy were purely volumic. To shift from volume anisotropy to surface anisotropy, it is sufficient to multiply the previous relation by the surface-to-volume ratio of the particles. In this case, the anisotropy field is linked to the surface anisotropy constant K_S by [Gazeau, 1997]:

$$H_a = \frac{12K_S}{Md_{mp}} \qquad [12.15]$$

The value of K_S deduced from this relation can be verified to be practically independent of d_{mp} (figure 12.13b). We obtain $K_S \cong 2.8 \times 10^{-5}\,\mathrm{J\,m^{-2}}$ at 3.5 K.

FMR experiments thus unambiguously demonstrate that, despite their cubic crystal structure, nanoparticles present uniaxial anisotropy mainly due to surface effects. The notion of surface anisotropy was originally introduced by L. Néel [Néel, 1954]. Its origin is due to the fact that interactions between the spins of ions located at the surface of the particles are different to those existing inside the particle. This effect is particularly significant in nanoparticles. L. Néel described surface anisotropy by associating it with a surface density K_{SR} and a local axis perpendicular to any point at the surface. This anisotropy is null for a perfectly spherical particle, but it becomes significant when the particle has a different shape. For an ellipsoid of revolution with a low-ellipticity e, the anisotropy constant K_S is linked to K_{SR} by the following relation:

$$K_S = \frac{4}{15} e^2 K_{SR}$$

According to L. Néel, K_{SR} should be around 10^{-4} to $10^{-3}\,\mathrm{J\,m^{-2}}$ [Néel, 1954]. Our results confirm this prediction. Indeed, for $e = 0.4$ (which corresponds to a 10 % difference relative to a spherical shape), the value $K_S = 2.8 \times 10^{-5}\,\mathrm{J\,m^{-2}}$ measured by FMR results in $K_{SR} = 0.6 \times 10^{-3}\,\mathrm{J\,m^{-2}}$.

12.6.3 - Evidence for a superparamagnetic component

At higher temperatures, typically between 200 K and 300 K, the width and position of the spectrum are highly dependent on the size of the particles (figures 12.14, 12.15):

▷ For larger particles, the line is broad ($\Delta H_{pp} \cong 60$ mT) and the resonance field is weaker than the reference field $H_R = 330$ mT. When the sample is cooled in the field $\mathbf{H}_{FC}$, the resonance field depends on the direction of the field $\mathbf{H}$ relative to $\mathbf{H}_{FC}$, like in figure 12.10.

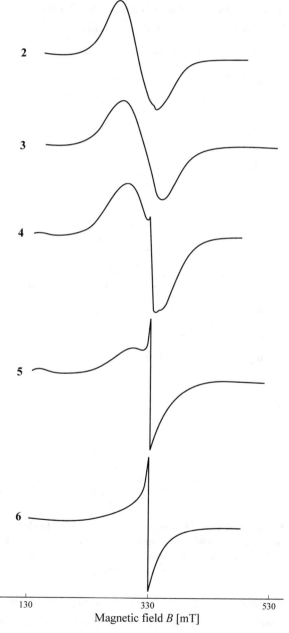

Figure 12.14 - X-band spectra for samples 2 to 6 recorded at room temperature.

▷ For the smallest particles, the line is narrow ($\Delta H_{pp} \cong 6$ mT) and centred on the H_R field. When the sample is cooled in the field $\mathbf{H}_{FC}$, the resonance field remains equal to H_R whatever the direction of the field $\mathbf{H}$. The particles producing this line are thus not affected by an anisotropy field.

▷ For medium-sized particles, the spectrum contains both components, with the same characteristics as those described above.

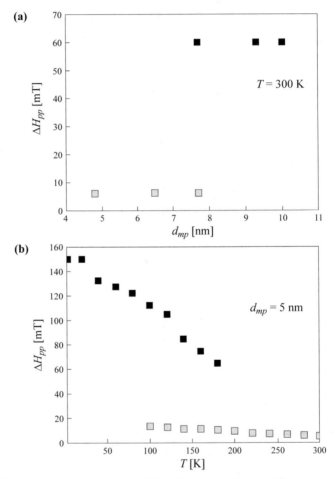

Figure 12.15 - Peak-to-peak widths ΔH_{pp} of resonance lines: **(a)** at room temperature as a function of the particle diameter, **(b)** as a function of temperature for sample 6 (d_{mp} = 4.8 nm). The black and grey squares correspond to the broad and narrow components, respectively.

The narrow component of the spectrum might be due to paramagnetic impurities present in the carrier liquid. To test this hypothesis, we compared its intensity to that of a "standard" ruby sample, an alumine crystal doped with a known number of Cr^{3+} ions. The measurements show that the intensity of the narrow component corresponds to a number of centres three orders of magnitude greater than that for the Fe^{3+} ions present in the sample. We can thus eliminate the hypothesis of a paramagnetic signal and attribute the narrow component to a *superparamagnetic* state of the nanoparticle system.

To interpret these results, we must call on a model which describes the effects of anisotropy and thermal fluctuations on the width of the resonance lines for a system of monodomain particles. This model relies on the Landau-Lifshitz phenomenological equation (section 12.2.4) and on an equation describing the rotational diffusion of the moment of a particle in the presence of fluctuations [Brown, 1963; Raikher *et al.*, 1974]. It accounts for most of the experimental observations, but only qualitatively [Gazeau *et al.*, 1999].

12.7 - Conclusion

Today, the field of application for FMR is very diverse. The nature of the objects studied has changed considerably since the first studies, as shown by the examples presented in this chapter. Most studies now focus on ultrathin films, multilayers or nano-objects within which new magnetic effects are observed. This research is often motivated by potential applications in spin electronics. For example, we can mention the case of ultrathin bilayers of GaMnAs/Fe in which a significant increase in the Curie temperature for GaMnAs near the interface was observed [Song *et al.*, 2011], or the effects of spin pumping at the magnetic insulator/metal (YIG/Au) interface [Heinrich *et al.*, 2011]. FMR also makes a significant contribution to our understanding of the magnetic properties of Heusler-type alloys, such as thin CoMnGe films. These materials are of interest in spintronics [Belmeguenai *et al.*, 2010]. Similarly, the dynamics of the magnetisation of thin layers of $La_{0.7}Sr_{0.3}MnO_3$, a material in the manganese perovskites family, can be better understood by FMR analysis [Belmeguenai *et al.*, 2011]. Very recently, electrically-detected FMR was used to explore the inverse Hall effect of spins [Ando *et al.*, 2009] and to generate microwave fields within single nano-objects through spin-orbit coupling [Fang *et al.*, 2011]. These very varied studies illustrate the relevance of FMR in the field of nanosciences.

References

ALLOUL, H. (2011) *Introduction to the Physics of Electrons in Solids*, Springer, Heidelberg, London, New York.

ANDO K. *et al.* (2009) *Applied Physics Letters* **94**: 262505-26507.

ARTMANN J.O. (1957) (a) *Physical Review* **105**: 62-73.

ARTMANN J.O. (1957) (b) *Physical Review* **105**: 74-84.

BELMEGUENAI M. *et al.* (2010) *Journal of Applied Physics* **108**: 63926-63932.

BELMEGUENAI M. *et al.* (2011) *Journal of Applied Physics* **109**: 7C120-7C123.

BICKFORD L.R. Jr. (1950) *Physical Review* **78**: 449-454.

BLOEMBERGEN N. (1950) *Physical Review* **78**: 572-580.

BROWN W.F. Jr. (1963) *Physical Review* **130**: 1677-1686.

CALLEN H.B. & CALLEN E. (1996) *Journal of Physics and Chemistry of Solids* **27**: 1271-1285.

CHEN L. *et al.* (2009*) Applied Physics Letters* **95**: 182505-182508.

CUBUKCU M. *et al.* (2010) *Physical Review B* **81**: 41202-41206.

DE BIASI R.S. & DEVEZAS (1978) *Journal of Applied Physics* **49**: 2466-2469.

DZIATKOWSKI K. *et al.* (2004) *Physical Review B* **70**: 115202-115214.

FANG D. *et al.* (2011) *Nature Nanotechnology* **6**: 413-417.

FARLE M. (1998) *Reports on Progress in Physics* **61**: 755-826.

FRASCA G *et al.* (2011) *Reflets de la Physique* revue de la Société Française de Physique **23**: 6-10.

GAZEAU F. (1997) PhD thesis, *Magnetic and Brownian dynamics of nanoparticles in a ferrofluid*, Université Paris 7 Denis Diderot.

GAZEAU F. *et al.* (1998) *Journal of Magnetism and Magnetic Materials* **186**: 175-187.

GAZEAU F. *et al.* (1999) *Journal of Magnetism and Magnetic Materials* **202**: 535-546.

GILBERT T.L (1956) *Ph.D thesis " Formulation, Foundations and Applications of the Phenomenological Theory of Ferromagnetism "* Illinois Institut of Technology.

GILBERT T.L. (2004) *IEEE Transactions on Magnetics* **40**: 3443-3449.

GRIFFITHS J.H.E (1946) *Nature* **158**: 670-671.

HEALY D.W Jr (1952) *Physical Review* **86**: 1009-1013.

HEINRICH B. & BLAND J.A.C. (2005) *Ultrathin Magnetic Structures II-III*, Springer, Heidelberg.

HEINRICH B. *et al.* (2011) *Physical Review Letters* **107**: 66604-66608.

HERPIN A. (1968) *Théorie du magnétisme,* Presses Universitaires de France, Paris.

JUNGWIRTH T. *et al.* (2006) *Review of Modern Physics* **78**: 809-864.

KHAZEN Kh. *et al.* (2008) (a) *Physical Review B* **77**: 165204-165219.

KHAZEN Kh. *et al.* (2008) (b) *Physical Review B* **78**: 195210-195218.

KIP A.F. & ARNOLD R.D. (1949) *Physical Review* **75**: 1556-1560.

KITTEL C. (1947) *Physical Review* **71**: 270-271.

KITTEL C. (1948) *Physical Review* **73**: 155-161.

KOPEC M. *et al.* (2009) *Chemical Physics* **21**: 2525-2533.

KUANR B.K. *et al.* (2003) *Journal of Applied Physics* **93**: 7723-7726.

LANDAU L. & LIFSHITZ E. (1935) *Physikalische Zeitschrift den Sovjetunion* **8**: 153-169.

LEMAITRE A. *et al.* (2009) *Applied Physics Letters* **93**: 21123-21126.

LÉVY M. *et al.* (2010) *Nanotechnology* **21**: 395103-395114 and references therein.

LIU X. *et al.* (2003) *Physical Review B* **67**: 205204-205216.

LIU X. & FURDYNA J.K. (2006) *Journal of Physics: Condensed Matter* **18**: R245-R279.

DU TRÉMOLET DE LACHEISSERIE E. (2004) *Magnetism I: fundamentals*, Springer, Heidelberg, London, New York.

MOTT N. (1982) *Proceedings of the Royal Society* (London) **1382**: 1-24.

NÉEL L. (1949) *Comptes Rendus de l'Acad*émie des *Sciences* **228**: 664-668.

NÉEL L. (1954) *Journal de Physique et le Radium* **15**: 225-239.

OHNO H. *et al.* (1992) *Physical Review Letters* **68**: 2664-2667.

RAIKHER Y.L. & SHLIOMIS M.I. (1974) *Sovjet Physics JETP* **40**: 526-532.

RAIKHER Y.L. & STEPANOV V.I. (1994) *Physical Review B* **50**: 6250-6259.

SMIT J. & BELJERS H.G. (1955) *Philips Research Reports* **10**: 113-130.

SONG C. *et al.* (2011) *Physical Review Letters* **107**: 56601- 56605.

TANNEWALD P.E. (1955) *Physical Review* **100**: 1713-1719.

THEVENARD L. *et al.* (2005) *Applied Physics Letters* **87**: 182506-182509.

VAN VLECK J.H. (1950) *Physical Review* **78**: 266-274.

Volume 1: BERTRAND P. (2020) *Electron Paramagnetic Resonance Spectroscopy - Fundamentals*, Springer, Heidelberg.

WASTLBAUER G. *et al.* (2005) *Advances In Physics* **54**: 137-219.

WILHELM C. *et al.* (2007) *Physical Review E* **75**: 41906-41912.

YAGER W.A. *et al.* (1950) *Physical Review* **80**: 744-748.

ZAKERI Kh. *et al.* (2006) *Phase Transitions*, Magnetic anisotropy in nanoscaled materials probed by ferromagnetic resonance, Taylor & Francis Ltd, **79**: 793- 813.

ZAKERI Kh. (2007) *Magnetic monolayers on semiconducting substrates: an in situ FMR study of Fe-based heterostructures*, PhD thesis, Universität Duisburg-Essen.

Principles of magnetic resonance: Bloch equations and pulsed methods

Belorizky E. and Fries P.H.

In this appendix, we describe the general principles of magnetic resonance using the simple example of proton NMR, which is used in chapter 11. Further details can be found in [Abragam, 1983; Canet, 1996]. These principles are directly transposable to EPR. The transposition is made in complement 1 to chapter 10 for *continuous wave EPR*, and in appendix 2 to this volume for *pulsed EPR.*

1 - Proton NMR

Interaction of a magnetic field $\mathbf{B}_0 = B_0\mathbf{k}$ – oriented along Oz – and a proton of spin $I = \frac{1}{2}$ which has a magnetogyric ratio $\gamma_I = \dfrac{g_p\beta_N}{\hbar} = 2.68 \times 10^8 \text{ rad s}^{-1}\text{ T}^{-1}$

and a magnetic moment $\qquad\qquad \boldsymbol{\mu} = \gamma_I\hbar\mathbf{I}$ [1]

is described by the Hamiltonian $\hat{H}_Z = -\boldsymbol{\mu}\cdot\mathbf{B}_0$ [2]

Two energy levels emerge $E_m = -m\gamma_I\hbar B_0$, where $m = \pm\frac{1}{2}$ is the eigenvalue for the $\hat{I}_z$ operator (figure A1.1). Their difference is equal to $\Delta E = \gamma_I\hbar B_0 = \hbar\omega_0 = h\nu_0$.

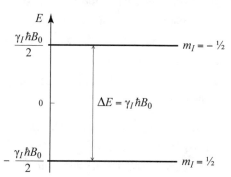

Figure A1.1 - Energy levels for a proton in a field $\mathbf{B}_0$.

For example, we obtain $v_0 = 42.577$ MHz for $B_0 = 1$ T. The magnetic moment for a proton is 657-fold smaller than that for a free electron. Thus, for the same value of magnetic field, the Zeeman separations are three orders of magnitude weaker than those involved in EPR.

At thermodynamic equilibrium, the ratio between the populations of the two levels can be written:

$$\frac{N_{-1/2}}{N_{1/2}} = \exp\left(-\frac{\Delta E}{k_B T}\right) \qquad [3]$$

For $B_0 = 1$ T, we obtain $\dfrac{N_{1/2}}{N_{-1/2}} = 1 + 6.87 \times 10^{-6}$ at room temperature ($T = 298$ K).

If we consider a system with N_0 protons at equilibrium, the population difference is written [Volume 1, section 1.4.4]:

$$N_{1/2} - N_{-1/2} = N_0 \tanh\left(\frac{\Delta E}{2k_B T}\right) \approx N_0 \frac{\Delta E}{2k_B T} \qquad [4]$$

In these conditions, the mean value at equilibrium M_{eq} for the magnetisation M_z of protons in the direction of the field is given by Curie's law:

$$M_{eq} = \frac{N_0 \gamma_I^2 \hbar^2}{4k_B T} B_0 \qquad [5]$$

The mean magnetic moment for a proton, which can be written $\dfrac{\gamma_I \hbar}{2} \dfrac{\Delta E}{2k_B T}$, is thus very small relative to its intrinsic moment $\mu = \gamma_I \hbar / 2$. For resonance to be observed, transitions must be induced between the two Zeeman levels. This can be achieved by subjecting the protons to a magnetic field $\mathbf{B}_1$ with a direction perpendicular to $\mathbf{B}_0$, an amplitude B_1 much smaller than B_0, and which rotates with an angular velocity $-\omega$. Resonance occurs when ω is close to $\omega_0 = \gamma_I B_0$. In practice, a coil of axis Ox, in which a sinusoidal current of angular frequency ω circulates, creates an oscillating field $B_1^{Ox} = 2B_1 \cos \omega t$ along this axis. This field can be considered to be the sum of two fields rotating in opposing directions with angular velocities $\pm \omega$ in the xy plane, of which only the component rotating at the angular frequency $-\omega$ is useful.

Interaction with the variable field $\mathbf{B}_1$ produces transitions between the two Zeeman levels. The probability of these transitions per unit of time, which is the same in both directions, is given by time-dependent perturbation theory:

$$w = \frac{\pi}{2} \gamma_I^2 B_1^2 f(\omega - \omega_0) \qquad [6]$$

where $f(\omega - \omega_0)$ is a normalised function representing the shape of the resonance line. As the population of the lower level is slightly greater than that of the higher level, the system of N_0 protons absorbs from the radiofrequency field a power P_a given by:

$$P_a = w(N_{1/2} - N_{-1/2})\Delta E$$

Thus, according to equation [4]

$$P_a \cong wN_0 \frac{(\hbar\omega_0)^2}{2k_{\mathrm{B}}T} \tag{7}$$

2 - Motion of the macroscopic magnetisation

The magnetisation $\mathbf{M}$ for a sample containing a large number of protons is obtained by adding the microscopic moments $\boldsymbol{\mu}_i$ of the different protons: $\mathbf{M} = \sum_i \boldsymbol{\mu}_i$. When the sample is placed in a field $\mathbf{B}_0$, the movement of the magnetisation, which can be considered a classical vector (see for example the F_{IV} complement in [Cohen-Tannoudji *et al.*, 2015]), is determined by equation

$$\frac{d\mathbf{M}}{dt} = \gamma_I \mathbf{M} \times \mathbf{B}_0$$

By projecting this equation on the three axes of coordinates, we obtain the system:

$$\begin{cases} \dfrac{dM_x}{dt} = \gamma_I M_y B_0 \\ \dfrac{dM_y}{dt} = -\gamma_I M_x B_0 \\ \dfrac{dM_z}{dt} = 0 \end{cases} \tag{8}$$

We will first assume that at $t = 0$ the magnetisation $\mathbf{M}(0)$ takes its equilibrium value $M_{eq}\mathbf{k}$, where $\mathbf{k}$ is the unit vector along Oz. In this case, the solution to this system is time-independent and is $M_x = M_y = 0$, $M_z = M_{eq} = $ constant. We will now assume that at $t = 0$ the magnetisation is separated from its equilibrium position by an angle α, for example:

$$M_x(0) = 0, M_y(0) = M_{eq}\sin\alpha, M_z(0) = M_{eq}\cos\alpha \tag{9}$$

The solution to system [8] is then written

$$\begin{cases} M_x(t) = M_{eq}\sin\alpha\sin\omega_0 t \\ M_y(t) = M_{eq}\sin\alpha\cos\omega_0 t \\ M_z(t) = M_{eq}\cos\alpha \end{cases} \tag{10}$$

where $\omega_0 = \gamma_I B_0$. The magnetisation **M** precesses around Oz with an angular velocity $-\omega_0$, i.e., at a velocity of ω_0 in the clockwise direction.

We now apply the radiofrequency field $B_1^{Ox} = 2B_1 \cos \omega t$ created by a coil of axis Ox in which a sinusoidal current of angular frequency ω, close to ω_0, flows. Transitions between the energy levels for the protons characterised by $m = \pm \frac{1}{2}$ are induced by the component of B_1^{Ox} which rotates about Oz at a frequency $-\omega$. The projections of this component onto the (x, y, z) axes can be written:

$$B_{1x} = B_1 \cos \omega t, \; B_{1y} = -B_1 \sin \omega t, \; B_{1z} = 0 \qquad [11]$$

To determine the motion of the magnetisation, it is convenient to introduce a system of coordinates $R = (x', y', z' = z)$ which rotates about z at an angular velocity $-\omega$, and thus follows the rotation of the field **B**$_1$ defined by relations [11]. The reference frames (x', y', z) and (x, y, z) are respectively known as *the rotating reference frame* (R) and *the laboratory's reference frame* (L). In the rotating reference frame (R), **B**$_1$ appears to be fixed (figure A1.2).

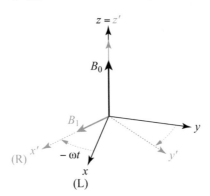

Figure A1.2 - The laboratory's reference frame (L) = (x, y, z) and the rotating reference frame (R) = $(x', y', z' = z)$ rotating about the field **B**$_0$ $//$ z at an angular velocity $-\omega$.

In reference frame (L), movement of **M** in the presence of **B**$_0$ and **B**$_1$ is determined by the following equation:

$$\left(\frac{d\mathbf{M}}{dt}\right)_L = \gamma_I \mathbf{M} \times (\mathbf{B}_0 + \mathbf{B}_1) \qquad [12]$$

In reference frame (R), this movement is determined by $\left(\frac{d\mathbf{M}}{dt}\right)_R$, which is linked to $\left(\frac{d\mathbf{M}}{dt}\right)_L$ by:

$$\left(\frac{d\mathbf{M}}{dt}\right)_L = \left(\frac{d\mathbf{M}}{dt}\right)_R + \boldsymbol{\Omega} \times \mathbf{M} \qquad [13]$$

where $\boldsymbol{\Omega} = -\omega\mathbf{k}$ is the rotation vector of (R) relative to (L). By replacing this expression in equation [12], we obtain:

$$\left(\frac{d\mathbf{M}}{dt}\right)_R = \gamma_I \mathbf{M} \times \mathbf{B}_{eff} \tag{14}$$

where $\mathbf{B}_{eff}$ is the "effective field" defined by

$$\mathbf{B}_{eff} = \mathbf{B}_0 + \mathbf{B}_1 + \frac{\boldsymbol{\Omega}}{\gamma_I} \tag{15}$$

The motion of the magnetisation is much easier to determine in reference frame (R) than in reference frame (L) as $\mathbf{B}_{eff}$ is *constant* in (R). Note that $\boldsymbol{\Omega}$ and $\mathbf{B}_0$ have opposite directions. In the specific case where $\omega = \omega_0 = \gamma_I B_0$ (resonance), $\mathbf{B}_{eff}$ is equal to $\mathbf{B}_1$ such that in the rotating reference frame, the magnetisation rotates about $\mathbf{B}_1$ at an angular velocity $\omega_1 = -\gamma_I B_1$. If $\mathbf{B}_1$ is applied for a duration τ, $\mathbf{M}$ rotates by an angle $\alpha = \omega_1\tau = -\gamma_I B_1\tau$ from its original position. For example, if we choose a short τ (a few µs) such that $\omega_1\tau = \pi/2$ or π, this is termed *applying a radiofrequency (rf) pulse of $\pi/2$ or π*, or 90 ° or 180 °. If the magnetisation initially has its equilibrium value $\mathbf{M} = M_{eq}\mathbf{k}$, pulses of $\pi/2$ and π will align it with the axis Oy' and opposite to the $Oz' = Oz$ axis of the rotating reference frame, respectively. After suppression of $\mathbf{B}_1$, $\mathbf{M}$ generally has a non-null component in the xy plane, $M_{xy} = M_{eq}\sin\alpha$. In the laboratory's reference frame, this component rotates about $\mathbf{B}_0$ with a frequency of $-\omega_0$. As a result, a current is induced in the detector coil placed near to the sample. The same coil is often used for both excitation and detection.

3 - Relaxation phenomena

In the previous section we saw that an rf pulse applied for a short duration can cause the direction of the magnetisation $\mathbf{M}$ to shift from its equilibrium direction Oz. We assumed that after suppression of $\mathbf{B}_1$, the rotation of $\mathbf{M}$ around $\mathbf{B}_0$ continues indefinitely. In fact, the magnetisation will progressively return to its equilibrium value along the z axis, and the NMR signal will disappear. This return of the magnetisation to equilibrium is known as *relaxation*.

After an rf pulse, the M_z and M_{xy} components of the magnetisation $\mathbf{M}$, which are parallel and perpendicular to the field $\mathbf{B}_0$, respectively, progressively return towards their equilibrium values: $M_z \rightarrow M_{eq}$, where M_{eq} is given by equation [5], and $M_{xy} \rightarrow 0$. This relaxation is due to the fluctuating magnetic fields acting on the protons. These magnetic fields mainly originate from the dipole-dipole interactions between the proton's magnetic moment and the magnetic moments of neighbouring molecules, whether of nuclear or electronic origin. They can also result from hyperfine interactions or a rotational-spin effect, or even chemical displacement anisotropy interactions. The fluctuations of these magnetic fields are created by random movements of the molecules near to the protons studied. The return to equilibrium of M_z is the result of energy exchanges between Zeeman levels for the protons. This phenomenon is known as *spin-lattice relaxation* or simply T_1 *relaxation* as it often follows an exponential law characterised by a T_1 time constant. When several relaxation mechanisms come into play, the longitudinal nuclear relaxation rate $R_1 = 1/T_1$ is, most often, the sum $R_1 = \sum_m R_{1m}$ of the different R_{1m} contributions. Similarly, the return to equilibrium of M_{xy} is known as *spin-spin relaxation or T_2* relaxation, where T_2 is the time constant characterising this phenomenon. The changes to the components M_z and $M_{x'y'}$ in (R) can be described by the Bloch equations

$$\frac{dM_z}{dt} = -\frac{1}{T_1}(M_z - M_{eq}) \qquad [16]$$

$$\frac{dM_{x'y'}}{dt} = -\frac{1}{T_2}M_{x'y'} \qquad [17]$$

At the instant when the rf pulse is completed – which is chosen as time $t = 0$ – the values of components M_z and $M_{x'y'}$ are $M_z(0)$ and $M_{x'y'}(0)$. Resolution of the Bloch equations then gives

$$M_z(t) = M_{eq}[1 - \exp(-t/T_1)] + M_z(0)\exp(-t/T_1) \qquad [18]$$

$$M_{x'y'}(t) = M_{x'y'}(0)\exp(-t/T_2) \qquad [19]$$

We will now consider the important specific cases of 180 ° and 90 ° pulses around the Ox axis.

▷ Just after a 180 ° pulse, we get $M_z(0) = -M_{eq}$ and $M_{y'}(0) = 0$. We deduce that

$$\begin{cases} M_z(t) = M_{eq}[1 - 2\exp(-t/T_1)] \\ M_{xy}(t) = 0 \end{cases} \qquad [20]$$

in the rotating reference frame and the laboratory's reference frame. This experiment can be used to measure T_1.

▷ Just after a 90 ° pulse, we get $M_z(0) = 0$ and $M_{y'}(0) = M_{eq}$, thus

$$\begin{cases} M_z(t) = M_{eq}[1 - \exp(-t/T_1)] \\ M_{y'}(t) = M_{eq}\exp(-t/T_2) \end{cases} \qquad [21]$$

in the rotating reference frame and

$$\begin{cases} M_z(t) = M_{eq}[1 - \exp(-t/T_1)] \\ M_x(t) = M_{eq} \sin(\omega_0 t)\exp(-t/T_2) \\ M_y(t) = M_{eq} \cos(\omega_0 t)\exp(-t/T_2) \end{cases} \qquad [22]$$

in the laboratory's reference frame. A receiver coil aligned with Oy then receives a *free induction decay signal*, which is an induced current proportional to $M_y(t)$ with the form $A\cos(\omega_0 t)\exp(-t/T_2)$, that can be recorded. In addition, using a second receiver coil aligned with Ox, we can construct the complex signal

$$M_y(t) - iM_x(t) = M_{eq}\exp(-i\omega_0 t)\exp(-t/T_2) \quad \text{for } t \geq 0$$

If we set $M_y(t) - iM_x(t) = 0$ for $t < 0$, the real part of the Fourier transform of this complex signal is proportional to

$$F(\omega) = \frac{T_2}{1 + (\omega - \omega_0)^2 T_2^2} \qquad [23]$$

This is the NMR absorption line, which takes the form of a Lorentzian of full width at half maximum, $\Delta\omega = 2\pi\Delta\nu = 2/T_2$.

In practice, if the field B_0 is not perfectly homogeneous, the various spins in the sample are subjected to slightly different fields depending on their position in the gap of the electromagnet or inside the superconductor solenoid of the cryomagnet. Their resonance frequencies are thus slightly different. This leads to a faster attenuation of the transverse magnetisation, and T_2 is replaced by the effective time T_2^* defined by

$$\frac{1}{T_2^*} = \frac{1}{T_2} + \frac{1}{T_{2(B_0)}} \qquad [24]$$

In figure A1.3 we have represented the return to equilibrium of $M_z(t)$ after a 180 ° pulse which can be used to measure T_1, and in figure A1.4 we show the free induction decay signal after a 90 ° pulse, and its Fourier transform $F(\omega) = F(2\pi\nu)$, which can be used to determine T_2^*.

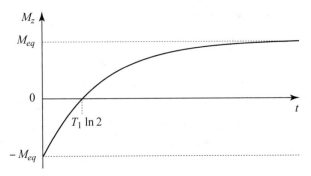

Figure A1.3 - Evolution of M_z after a 180 ° pulse.

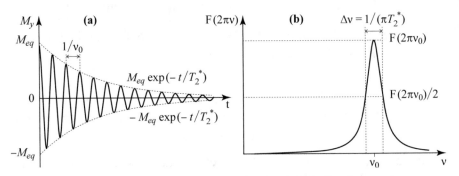

Figure A1.4 - (a) Free induction decay signal and **(b)** its Fourier transform $F(\omega) = F(2\pi\nu)$ after a 90 ° pulse.

Note. If equations [14] and [15] are modified to take relaxation phenomena into account, a system of differential equations is obtained which determines the time-dependence of the magnetisation in the rotating reference frame (R). This system can be used to determine the absorption signal detected by a continuous wave spectrometer, in the steady state regime where the magnetisation is constant in the reference frame (R), but also in the transient regime (see complement 1 to chapter 10).

References

ABRAGAM A. (1983) *The Principles of Nuclear Magnetism*, Clarendon Press, Oxford.

CANET D. *et al.* (1996) *Nuclear Magnetic Resonance: Concepts and Methods*, Wiley, New York.

COHEN-TANNOUDJI C., DIU B. & LALOE F. (2015) *Quantum Mechanics*, John Wiley & Sons, New York.

Introduction to pulsed EPR: ESEEM, HYSCORE and PELDOR experiments

Dorlet P. and Bertrand P.

Pulsed EPR is mainly used to measure weak interactions, whether hyperfine interactions between a paramagnetic centre and distant nuclei (ESEEM and HYSCORE experiments) or dipolar interaction between two paramagnetic centres separated by a large distance (PELDOR experiments). The general principle is as follows: when two systems of spin A and B are weakly coupled, a very brief excitation of A induces transitions in B which, in some conditions, modulate the amplitude of a temporal signal produced by A. These experiments can be rigorously interpreted by applying the *density operator* formalism, which is described in monographs [Schweiger and Jeschke, 2001], and well-written chapters of theses [Junk, 2012]. In this appendix, we will provide a simplified presentation based on a classical model.

In continuous wave EPR, the spectrum is recorded in the steady-state regime [Volume 1, section 5.3.1], where the magnetic moments of the sample interact continuously with the magnetic component $\mathbf{B}_1(t)$ of the microwave radiation. In pulsed EPR, the magnetic moments are subjected to sequences of very brief microwave pulses (a few nanoseconds to a few tens of nanoseconds) which can be used to "handle" them. Between two pulses, the magnetic moments precess around the static field $\mathbf{B}_0$. In this appendix, we describe the ESEEM, HYSCORE and PELDOR sequences used in some chapters of this volume. We will assume that the continuous wave EPR spectrum for the sample recorded at a frequency ν consists of a unique *symmetric inhomogeneous line centred* at

© Springer Nature Switzerland AG 2020
P. Bertrand, *Electron Paramagnetic Resonance Spectroscopy*,
https://doi.org/10.1007/978-3-030-39668-8

$B_0 = 2\pi v / \gamma_0$. This line results from the superposition of *homogeneous lines* produced by "spin packets" [Volume 1, complement 3 to chapter 5], each packet p is characterised by its magnetogyric ratio γ_p. In a pulsed EPR experiment, the sample is placed in a static field $\mathbf{B}_0$ and we are interested in the motion of the moments created by the spin packets and the motion of the total magnetisation $\mathbf{M}$ of the sample. The notations from appendix 1 are used: the movements are either described in the *laboratory's reference frame* (x, y, z), where z is the direction of $\mathbf{B}_0$, or in the *rotating reference frame* $(x', y', z' = z)$, where x' is the direction of the field $\mathbf{B}_1(t)$ perpendicular to $\mathbf{B}_0$, which rotates about $\mathbf{B}_0$ with an angular velocity $\omega = 2\pi v$. We assume that the spectrometer frequency is such that $\omega = \omega_0 = \gamma_0 B_0$.

1 - Spin echo and ESEEM experiments

1.1 - Spin echo

The simplest sequence generating a spin echo is the two-pulse sequence, which was first proposed by Hahn for nuclear spins [Hahn, 1950] (figure A.2.1).

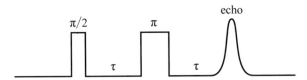

Figure A2.1 - Generating a spin echo by a 2-pulse sequence.

We assume that the moments of the spin packets are initially aligned in their equilibrium direction z, creating a resulting magnetisation $\mathbf{M}$ parallel to z.

▷ A $\pi/2$-pulse (see section 2 in appendix 1) causes all the magnetic moments to rotate 90° about x'. As a result, they all align in the $-y'$ direction, as does their resultant $\mathbf{M}$ (figure A2.2a). After suppression of $\mathbf{B}_1$, the moments precess about z in the (x', y') plane. If all the packets had the same magnetogyric factor γ_0, their moments would rotate in-phase at a velocity ω_0, the magnetisation $\mathbf{M}$ would be maximal and its movement would generate a *free induction decay signal* in the detector (see section 3 in appendix 1). In the present case, the moment of each packet p rotates at a velocity $\omega_p = \gamma_p B_0$ which is greater than or less than ω_0. In the rotating reference frame, this motion corresponds to precessing with a positive or negative velocity $\omega_p - \omega_0$: precession takes place in one or other direction, and the resulting magnetisation **M** decreases rapidly (figure A2.2b).

▷ At time τ, a π-pulse is applied (figure A2.1) causing all the moments to pivot by 180 ° about x'. The moments continue to precess at the same speed and in the same direction (figure A2.2c). As a result, after time τ, they all align in the y' direction. At this moment, the magnetisation **M** becomes large once again and generates a "spin echo" signal in the detector (figure A2.2d).

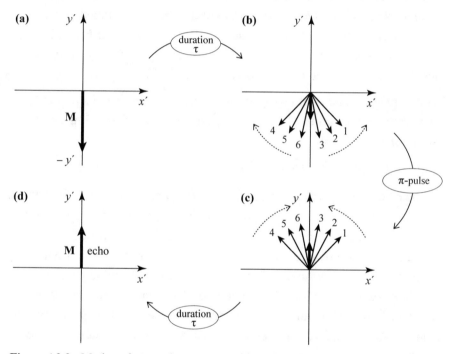

Figure A2.2 - Motion of magnetic moments subjected to the sequence from figure A2.1. **(a)** The moments initially oriented along z all align along $-y'$ after the $\pi/2$-pulse. **(b)** Within the duration τ, the moments are dephased and **M** decreases. **(c)** The moments which have been rotated 180 ° about x' by the π-pulse, continue to precess in the same direction. **(d)** After time τ, all the moments are aligned with y', and an echo is observed.

A decrease in magnetisation produced by transversal relaxation superposes on these phenomena. It is characterised by the electronic "phase memory time" T_m^e. This decrease is due to spin-spin relaxation (characteristic time T_2^e, see section 3 in appendix 1) and to other processes, such as spectral diffusion and instantaneous diffusion. Generation of a spin echo resolves the "dead time" problem which delays signal acquisition after the end of the $\pi/2$-pulse, and allows the EPR spectrum to be obtained by recording the amplitude of the echo when sweeping B_0.

1.2 -2-pulse ESEEM experiments

If we vary the *time* τ between the two pulses (figure A.2.1) and record the *amplitude of the echo* as a function of τ, the signal obtained decreases exponentially due to relaxation. When the EPR line is inhomogeneous due to unresolved hyperfine interactions, the anisotropic components of these interactions create *modulations* which are superposed on this decrease. By determining the Fourier transform of the modulated signal, the ESEEM (*Electron Spin Echo Envelope Modulation*) spectrum is obtained, where peaks appear at the resonance frequencies for the coupled nuclei [Rowan *et al.*, 1965].

To determine the nature of this spectrum, consider a nucleus of spin I coupled to a paramagnetic centre of spin S. To first order in perturbation theory, its Hamiltonian is written:

$$\hat{H}_I = -v_N \hat{I}_z + (a_{iso} + T_{ZZ})\hat{S}_Z \hat{I}_Z + T_{ZX}\hat{S}_Z \hat{I}_X + T_{ZY}\hat{S}_Z \hat{I}_Y + \kappa[3\hat{I}_{Z'}^2 + \eta(\hat{I}_{X'}^2 - \hat{I}_{Y'}^2)]$$

The energies are expressed here in units of frequency. $v_N = g_N \beta_N B_0 / h$ is the Larmor frequency of the nucleus, a_{iso} is the isotropic hyperfine constant and the T_{IJ} are anisotropic components due to the dipolar interactions [Volume 1, appendix 3]. κ is the quadrupole constant of the nucleus and η is the asymmetry parameter. The axis z is the direction of $\mathbf{B}_0$, (X, Y, Z) are the principal axes of the hyperfine matrix and (X', Y', Z') are those of the quadrupole matrix. Consider the case where $S = \frac{1}{2}$ and $I = \frac{1}{2}$. The quadrupole terms disappear from the Hamiltonian. When $\mathbf{B}_0$ is parallel to axis Z of the hyperfine matrix, the signal amplitude is modulated according to the following expression:

$$V_{2p}(\tau) = 1 - \frac{k}{4} \{2 - 2\cos(2\pi v_\alpha \tau) - 2\cos(2\pi v_\beta \tau)$$

$$+ \cos[2\pi(v_\alpha + v_\beta)\tau] + \cos[2\pi(v_\alpha - v_\beta)\tau]\} \quad [1]$$

The frequencies v_α and v_β are those of the transitions labelled v_{NMR1} and v_{NMR2} in figure A3.1 in appendix 3. They are given by:

$$v_{\alpha,\beta} = \sqrt{\left(v_N \pm \frac{a_{iso} + T_{ZZ}}{2}\right)^2 + \frac{T_{ZX}^2 + T_{ZY}^2}{4}} \quad [2]$$

The quantity k, which determines the *amplitude* of the modulation, is known as the "modulation depth". It is given by:

$$k = \frac{v_N^2(T_{ZX}^2 + T_{ZY}^2)}{v_\alpha^2 v_\beta^2} \qquad [3]$$

The Fourier transform of the amplitude of the modulated signal contains peaks at frequencies v_α, v_β and at the *combination frequencies* $v_\alpha + v_\beta$, $v_\alpha - v_\beta$, which can be used to determine v_N and the T_{IJ}. When the hyperfine coupling constants are much weaker than v_N, equations [1] and [2] show that the peaks form at the Larmor frequency v_N and at $2v_N$. These frequencies are easy to identify on the spectrum. In this case, the position of the peaks provides no information on the coupling, but the modulation amplitude depends on it (equation [3]).

Expressions similar to equations [1] and [2] exist for any values of S and I, and in the case where several nuclei are coupled [Mims, 1972a, b]. Analysis of the modulation can generally be used to determine the nature and number of nuclei coupled to the paramagnetic centre [Dikanov and Tsvetkov, 1992].

1.3 - 3-pulse ESEEM experiments

The main limitation of 2-pulse ESEEM is the rapid decrease in the amplitude of the echo (characteristic time T_m^e) which leads to *broadening* of the spectral peaks. To overcome this limitation, a 3-pulse ESEEM sequence can be used to generate a *stimulated echo*.

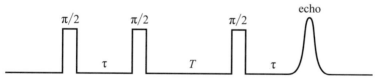

Figure A2.3 - Generation of a simulated spin echo by a 3-pulse sequence.

In this experiment, τ *is fixed* and the time T between the 2nd and 3rd pulses is varied. For $S = \frac{1}{2}$ and $I = \frac{1}{2}$, expression [1] is replaced by:

$$V_{3p}(\tau,T) = 1 - \frac{k}{4}([1 - \cos(2\pi v_\alpha \tau)]\{1 - \cos[2\pi v_\beta(T+\tau)]\}$$
$$+ [1 - \cos(2\pi v_\beta \tau)]\{1 - \cos[2\pi v_\alpha(T+\tau)]\})$$

The Fourier transform for the modulated signal only contains peaks at frequencies v_α and v_β. These peaks disappear for certain values of τ, an effect known as "τ suppression effect". To more easily interpret the spectrum, it is generally recorded for several values of τ. With the 3-pulse sequence, the decrease in amplitude of the echo is determined by the *nuclear* phase memory time T_m^n, which is much longer than T_m^e and is of the same order of magnitude as T_1^e. The spectral lines are narrower than in the 2-pulse experiment. Sequences with more than two pulses generate multiple echos. Undesirable echos can be eliminated by *phase cycling*: repeating the sequence several times by varying the phase of some pulses and summing the echos obtained.

When the coupled nucleus has a spin $I = 1$ (as for ^2H or ^{14}N for example), the 3-pulse ESEEM spectrum has a remarkable appearance when $|a_{iso}|/2 = v_N$ (known as the "perfect compensation" situation). Splitting of a multiplicity M_S in this case only results from *the quadrupole interaction*, and only three narrow peaks are observed in the low-frequency part of the spectrum. Their positions are given by:

$$v_0 = 2\,\kappa\eta, \; v_- = \kappa\,(3 - \eta), \; v_+ = \kappa\,(3 + \eta) = v_0 + v_-$$

The "$\Delta M_I = \pm 2$" transition (double quantum) of the other multiplicity produces a broad peak at frequency:

$$v_{dq} \approx 2\,[(v_N + |a_{iso}|/2)^2 + \kappa^2\,(3 + \eta^2)]^{1/2} \approx 4v_N$$

In complexes where a Cu(II) ion is coordinated to an imidazole group in the equatorial position, perfect compensation is produced at X-band for the ^{14}N nucleus of the distal nitrogen (figure A2.4).

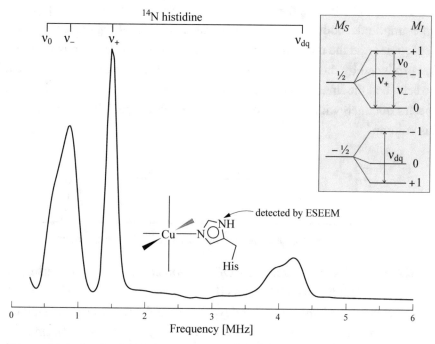

Figure A2.4 - 3-pulse ESEEM spectrum at X-band for a protein where a Cu(II) ion is coordinated to the imidazole group of a histidine in the equatorial position.
The peak at v_0 appears here as a shoulder. The inset shows the energy level diagram for a centre of spin $S = \frac{1}{2}$ coupled to a nucleus of spin $I = 1$. The nuclear frequencies which appear on the ESEEM spectrum in the "perfect compensation" situation are indicated.

Alternative techniques must be used to study other hyperfine interactions existing in these complexes: continuous wave EPR for interactions with ^{63}Cu and ^{65}Cu nuclei [Volume 1, section 4.4.3] and ENDOR spectroscopy for interactions with the nucleus of the proximal nitrogen, ligand of the histidine residue.

ESEEM spectra are represented in figure 4.16 of this volume (complement 1 to chapter 4). A description of numerical simulation software can be found in [Shane *et al.*, 1998; Madi *et al.*, 2002].

2 - 4-pulse ESEEM experiments and HYSCORE experiments

Four-pulse ESEEM is based on a 3-pulse sequence with an additional π-pulse applied between the second and third $\pi/2$-pulses. The final $\pi/2$-pulse generates the stimulated echo. In the 4-pulse sequence represented in figure A2.5, τ *is constant.* The ESEEM spectrum is obtained by recording the amplitude of the echo as a function of $t_1 = t_2$. For some values of τ, this sequence produces large-amplitude modulations which generate marked peaks at the basic nuclear frequencies and the combination frequencies. The spectrum is thus similar to that produced by the 2-pulse sequence, but the decrease in the amplitude of the echo and the linewidth are determined by the *nuclear* relaxation time T_m^n. Spectra obtained with this sequence are shown in figure 7.10, in chapter 7 to this volume.

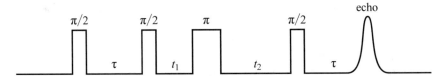

Figure A2.5 - Diagram representing a 4-pulse ESEEM experiment.

HYSCORE (*HYperfine Sublevel CORrElation spectroscopy*) experiments, developed by Höfer and collaborators [Höfer *et al.*, 1996], are 4-pulse ESEEM experiments in which the times t_1 and t_2 are independently varied. The 2-dimensional Fourier transform of the modulated signal produces a spectrum in the (ν_1, ν_2) plane, the amplitude of which is represented by contour lines or colours:

▷ The diagonal $(\nu_1 = \nu_2)$ is the ESEEM spectrum.

▷ *Correlation peaks*, symmetric about the diagonal or the antidiagonal, appear for some values of (ν_1, ν_2). Quantitative study of these peaks can be used to determine the isotropic and anisotropic components of hyperfine interactions of the coupled nuclei. A 6-pulse version also exists which has the advantage of overcoming time τ suppression effects [Kasumaj and Stoll, 2008]. HYSCORE experiments are very useful for analysing complex ESEEM spectra, in particular those produced by systems containing several coupled nuclei. HYSCORE spectra are represented in figures 4.17 (complement 1 to chapter 4) and 7.11, 7.19 (chapter 7) in this volume.

3 - PELDOR experiments

PELDOR (*Pulsed ELectron electron DOuble Resonance*) experiments aim to measure the dipolar interaction between two paramagnetic centres, A and B, so as to determine the *distance r* separating them, which is assumed to be large enough for exchange coupling to be negligible. These measurements are obtained by applying the 4-pulse DEER (*Double Electron Electron Resonance*) sequence. In most applications, these centres (most often nitroxides) are bound to biological macromolecules and the sample is a frozen solution. First consider molecules in which the angle between the intercentre axis and the field $\mathbf{B_0}$ has a given value θ, between 0 and 180 °. The DEER sequence relies on selective excitation of the A and B centres (figure A2.6a). Excitation of the A centres at a frequency v_{obs} by the sequence $\pi/2 - \tau_1 - \pi - \tau_1$ generates a *primary echo*, and the sequence $\tau_2 - \pi - \tau_2$ generates a *secondary echo*. At time t after formation of the primary echo, the B centres are excited by a pulse π at the v_{pump} frequency. Because of dipolar interactions between A and B, the amplitude of the secondary echo depends on the value of t. The decrease in this amplitude as a function of t is modulated at the angular frequency $\omega_{dip}(r, \theta)$ given by:

$$\omega_{dip}(r,\theta) = \frac{\mu_0 \beta^2 g_A g_B}{4\pi\hbar r^3}(3\cos^2\theta - 1) \qquad [4]$$

In a frozen solution of molecules, the angle θ, and thus the angular frequency ω_{dip}, are *random* quantities. The Fourier transform of the modulated amplitude (figures A2.6b and A2.6c) produces the *distribution* of ω_{dip}. When all directions are equally probable, the probability density of θ is $p(\theta) = \sin\theta$ [Volume 1, complement 3 to chapter 4] and the distribution of ω_{dip} is dominated by the $\omega_{dip}(r, 90°)$ component. The density of ω_{dip}, calculated in section 7.3.1 of [Volume 1] and represented in figure 6.5 of that volume, is often known as "Pake's doublet". The distance r can be deduced from the value $\omega_{dip}(r, 90 °)$. When the value of r is distributed over the sample, a method known as "Tikhonov regularisation" can be used to determine its distribution (figure A2.6d) [Jeschke *et al.*, 2006].

In principle, this method can be used to measure intercentre distances between about 1.5 and 8 nm. In practice, the precision of measurements is highly dependent on the nature of the paramagnetic centres and the sample concentration [Pannier *et al.*, 2000; Jeschke, 2002].

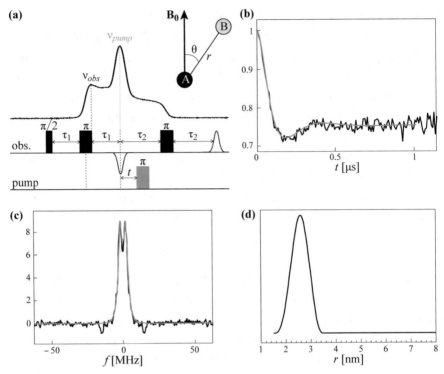

Figure A2.6 - PELDOR experiment performed on a pair of nitroxide radicals
attached to a protein (see section 8.4.3 in chapter 8).

(a) 4-pulse DEER sequence. The irradiation frequencies ν_{obs} and ν_{pump} are identified
on the absorption spectrum for a frozen solution of proteins.

(b) Decrease of the echo amplitude as a function of t, modulated by the dipolar interaction.
(c) Distribution of the frequency $f = \omega_{dip}/2\pi$ obtained by Fourier transform of the signal
represented in **(b)** (Pake's doublet). The maxima correspond to $\theta = 90\,°$. **(d)** Distribution
of the distance r obtained after Tikhonov regularisation (*DeerAnalysis* software
[Jeschke *et al.*, 2006]). $T = 60$ K, $\Delta\nu = \nu_{pump} - \nu_{obs} = 72.12$ MHz, pulse duration
16 ns for $\pi/2$ and 32 ns for π, $\tau_1 = 200$ ns, $\tau_2 = 1200$ ns.

References

Dikanov S.A. & Tsvetkov Y.D. (1992) *Electron Spin Echo Envelope Modulation (ESEEM) Spectroscopy*, CRC Press, Boca Raton.

Hahn E.L. (1950) *Physical Review* **80**: 580-594.

Höfer P. *et al.* (1996) *Bruker Report* **142**: 15

Jeschke G. (2002) *Chem Phys Chem* **3**: 927-932.

Jeschke G. *et al.* (2006) *Applied Magnetic Resonance* **30**: 473-498.

Junk M. (2012) *Assessing the Functional Structure of Molecular Transporters by EPR Spectroscopy*, Springer Verlag, Berlin.

Kasumaj B. & Stoll S. (2008) *Journal of Magnetic Resonance* **190**: 233-247.

Madi B.L., Van Doorslaer S. & Schweiger A. (2002) *Journal of Magnetic Resonance* **154**: 181-191.

Mims W.B. (1972) (a) *Physical Review* **B5**: 2409-2419.

Mims W.B. (1972) (b) *Physical Review* **B6**: 3543-3545.

Pannier M. *et al.* (2000) *Journal of Magnetic Resonance* **142**: 331-340.

Rowan L.G, Hahn E.L. & Mims W.B. (1965) *Physical Review* A**137**: 61-71.

Schweiger A. & Jeschke G. (2001) *Principles of Pulse Electron Paramagnetic Resonance*, Oxford University Press, New York.

Shane J.J., Liesum L.P. & Schweiger A.. (1998) *Journal of Magnetic Resonance* **134**: 72-75.

Volume 1: Bertrand P. (2020) *Electron Paramagnetic Resonance Spectroscopy - Fundamentals*, Springer, Heidelberg.

Principle of continuous wave ENDOR spectroscopy

Grimaldi S.

*Laboratory of Bioenergetics and Protein Engineering, UMR 7281,
Institute of Microbiology of the Mediterranean,
CNRS & Aix-Marseille University, Marseille.*

1 - Introduction

Electron Nuclear Double Resonance (ENDOR) spectroscopy is mainly used to quantitatively study magnetic interactions between unpaired electrons and the surrounding paramagnetic nuclei. It is particularly useful when seeking to detect and measure weak hyperfine couplings which can be difficult or even impossible to study by continuous wave EPR. This technique was initially developed by G. Feher [Feher, 1956a, b] to polarise nuclei in paramagnetic substances. By appropriately exciting EPR and NMR transitions through the application of double microwave and radiofrequency radiation, he obtained nuclear polarisation with the same order of magnitude as electronic polarisation, which allows the NMR signal to be detected through variations in the intensity of the EPR lines [Feher, 1956c]. ENDOR was initially used in solid-state physics to probe the structure of doped semi-conductors, and was then extended to chemistry applications thanks to the first in-solution ENDOR experiments [Cederquist, 1963; Hyde and Maki, 1964], and subsequently to biological systems [Ehrenberg *et al.,* 1968]. Several methodological developments followed the first continuous wave ENDOR experiments, these included *triple resonance* which can be used to measure the *relative sign* of hyperfine coupling constants [Dinse *et al.,* 1974], and pulsed ENDOR [Mims, 1965] which considerably extended the

© Springer Nature Switzerland AG 2020
P. Bertrand, *Electron Paramagnetic Resonance Spectroscopy*,
https://doi.org/10.1007/978-3-030-39668-8

technique's field of application. In this appendix we will present the principle of continuous wave ENDOR and the information that can be gleaned from it. For further details, the reader is invited to consult [Kevan and Kispert, 1976; Kurreck *et al.*, 1988; Lowe, 1995] or review articles [Murphy and Farley, 2006; Kulik and Lubitz, 2009] dealing with methodological developments and applications of ENDOR spectroscopy.

2 - ENDOR spectrum for radicals in the isotropic regime

2.1 - Positions of ENDOR lines

Consider a radical ($S = \frac{1}{2}$) with an unpaired electron which interacts with a nucleus of magnetic moment $g_N \beta_N \mathbf{I}$ with $I = \frac{1}{2}$, for example a proton. When this radical is placed in a magnetic field $\mathbf{B}$, its spin Hamiltonian in the isotropic regime is written [Volume 1, section 2.3.1]:

$$\hat{H}_S = g_{iso}\beta\mathbf{B}\cdot\mathbf{S} + A_{iso}\mathbf{S}\cdot\mathbf{I} - g_N\beta_N\mathbf{B}\cdot\mathbf{I}$$

To first order in perturbation theory, the energy levels are given by:

$$E(M_S, M_I) = g_{iso}\beta B M_S + A_{iso}\, M_S\, M_I - g_N\beta_N\, B M_I \qquad [1]$$

It is convenient to express these energies in units of frequency. By dividing the two sides of equation [1] by the Planck constant h, we obtain:

$$E(M_S, M_I)/h = \nu_e\, M_S + A'_{iso}\, M_S\, M_I - \nu_N\, M_I \qquad [2]$$

where $\nu_e = g_{iso}\beta B/h$ and $A'_{iso} = A_{iso}/h$. The quantity $\nu_N = g_N\beta_N B/h$ is the Larmor frequency of the nucleus in the field B. In the case of the proton, for example, $\nu_N \approx 14\,\text{MHz}$ for the typical value $B = 0.34\,\text{T}$. The energy levels given by equation [2] are schematically represented in figure A3.1 where we have assumed $A_{iso} > 0$ and we distinguished between the situations $\nu_N > A'_{iso}/2$ and $A'_{iso}/2 > \nu_N$.

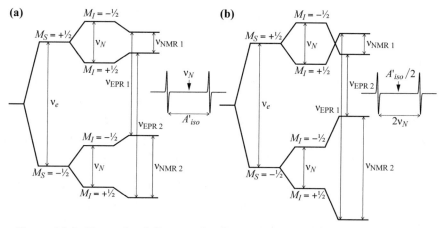

Figure A3.1 - Energy level diagram given by equation [2]. We assumed $A_{iso} > 0$ and $g_N > 0$, and distinguished between the cases **(a)** $A'_{iso}/2 < v_N$ and **(b)** $A'_{iso}/2 > v_N$. The frequencies of the two ENDOR lines are given by $v_{NMR} = v_N \pm A'_{iso}/2$ in case **(a)** and $v_{NMR} = A'_{iso}/2 \pm v_N$ in case **(b)**. The appearance of the ENDOR spectrum is shown for both cases.

▷ The energies of the EPR transitions, which verify the selection rule $\Delta M_I = 0$, $\Delta M_S = \pm 1$, are unaltered by the nuclear Zeeman term. If v_{EPR} is the frequency of the spectrometer, the position of the lines is given by [Volume 1, section 2.3.1]:

$$\text{EPR transition 1 } (M_I = -\tfrac{1}{2}): B_1 = hv_{EPR}/g_{iso}\beta + A_{iso}/2g_{iso}\beta$$

$$\text{EPR transition 2 } (M_I = \tfrac{1}{2}): B_2 = hv_{EPR}/g_{iso}\beta - A_{iso}/2g_{iso}\beta \qquad [3]$$

▷ The NMR transitions are determined by the selection rule $\Delta M_I = \pm 1$, $\Delta M_S = 0$. Resonance occurs at the following frequencies:

$$v_{NMR\ 1} = |v_N - A'_{iso}/2|$$

$$v_{NMR\ 2} = |v_N + A'_{iso}/2|$$

We will see in the next section that NMR transitions induce variations in the amplitude of the EPR signal. The ENDOR spectrum, which represents this variation in amplitude when the frequency of the radiofrequency wave is scanned, is thus composed of a pattern of two lines, the positions and splitting of which depend on the relative values of v_N and $|A'_{iso}|/2$:

a) If $v_N > |A'_{iso}|/2$, the pattern is centred on the nuclear Larmor frequency v_N and its splitting is $|A'_{iso}|$ (figure A3.1a).

b) If $|A'_{iso}|/2 > v_N$, the pattern is centred on $|A'_{iso}|/2$ and its splitting is $2v_N$ (figure A3.1 b).

In both cases, the ENDOR spectrum provides the absolute value of the hyperfine constant A_{iso} and the value of v_N, which can be used to identify the coupled nucleus. When this nucleus is a proton, the high value of g_N and thus of v_N means that situation a) is the most common. To record the ENDOR spectrum, a weak-amplitude frequency *modulation* is generally superposed on the frequency scan. The ENDOR lines therefore take the shape of derivatives, like in EPR.

We will now examine the case of an unpaired electron interacting with several nuclei [Volume 1, section 2.4]. Figure A3.2 shows the energy level diagram when there are four *equivalent nuclei* of spin $I = \frac{1}{2}$. This is the case, for example, of the anionic radical *p*-benzosemiquinone which contains four equivalent protons. In the isotropic regime, the EPR spectrum consists of five equidistant lines separated by $A_{iso}/g_{iso}\beta$, the relative intensities of which (1:4:6:4:1) are given by Pascal's triangle [Volume 1, chapter 2, complement 2]. As all the NMR transitions associated with a given value of M_S have the same frequency, the ENDOR spectrum contains only two lines at $v_{NMR\ 1}$ and $v_{NMR\ 2}$ (figure A.3.2).

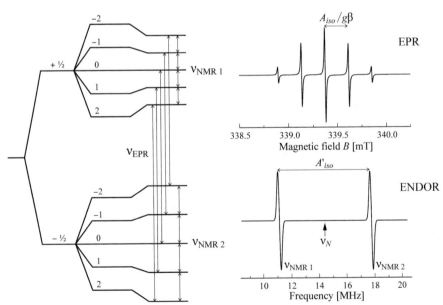

Figure A3.2 - Energy levels for a centre of spin $S = \frac{1}{2}$ coupled to four equivalent protons ($I = \frac{1}{2}$). We assumed $A_{iso} > 0$ and $v_N > A'_{iso}/2$. Both the EPR and ENDOR spectra are represented.

In the general case, where there exist p groups of n_i equivalent coupled nuclei of spin I_i, the total number of hyperfine lines in the EPR spectrum is given by [Volume 1, equation 2.12]:

$$N(p,n_i,I_i) = \prod_{i=1}^{p} (2n_iI_i + 1)$$

In contrast, the ENDOR spectrum contains only p pairs of lines. In a homonuclear ENDOR experiment (i.e., where we are only interested in a single type of nucleus), the frequency range covered by this spectrum is equal to $|A'_{max}|$ where A'_{max} is the highest hyperfine constant. To illustrate this, consider for example the phenalenyl radical which has a group of 3 equivalent protons and a group of 6 equivalent protons (figure A3.3). In the isotropic regime, the EPR spectrum has $4 \times 7 = 28$ lines, whereas the ENDOR spectrum contains only 2 pairs of lines centred on the Larmor frequency for the proton. Examination of the ENDOR spectrum straightforwardly gives the number of groups of equivalent protons. The number of equivalent protons in each group can be determined by triple resonance experiments [Murphy and Farley, 2006].

The hyperfine structure of the EPR spectrum in figure A3.3 is well resolved and can be directly interpreted. But the EPR spectrum for many radicals composed of a very large number of more or less strongly-coupled protons is composed of a compact cluster of hyperfine lines which cannot be analysed. In this case, ENDOR spectroscopy must be used.

To conclude this section, we indicate that nuclei of spin $I > \frac{1}{2}$ produce $2I$ pairs of ENDOR lines in the isotropic regime. In this regime, quadrupole interactions are not involved.

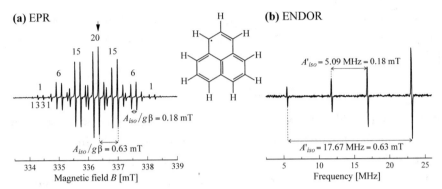

Figure A3.3 - EPR and ENDOR spectra for the phenalenyl radical obtained in derivative mode. On the EPR spectrum, the numbers represent the relative intensities of the hyperfine lines for the 2 groups of equivalent protons and the arrow indicates the field at which the ENDOR spectrum was recorded.

2.2 - Effect of NMR transitions on the amplitude of EPR lines

◇ The ENDOR phenomenon

We will now return to the four energy levels shown in figure A.3.1. At thermal equilibrium, their relative populations are determined by the Boltzmann distribution. This figure (and figure A.3.2) is not to-scale. Indeed, the separation between energy levels identified by $M_S = \frac{1}{2}$ and $M_S = -\frac{1}{2}$ is equal to ν_e, whereas their splitting is around ν_N which is three orders of magnitude smaller. As a result, if we label these energy levels a, b, c, d (figure A3.4a), their populations verify the following relations with a very good approximation [Volume 1, section 1.4.4]:

$$n(a) = n(d), \quad n(b) = n(c)$$

$$n(b)/n(a) = \exp(-h\nu_e/k_B T) = 1 - h\nu_e/k_B T$$

The relative populations of the four levels can thus be written as indicated in figure A3.4a, with

$$\varepsilon = \frac{h\nu_e}{2k_B T} \ll 1$$

If we set the magnetic field equal to the value B_2 of the resonance field for the EPR2 transition (equation [3]), the microwave radiation induces the transitions $a \leftrightarrow b$ and the intensity of the resulting EPR line is proportional to the population difference [Volume 1, chapter 5], which is 2ε at thermal equilibrium (figure A3.4a). If the transition is completely *saturated*, we obtain $n(a) = n(b)$ (figure A3.4b) and the line disappears. It can partially reappear if the nuclear moments interact with a powerful radiowave at the appropriate frequency. Indeed, the transitions of frequency $v_{NMR\,1}$ between energy levels b and c tend to equalise their populations and thus re-establish population difference between a and b. If the radiowave power is sufficient to saturate the NMR transition, we obtain the relative populations indicated in figure A3.4c. The population difference between levels a and b becomes equal to $\varepsilon/2$ and the EPR line recovers 25 % of its intensity. Similarly, the frequency transitions $v_{NMR\,2}$ between levels a and d tend to equalise their populations, which leads to the relative populations in figure A3.4d, and the same recovery of 25 % of the EPR line's initial intensity. We thus observe one of the ENDOR spectra from figure A3.1, where v_N is the Larmor frequency corresponding to the field B_2.

The same reasoning applies if the magnetic field is set to the value B_1 of the resonance field for the EPR1 transition (equation [3]). The same ENDOR spectrum is obtained, but this time with the value of v_N corresponding to B_1.

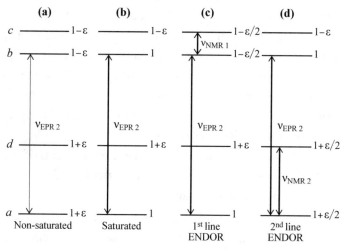

Figure A3.4 - Effect of NMR 1 and NMR 2 transitions on the populations of the a–d levels from figure A3.1a. The magnetic field is assumed to have the value B_2 corresponding to the resonance field for EPR transition 2. The saturating transitions are represented by thick lines.

In fact, the fraction of EPR signal recovered thanks to the NMR transitions is highly dependent on the characteristic times of the multiple relaxation processes through which the system returns to thermal equilibrium. These processes are dominated by electronic transitions, but they also include nuclear transitions and mixed electronic/nuclear transitions. The fraction recovered also depends on the temperature and microwave and radiofrequency powers used. This explains why the relative intensities of the ENDOR lines do not necessarily reflect the number of coupled nuclei. In practice, the intensity of the ENDOR lines only represents a few percent of that of the EPR lines.

3 - ENDOR spectrum of a polycrystalline powder or frozen solution

The ENDOR spectrum for molecules in liquid solution is simple to interpret, but it only provides the *isotropic* component of the hyperfine interactions. The ENDOR spectra for polycrystalline powders and frozen solutions are more complex to analyse, but they can be used to determine the principal values of the $\tilde{A}$ matrix, to deduce the dipolar matrix and consequently the distance to the coupled nuclei [Volume 1, appendix 3].

The EPR spectrum for polycrystalline powders and frozen solutions results from the superposition of the lines produced by all the paramagnetic centres in the sample, which are randomly oriented relative to the field **B**. This superposition generally causes the line splitting due to weak interactions to disappear. This is particularly the case when the resonance lines are broadened by g-strain or A-strain type phenomena [Volume 1, section 9.5.3]. The features produced by weak hyperfine interactions are thus rarely visible on EPR spectra for this type of sample. However, there are exceptions, like for example the superhyperfine structure observed on spectra of certain Cu(II) centres (see chapter 4 in this volume). To understand the nature of the ENDOR spectrum produced by this type of sample for a given value of the magnetic field $B = B_{res}$, the following phenomena must be taken into account:

> ▷ The EPR lines for solid samples are often *inhomogeneous*, which means that they result from the superposition of *homogeneous lines* due to "spin packets" [Volume 1, complement 3 to chapter 5]. In the ENDOR experiment, the spin packet centred at B_{res} is saturated, as well as, to a lesser extent, adjacent packets. It is thus this packet that is mainly "desaturated" by ENDOR transitions. As a result, the width of the ENDOR lines is of the same order of magnitude as that of the homogeneous lines for spin packets, and may be increased by A strain effects.

> ▷ The ENDOR spectrum is produced by paramagnetic centres with an orientation relative to **B** such that their resonance line forms at B_{res}. By choosing specific values of B_{res}, sub-groups of paramagnetic centres with a well

defined orientation relative to **B** can be selected. This selection simplifies interpretation of the spectra.

These different points can be illustrated through the example of a molybdenum centre. As the Mo^{5+} ion has a $(4d^1)$ configuration, Mo(V) complexes are characterised by $S = \frac{1}{2}$ and three principal values of the $\tilde{g}$ matrix slightly less than g_e [Volume 1, section 4.2.2]. In figure A3.5a, the X-band EPR spectrum for the Mo(V) centre of nitrate reductase from the bacterium *Escherichia coli* is shown. Around 25 % of the molybdenum nuclei are paramagnetic with $I = \frac{5}{2}$ [Volume 1, table 2.2]. These nuclei produce a hyperfine structure which is manifested by satellite signals to the main group of peaks, identified by asterisks in the figure. Numerical simulation produces $g_x = 1.962$, $g_y = 1.981$, $g_z = 1.987$. The ENDOR spectrum for the protons was recorded for the corresponding canonical values B_x, B_y and B_z (figure A3.5b). It displays well resolved structures (peak-to-peak width around 0.1 MHz), which are practically symmetric relative to the Larmor frequency for the proton, and which vary with the value of the resonance field.

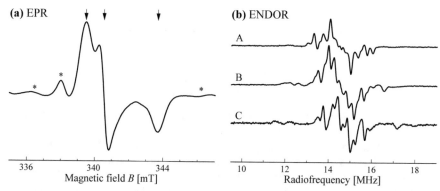

Figure A3.5 - EPR and ENDOR spectra for the Mo(V) centre in nitrate reductase A from *Escherichia coli*. **(a)** EPR spectrum. Temperature 50 K, microwave frequency 9.424 GHz, power 4 mW, modulation amplitude 0.4 mT at 100 kHz. **(b)** ENDOR spectra recorded for resonance fields corresponding to the three principal values for the $\tilde{g}$ matrix, identified by the arrows in figure A3.5a: A: $g_z = 1.987$, B: $g_y = 1.981$, C: $g_x = 1.962$. Temperature 25 K, microwave frequency 9.471 GHz, microwave power 25 mW, radiofrequency power ~ 100 W, depth of radiofrequency modulation 100 kHz.

To understand the origins of these structures, we consider the case where *a single proton* interacts with the Mo(V) centre. We assume that the interaction takes place at a large distance, such that the hyperfine matrix is axial [Volume 1, Appendix 3], and that its principal axes differ from those of the $\tilde{g}$ matrix. The ENDOR spectrum resulting from this interaction was calculated for the values

B_x, B_y, B_z of the field (figure A3.6). We consider for example the spectrum calculated for $B = B_x$. The EPR line at B_x is produced by the paramagnetic centres for which the magnetic axis x *is parallel* to the field **B**. These centres are related by rotations about x [Volume 1, complement 2 to chapter 4]. As the hyperfine matrix is *anisotropic*, this rotation causes the energy levels to vary, and consequently the frequency of the ENDOR lines. The ENDOR spectrum calculated for $B = B_x$ is thus composed of an *ensemble of pairs of lines*, which superpose to produce the structures shown in figure A3.6. This is also the case for the spectra calculated for $B = B_y$ and $B = B_z$.

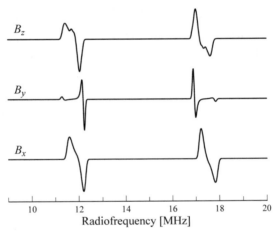

Figure A3.6 - Calculation of the ENDOR spectrum resulting from coupling between Mo(V) and a single proton. We assumed that the Mo(V) producing the EPR spectrum in figure A3.5a interacts with a single proton, and that the matrix $\tilde{\mathbf{A}}$ is axial with $A_\perp = 4.7$ MHz, $A_{//} = 6.6$ MHz. The orientation of the principal axes of this matrix relative to the (x, y, z) axes of the $\tilde{\mathbf{g}}$ matrix is defined by the Euler angles $(0°, 45°, 0°)$. The spectra were calculated for the values B_x, B_y, B_z of the magnetic field, which correspond to (g_x, g_y, g_z), using *Easyspin* software [Stoll and Schweiger, 2006].

The paramagnetic centres which form their resonance lines at B_y have more varied orientations than those forming lines at B_x and B_z [Volume 1, complement 2 to chapter 4]. We thus expect the frequency range covered by the ENDOR spectrum to be broader for this field value (figure A3.6). With the orientation of the principal axes of the $\tilde{\mathbf{A}}$ matrix specified in the legend to figure A3.6, separation of the external peaks of the spectrum calculated for $B = B_y$ is equal to 6.6 MHz, which is the value of $A_{//}$, and that of the internal peaks is equal to 4.7 MHz, which is the value of $A_\perp$. In general, the spectra must be *simulated* for a series of values of B to determine the principal values of the $\tilde{\mathbf{A}}$ matrix and the orientation of its principal axes relative to those of the $\tilde{\mathbf{g}}$ matrix.

4 - Comparison with pulsed ENDOR and other high-resolution spectroscopic techniques

Pulsed ENDOR experiments are mainly based on sequences developed by W. B. Mims [Mims, 1965] and E. R. Davies [Davies, 1974]. They consist in measuring the variation in intensity of an electronic spin echo generated by *selective* microwave pulses (i.e., acting only on one of the transitions such as $v_{EPR\,1}$ or $v_{EPR\,2}$ in figure A3.1) (Davies ENDOR), or non-selective pulses (Mims ENDOR), when the frequency of a radiofrequency pulse inserted in the sequence of microwave pulses is varied (see appendix 2). The two techniques are *complementary*: Davies ENDOR is well suited to the detection of strongly coupled nuclei, whereas Mims ENDOR is mainly used to study weak interactions. Pulsed ENDOR techniques are less sensitive to experimental artefacts due to the simultaneous double irradiation required in the continuous mode. They are also more generally applicable: a signal can be detected when a) the electronic phase memory time T_m^e is long enough to be able to observe a spin echo (see section 1.1 in appendix 2) and b) the spin-lattice electronic relaxation time T_1^e is greater than, or at least comparable to, the duration of the radiofrequency pulse. In liquid solution, the EPR lines are generally homogeneous, and only continuous wave ENDOR spectroscopy can be used. In contrast, pulsed ENDOR is often better adapted to the study of solid samples. Other pulsed ENDOR techniques have been developed [Schweiger and Jeschke, 2001].

ESEEM (*Electron Spin Echo Envelope Modulation*) and ENDOR experiments are *complementary* as they are suited for use in different situations. ESEEM spectroscopy is better adapted to detecting low-frequency nuclear transitions which appear in particular in the "perfect compensation" situation, in an intermediate regime between the strong and weak coupling situations (section 1.3 in appendix 2). However, the intensity of the experimentally-accessible microwave pulses limits the spectral window which can be covered by this method operating in the time domain. A significant advantage of ENDOR experiments over ESEEM experiments is that their intrinsic sensitivity increases with the microwave frequency used. In addition, when the anisotropy of the $\tilde{\mathbf{g}}$ matrix is not resolved on the X-band EPR spectrum (as is frequently the case for radical species), the ENDOR spectra recorded at frequencies higher than X-band can be used to study the anisotropy of the hyperfine interactions and any quadrupole interactions of the coupled nuclei. Finally, by using a higher EPR frequency is it possible to better separate the ENDOR lines produced by nuclei of different natures.

References

ABRAGAM A. & BLEANEY B. (1986) *Electron Paramagnetic Resonance of Transition Ions*, Dover Publications, New York.

CEDERQUIST A.L. (1963) PhD thesis, Washington University, St. Louis.

DAVIES E.R. (1974) *Physics Letters* A **47**: 1-2.

DINSE K.P. *et al.* (1974) *Journal of Chemical Physics* **61**: 4335-4341.

EHRENBERG *et al.* (1968) *Biochimica et Biophysica Acta* **167**: 482-484.

FEHER G. (1956) (a) *Physical Review* **103**: 500-501.

FEHER G. (1956) (b) *Physical Review* **103**: 834-835.

FEHER G. (1956) (c) *Physical Review* **103**: 501-503.

HYDE J. S. & MAKI A.H. (1964) *Journal of Chemical Physics* **40**: 3117-3118.

KEVAN L. & KISPERT L.D. (1976) *Electron Spin Double Resonance Spectroscopy*, Wiley-Interscience, New York.

KULIK L. & LUBITZ W. (2009) *Photosynthesis Research* **102**: 391-401.

KURRECK H. *et al.* (1988) *Electron Nuclear Double Resonance Spectroscopy of Radicals in Solution. Application to Organic and Biological Chemistry,* VCH Publishers, Inc., New York.

LOWE D. J. (1995) *ENDOR and EPR of Metalloproteins,* Springer, Heidelberg.

MIMS W. B. (1965) *Proceeding of the Royal Society London A* **283**: 452-457.

MURPHY D. M. & FARLEY R.D. (2006) *Chemical Society Review* **35**: 249-68.

SCHWEIGER A. & JESCHKE G. (2001) *Principles of Pulse Electron Paramagnetic Resonance*, Oxford University Press, Oxford.

STOLL S. & SCHWEIGER A. (2006) *Journal of Magnetic Resonance* **178**: 42-55.

Volume 1: BERTRAND P. (2020) *Electron Paramagnetic Resonance Spectroscopy - Fundamentals*, Springer, Heidelberg.

Macromolecules with very varied functions: proteins

Belle V. and Bertrand P.

Since the middle of the 19th century proteins have been recognised to be essential components of biological cells, but the vast diversity of their structures and functions only emerged during studies performed over the last fifty years. The description of the structure of these complex macromolecules provided here should facilitate reading of chapters 4, 6 and 8.

1 - From sequence to structure

A protein is composed of a "polypeptide" chain made of amino acids known as the *residues* of the protein. The *sequence* or **primary structure** of the protein, i.e., the succession of amino acids making up the chain, is determined by the *sequence* of the gene coding for the protein. The smallest proteins are composed of around sixty residues and have a molar mass of around 6 kDa ($1\ Da = 1\ g\,mol^{-1}$) (e.g. 5.8 kDa for insulin), the largest have several thousand residues and a molar mass of several hundred kDa (e.g. 480 kDa for myosin). The name "peptide" generally refers to short sequences of less than fifty residues. Each link in the chain is composed of amide N–H and carbonyl C=O groups and a HCR group, of which the carbon is the "alpha" carbon (C_α), and R represents the *side chain* of the amino acid (figure A4.1).

The polypeptide chain starts with an NH_2 amine group bound to the C_α of the first amino acid (N-terminal end) and ends with a carboxylate group COOH bound to the C_α of the last amino acid (C-terminal end). By convention, the amino acids are numbered from the N-terminal end. The proteins in all living

© Springer Nature Switzerland AG 2020
P. Bertrand, *Electron Paramagnetic Resonance Spectroscopy*,
https://doi.org/10.1007/978-3-030-39668-8

organisms are constructed from 20 amino acids which are differentiated based on the nature of their side chain (table A4.1). These amino acids are designated by a one-letter or three-letter code (first and second column in table A4.1). A 16-amino acid peptide is represented in chapter 4 (figure 4.13).

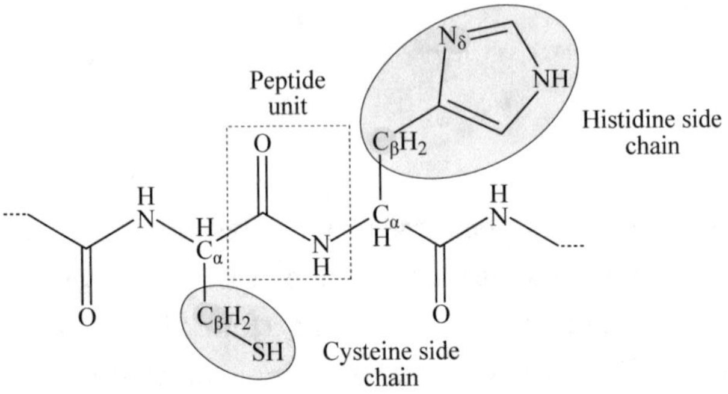

Figure A4.1 - Succession of a cysteine and a histidine in a polypeptide chain. The peptide unit is planar, but rotations around the C_α–CO and C_α–NH bonds allow the chain to adopt various geometries. The succession of $-C_\alpha$–CO–NH–C_α– patterns makes up the protein's « peptide backbone ».

The chemical and electrostatic properties of the side chains are very variable (last column in table A4.1) and they determine in large-part how the polypeptide chain will fold to form a more or less compact three-dimensional structure.

Table A4.1 - The 20 amino acids which make up proteins in living organisms. Some have ionisable side chains, and are thus charged at pH 7.

1-letter code	3-letter code	Amino acid (residue)	Nature of the side chain
A	Ala	Alanine	apolar, aliphatic
C	Cys	Cysteine	polar
D	Asp	Aspartic acid	acidic (−)
E	Glu	Glutamic acid	acidic (−)
F	Phe	Phenylalanine	apolar, aromatic
G	Gly	Glycine	apolar, aliphatic
H	His	Histidine	basic, aromatic
I	Ile	Isoleucine	apolar, aliphatic
K	Lys	Lysine	basic (+)
L	Leu	Leucine	apolar, aliphatic
M	Met	Methionine	apolar
N	Asn	Asparagine	polar
P	Pro	Proline	apolar
Q	Gln	Glutamine	polar
R	Arg	Arginine	basic (+)
S	Ser	Serine	polar
T	Thr	Threonine	polar
V	Val	Valine	apolar, aliphatic
W	Trp	Tryptophan	apolar, aromatic
Y	Tyr	Tyrosine	polar, aromatic

Protein sequences are determined by biochemical analysis and/or mass spectrometry. They can also be deduced from the DNA sequence of the gene coding for the protein.

Segments of the polypeptide chain often fold to form repetitive structures stabilised by hydrogen bonds between the C=O and N–H groups from different residues (figure A4.2). When the interacting residues are close together in the sequence, the peptide backbone takes the shape of an α helix with a pitch of 5.4 Å, with the side chains directed towards the exterior. In β sheets, segments positioned face to face interact *via* a network of hydrogen bonds, and the peptide backbone adopts a stretched structure. α helices and β sheets are elements of **secondary structure**.

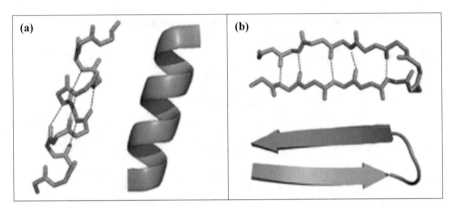

Figure A4.2 - Secondary structure elements and their diagrammatic representations.
(a) α helix, **(b)** β sheet. The dashed lines represent hydrogen bonds between the C=O and
N–H groups of different residues.

The spatial organisation of these elements and the loops linking them make up
the **tertiary structure** of the protein. The cohesion of this structure is ensured
by numerous non-covalent interactions involving the amino acid side-chains
and, sometimes, the peptide backbone (figure A4.3).

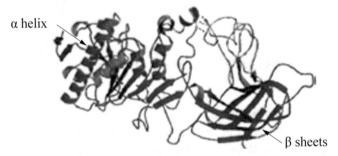

Figure A4.3 - Crystal structure of the complex formed by human pancreatic lipase
and colipase. The secondary structure includes α helices and β sheets.

In increasing order of strength, these interactions are: van der Waals interactions,
electrostatic bonds between charged groups, hydrogen bonds, hydrophobic
bonds between apolar groups. To fulfil their functions, proteins are sometimes
organised in *domains* which may display a certain level of independence.

Some proteins are formed through the association of several polypeptide chains,
the arrangement of which constitutes the **quaternary structure**. In this case,
each polypeptide chain is known as a *subunit* of the protein.

The three-dimensional structure of proteins can be determined by X-ray diffraction experiments on single crystals or nuclear magnetic resonance (particularly nitrogen and hydrogen nuclei) in solution. These structures have been gathered together in a data base, Brookhaven's *Protein Data Bank* (PDB), and they are accessible to the whole scientific community. Whereas only seven structures were known in 1971, almost 80,000 are currently available. The "mean structures" of the protein edifices produced by these techniques provide considerable information, but they can also be misleading. At physiological temperatures, proteins are mobile structures, and their complex dynamics are the result of combinations of vibrations and rotations at very variable timescales [Mittermaier and Kay, 2006]. This dynamicity gives the protein a certain flexibility which favours the conformational changes needed for some proteins to function. Small peptides and proteins or "intrinsically disordered" domains without a stable conformation in physiological conditions are particularly flexible. Their difficulty in folding is due to their sequence, which is rich in charged amino acids and poor in hydrophobic amino acids. When the polypeptide chain of these peptides or proteins includes coordination sites for metal ions, it can fold to form stable complexes with these ions. EPR can be used to study the complexation of copper by peptides and the disordered domains of proteins involved in some neurodegenerative diseases, as described in chapter 4.

2 - From structure to function

Each protein has a specific function linked to its structure. Some act in solution, others are membrane-bound. Proteins are involved in almost all cellular functions, and a few examples are sufficient to illustrate the diversity of their roles:

Catalysis: almost all chemical reactions occurring in biological cells are catalysed by specific proteins, known as *enzymes*. These proteins have an *active site* where reactants are activated (see chapter 6).

Transport: the haemoglobin in red blood cells and myoglobin in muscle cells transport dioxygen. These were the first proteins for which the crystal structure was resolved. [Kendrew *et al.*, 1958; Perutz *et al.*, 1960, shared the 1962 Nobel prize for chemistry]. Some proteins specialise in the transport and storage of metal ions. For example, mammalian ferritin contains 24 subunits, each of which is composed of 175 amino acids, and can accept up to 4500 Fe^{3+} ions, i.e., around one ion per amino acid. Proteins also ensure the transport of small molecules through pores in cellular membranes.

Motion: contractile proteins such as actin and myosin are responsible for muscle contraction.

Cellular signalling: hormone receptors are proteins that are generally inserted into membranes. Numerous hormones, such as insulin and growth hormone, are proteins.

Immunology: antibodies secreted by plasma cells are proteins.

Structural maintenance: some proteins such as collagen and keratin contribute to supporting tissues. Cytoskeletal proteins play an analogous role in cells.

Intrinsically-disordered proteins: currently, more than a third of proteins in eukaryotic cells are considered to have disordered domains composed of more than 30 amino acids. The great flexibility of these domains allows them to interact with multiple partners, and their role in numerous processes has been recognised: cellular processes (regulation, cell division, etc.), molecular recognition processes, key role in neurodegenerative diseases. These proteins often adopt a well-defined conformation when they interact with their physiological partners.

3 - Contribution of EPR to the study of proteins

EPR can be used to obtain structural information on proteins and to better understand their function at a molecular level. Its applications can be classed in two categories:

▷ The function of numerous proteins relies on the presence of "prosthetic groups" linked to the side chains by weak or covalent bonds, which can often be prepared in a paramagnetic oxidation state. Organic groups – such as flavins, quinones and chlorophylls – produce radical EPR signals at room temperature. Mononuclear or polynuclear metal centres produce a wide variety of spectra which are generally only observable at cryogenic temperatures. In particular, this type of cofactor is found in *redox enzymes* and examples of how EPR can be applied to study the structure and catalytic mechanisms of these enzymes are described in chapter 6.

▷ In other applications, one or more nitroxide radicals attached to well-defined sites on a protein act as *paramagnetic probes*: a detailed study of the shape of their EPR spectrum and their relaxation properties provides information on the environment of the binding site. This method can be used to study conformational changes which occur in some proteins following interaction with their physiological partners. Two examples are presented in chapter 8: conformational changes occurring in human pancreatic lipase (figure A4.3), a key enzyme in the human digestive system, and folding of an intrinsically disordered domain in the measles virus nucleoprotein.

All these studies benefit from the possibilities presented by "site-directed mutagenesis" techniques, which allow an amino acid to be replaced by another by modifying the sequence of the gene coding for the protein. For example, replacing an amino acid binding the prosthetic group of a protein by another residue modifies its spectroscopic properties and alters the protein's function. When this group is paramagnetic, EPR can be used to correlate its structural variations with variations in the protein's activity. Similarly, to affix a nitroxide radical to a well-defined position in the polypeptide chain, the amino acid at the target position is replaced by a cysteine. To conclude, EPR studies are generally performed in aqueous solutions using purified proteins, in capillary tubes or flat cells at room temperature, or in EPR tubes at cryogenic temperatures. The concentration of the samples is generally around 10 to 100 μM.

References

KENDREW J.C. *et al.* (1958) *Nature* **181**: 662-665.

MITTERMAIER A. & KAY L.E. (2006) *Science* **312**: 224-228.

PERUTZ M.F. *et al.* (1960) *Nature* **185**: 426-422.

Volume 1: BERTRAND P. (2020) *Electron Paramagnetic Resonance Spectroscopy - Fundamentals*, Springer, Heidelberg.

Index

U

V

Z

Printed in the United States
By Bookmasters